NEW CURING TECHNIQUES
IN THE PRINTING,
COATING AND PLASTICS INDUSTRIES

NEW CURING TECHNIQUES IN THE PRINTING, COATING AND PLASTICS INDUSTRIES

Maurice W. Ranney

NOYES DATA CORPORATION

Park Ridge, New Jersey   London, England

1973

Published in the United States of America by
Noyes Data Corporation
Noyes Building, Park Ridge, New Jersey 07656

FOREWORD

The detailed, descriptive information in this book is based on U.S. patents since 1971 relating to new curing techniques in the printing, coating and plastics industries. This book serves a double purpose in that it supplies detailed chemical information and can be used as a guide to the U.S. patent literature in this field. By indicating all the information that is significant and eliminating legal jargon and juristic phraseology, this book presents an advanced, commercially oriented review of new curing techniques which may be used to advantage by the industries cited above.

The U.S. patent literature is the largest and most comprehensive collection of technical information in the world. There is more practical, commercial, timely process information assembled here than is available from any other source. The technical information obtained from a patent is extremely reliable and comprehensive; sufficient information must be included to avoid rejection for "insufficient disclosure."

The patent literature covers a substantial amount of information not available in the journal literature. The patent literature is a prime source of basic commercially useful information. This information is overlooked by those who rely primarily on the periodical journal literature. It is realized that there is a lag between a patent application on a new process development and the granting of a patent, but it is felt that this may roughly parallel or even anticipate the lag in putting that development into commercial practice. Many of these patents are being utilized commercially. Whether used or not, they offer opportunities for technological transfer. Also, a major purpose of this book is to describe the number of technical possibilities available, which may open up profitable areas of research and development. One should have to go no further than this condensed information to establish a sound background before launching into research in this field.

Advanced composition and production methods developed by Noyes Data are employed to bring our new durably bound books to you in a minimum of time. Special techniques are used to close the gap between "manuscript" and "completed book." Industrial technology is progressing so rapidly that time-honored, conventional typesetting, binding and shipping methods are no longer suitable. We have bypassed the delays in the conventional book publishing cycle and provide the user with an effective and convenient means of reviewing up-to-date information in depth. The Table of Contents is organized in such a way as to serve as a subject index. Other indexes by company, inventor and patent number help in providing easy access to the information contained in this book.

15 Reasons Why the U.S. Patent Office Literature Is Important to You —

1. The U.S. patent literature is the largest and most comprehensive collection of technical information in the world. There is more practical commercial process information assembled here than is available from any other source.

2. The technical information obtained from the patent literature is extremely comprehensive; sufficient information must be included to avoid rejection for "insufficient disclosure."

3. The patent literature is a prime source of basic commercially utilizable information. This information is overlooked by those who rely primarily on the periodical journal literature.

4. An important feature of the patent literature is that it can serve to avoid duplication of research and development.

5. Patents, unlike periodical literature, are bound by definition to contain new information, data and ideas.

6. It can serve as a source of new ideas in a different but related field, and may be outside the patent protection offered the original invention.

7. Since claims are narrowly defined, much valuable information is included that may be outside the legal protection afforded by the claims.

8. Patents discuss the difficulties associated with previous research, development or production techniques, and offer a specific method of overcoming problems. This gives clues to current process information that has not been published in periodicals or books.

9. Can aid in process design by providing a selection of alternate techniques. A powerful research and engineering tool.

10. Obtain licenses — many U.S. chemical patents have not been developed commercially.

11. Patents provide an excellent starting point for the next investigator.

12. Frequently, innovations derived from research are first disclosed in the patent literature, prior to coverage in the periodical literature.

13. Patents offer a most valuable method of keeping abreast of latest technologies, serving an individual's own "current awareness" program.

14. Copies of U.S. patents are easily obtained from the U.S. Patent Office at 50¢ a copy.

15. It is a creative source of ideas for those with imagination.

CONTENTS AND SUBJECT INDEX

INTRODUCTION

With the increasing emphasis being placed on pollution-free coatings systems, much research effort has been directed towards the use of energy sensitive systems which can be cured by exposure to ultraviolet and electron beam sources. Low volatility, essentially 100% solids systems with minimal odor levels can be applied at relatively high rates using these techniques. The resulting coatings are thermoset in character, even though no heat, as such, has been applied. Thus, they can be formulated to have excellent hardness, abrasion and stain resistance and when pertinent, outstanding exterior durability.

High energy cured coatings are generally composed of cross-linked polymers, the reaction occuring "in situ" on the substrate. Free-radical vinyl polymerization of a low molecular weight unsaturated resin, dissolved in an appropriate reactive monomer, along with a sensitizer or photoinitiator is most commonly used.

Photocurable coatings are a recent development which largely started in Europe some ten years ago. While much of the effort continues to be directed towards unsaturated polyesters combined with styrene and acrylic monomers, many other photocurable systems, based on urethanes, acrylics and silicones, have been developed. Many light sensitive monomers and low molecular weight resins have been synthesized and studied as components for coatings, photo-resists and printing inks. Photoiniators are required for an efficient photocuring process. In addition to a number of energy sensitive additives such as benzophenone and the benzoin ethers, considerable effort has been directed to incorporating the initiator directly into the film-forming resin. Many new photo-resist formulations have appeared in the process literature, as this continues to be a high priority need

1

in the printing and electronics industries. Ultraviolet curing requires a
minimal capital investment compared to electron beam curing and many
conventional systems are designed for solvent removal. Thus, it is a practical route for small coaters and printers and is expected to gain rapid commercial acceptance.

Electron beam curing is receiving particular attention among the large consumers of coatings, particularly in the automotive, metal siding, textile
and plastics industries. Ford Motor Company has conducted extensive testing with many compounds and paint formulations and has some commercial
applications now underway. While electron beam curing has yet to reach
significant commercial status, the advantages of rapid curing (seconds),
elimination of curing ovens, and the use of essentially solvent-free vehicles
which minimize air pollution problems clearly indicate that this form of
radiation curing will become an integral part of our coatings technology
over the next decade.

Much of the reserach effort has focused on preparing reactive monomers,
particularly acrylic and vinyl, as well as on total paint systems. Formulations have been developed for various substituted resins, silicone modified
polyesters, polyurethanes, acrylics and ethers such as polyvinyl chloride
and ethylene-acrylic acid copolymers. This highly penetrating energy source
is also very useful for curing many plastic composites and in heating textiles.

This book describes over 120 processes involving the newest technology
available in the patent literature for photo and electron beam curing techniques as they apply to the coatings, printing ink and plastics industries.

Other books relating to new curing and application techniques for generally pollution-free processing include:

> Electrodeposition and Radiation Curing of Coatings,
>   Park Ridge, New Jersey, Noyes Data Corporation (1970)

> Powder Coatings and Fluidized Bed Techniques,
>   Park Ridge, New Jersey, Noyes Data Corporation (1971)

Part I.

Ultraviolet Curing

PRINTING AND PHOTORESIST COATINGS

LIGHT SENSITIVE POLYMERS AND MONOMERS

Maleic Anhydride Copolymers Modified with Amino Stilbene

A.A.R. Sayigh, F.A. Stuber and H. Ulrich; U.S. Patent 3,657,199; April 18, 1972; assigned to The Upjohn Company describe light sensitive ester-amide polymers. The polymers are characterized by the recurring unit:

$$\left[\text{CH}-\underset{\underset{R_1}{\overset{\mid}{C}O}}{\overset{\mid}{\underset{\mid}{}}}\text{CH}-\underset{\underset{R_2}{\overset{\mid}{C}O}}{\overset{\mid}{\underset{\mid}{}}}\text{CH}-\overset{R''}{\overset{\mid}{\underset{\mid}{}}}\text{CH}_2\right]$$

where one of R_1 and R_2 represents —O— (lower-alkyl) and the other of R_1 and R_2 represents —NHR where R is a stilbene residue having the formula:

$$\text{CH}=\text{CH} \quad (\text{lower-alkoxy})_n$$

n=1 to 3, and R''=lower-alkoxy or phenyl. The polymers have an average molecular weight within the range of 100,000 to 1,250,000. The above ester-amide polymers undergo cross-linking on exposure to ultraviolet light and can be used in photo-resist systems. The above ester-amide polymers

4

are prepared by reacting the appropriate vinyl ether–maleic anhydride or styrene–maleic anhydride copolymer with the appropriate amino–stilbene and esterifying the intermediate amide amine salt with a lower–aliphatic alcohol.

Example 1: A total of 10.2 g. (0.04 mol) of trans–2, 5–dimethoxy–4'–aminostilbene was added to a refluxing solution of 3.2 g. of a poly(maleic anhydride co–methylvinyl ether) (having an average molecular weight of 250,000; Gantrez AN 119) in 150 ml. of anhydrous tetrahydrofuran (previously dried by treatment with p–toluene sulfonyl isocyanate followed by distillation). The proportion of aminostilbene used represented 2 mol per anhydride group in the copolymer chain. The progress of the reaction was followed by infrared spectroscopic analysis of an aliquot.

When no further increase in absorption at the amide band (5.9u) of the spectrum was observed, 50 ml. of methanol was added to the reaction mixture and the latter was heated under reflux for a further 16 hours. At the end of this time the reaction mixture was cooled and carbon tetrachloride was added. The solid which separated was isolated by filtration and purified by dissolving in acetone followed by precipitation therefrom by addition of carbon tetrachloride. This procedure was repeated until no further aminostilbene starting material was detected in the solvent washings. The resulting brown solid was dried. There was thus obtained a photosensitive polymer having as the recurring unit a moiety of the following structure:

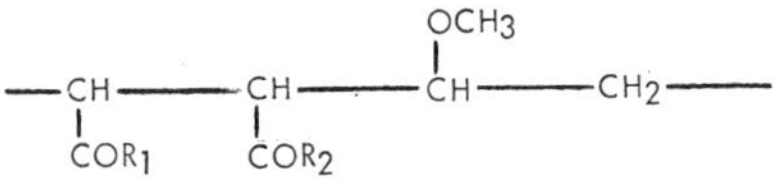

$$-CH-\underset{COR_1}{|}-CH-\underset{COR_2}{|}-\overset{OCH_3}{\underset{|}{CH}}-CH_2-$$

where one of R_1 and R_2 is methoxy and the other is

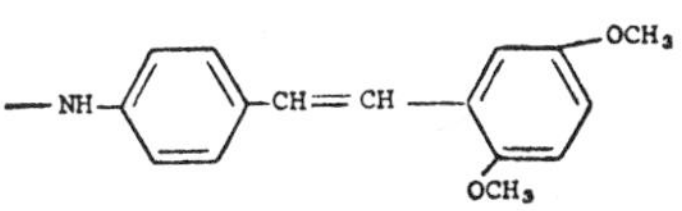

$$-NH-\langle\rangle-CH=CH-\langle\rangle\overset{OCH_3}{\underset{OCH_3}{}}$$

The stilbene residue in the above moiety is in transconfiguration. The ultraviolet absorption spectrum of the polymer (tetrahydrofuran) exhibited maxima at 302 to 347 nm.

Example 2: To a solution of 3.2 g. of a poly(maleic anhydride co–methylvinyl ether) (molecular weight average 250,000; Gantrez AN 119) in 75 ml. of methyl ethyl ketone was added 0.55 g. (0.002 mol) of trans–2, 5–dimethoxy–4'–aminostilbene and the resulting mixture was heated under reflux

for 80 minutes. (The proportion of amino-stilbene employed in the above
reaction was sufficient to react with 1 in 20 of the anhydride moieties in
the starting polymer.) At the end of this time 50 ml. of methanol and 1
drop of concentrated hydrochloric acid were added. The resulting mixture
was heated under reflux overnight (approximately 16 hours).

At the end of this time the bulk of the organic solvent was removed by evaporation and the residue was poured into carbon tetrachloride. The solid
which separated was isolated by filtration and was purified by dissolving in
a small quantity of methanol and pouring the methanol solution into carbon
tetrachloride. The resulting brown solid was isolated by filtration and dried.
There was thus obtained a modified copolymer of maleic anhydride and
methylvinyl ether in which approximately 1 in 20 of the recurring units in
the polymer chain had the structure:

$$
\begin{array}{c}
\text{----CH------CH------CH------CH}_2\text{----} \\
\quad\ |\qquad\quad |\qquad \overset{\displaystyle OCH_3}{|} \\
\quad COR_1 \qquad COR_2
\end{array}
$$

where one of R_1 and R_2 represents methoxy and the other is

The above procedure was repeated save that the amount of trans-2,5-dimeth-
oxy-4'-aminostilbene employed was increased to 2.1 g. (0.008 mol) corresponding to a proportion sufficient to react with 1 in 5 of the anhydride
moieties in the starting polymer. There was thus obtained a modified copolymer of maleic anhydride and methylvinyl ether in which approximately
1 in 5 of the recurring units in the polymer chain had the structure set forth
above. The infrared spectrum of the polymer (tetrahydrofuran solution)
shows the bands of the ester and amide groups from 5.85 to 5.95u.

Example 3: This example illustrates the use of the photosensitive polymers
as the active components of a typical photoresist system. A series of negatives were cast on quartz plates as follows: A 5% w./w. solution of the
ester-amide polymer obtained as described in Example 1 in a mixture of
acetone and Cellosolve (20 to 1 by volume) was cast on a series of quartz
plates (2 x 2 x 1/16 inches) to give films of thickness ranging from 0.0001
to 0.00001 in. (The film on any one plate was substantially uniform in
thickness.) The film so obtained was placed at a distance of 30 cm. from

the front of a mercury arc lamp (type GE H100 A4/T). A metal plate having a pattern of holes was employed as the negative to be reproduced and was mounted between the resist-coated plate and the light source. The resist-coated plate was exposed to light transmitted via the negative for a period of 2 to 5 minutes depending on the thickness of the film.

At the end of this time the image on the resist-coated plate was developed by washing with a mixture of methanol, acetone, Cellosolve (30:10:10) and the plate was found to have deposited thereon a solvent insoluble polymer in a pattern corresponding to that of the eight transmitting holes in the negative. The polymer coating was found to be resistant to acid and weak alkali solutions and to abrasive forces.

The above experiment was repeated using a film of photoresist deposited on a thin copper plate. After formation of the insoluble polymer image on the plate, the latter was exposed to 38% ferric chloride solution. The surface of the cooper plate etched readily but the polymer image was unaffected.

Maleic Anhydride Copolymers Modified with Dyestuff and Azidosulfonyl Groups

In a process described by A.A.R. Sayigh, F.A. Stuber and H. Ulrich; U.S. Patent 3,699,080; October 17, 1972; assigned to The Upjohn Company radiation (thermal and light)-sensitive polymers are provided and are characterized by (1) a recurring unit of the formula:

$$-CH-CH-\overset{\overset{\displaystyle R_1}{|}}{C}H-CH_2-$$
$$\underset{X}{|}\quad\underset{Y}{|}$$

where R_2 is lower-alkoxy or phenyl, one of X and Y is —COOR and the other is —COR$_1$, R is an azidosulfonylcarbanilyloxy alkylene group and R$_1$ is —O—dye, —NH—dye or hydroxyarylamino, or (2) a combination in the same polymer chain of recurring units having the formulas:

$$-CH-CH-\overset{\overset{\displaystyle R_1}{|}}{C}H-CH_2-$$
$$\underset{X_1}{|}\quad\underset{Z_1}{|}$$

where one of X_1 and Z_1 represents —COOH and the other represents —COOR as defined above, and

$$-CH-CH-\overset{\overset{\displaystyle R_1}{|}}{C}H-CH_2-$$
$$\underset{X_2}{|}\quad\underset{Z_2}{|}$$

where one of X_2 and Z_2 represents —COOH and the other represents —COR_1 as defined above. The polymers can be chemically bonded to substrates, such as those containing C—H bonds only (e.g. polyolefins), which are not normally dye receptive. The bonding is effected by exposure to appropriate radiation; the exposure can be carried out imagewise to produce an appropriate image on the substrate. Accordingly, the polymers find use in photoresist and printing.

The preparation of 2-hydroxyethyl 4-azido sulfonylcarbanilate, which is then used in the subsequent examples illustrating the process, is as follows. To 50.4 g. (0.8 mol) of ethylene glycol in 500 ml. of acetonitrile is added over a 10 min. period with stirring and cooling to 2° to 8°C., a solution of 43.2 g. (0.2 mol) of 4-isocyanatobenzenesulfonyl chloride (J. Polymer Science, 5, pp. 3212 to 3213, 1967). The mixture is then allowed to stand at room temperature until the NCO band stretching has disappeared in the infrared spectrum of an aliquot of the reaction mixture (approx. 30 minutes).

To the mixture so obtained is added 13 g. (0.2 mol) sodium azide and the resulting mixture is stirred for 1 hour at room temperature. The sodium chloride which has precipitated is removed by filtration and about 80% of solvent is evaporated from the filtrate under vacuum. Water is added to the remaining filtrate to precipitate the water insoluble product. The latter is separated by filtration, washed with water, and dried under vacuum at room temperature.

There is thus obtained 52 g. (91% theoretical yield) of a white crystalline powder identified by infrared and NMR spectrometric examination as 2-hydroxyethyl 4-azidosulfonyl carbanilate and having an MP of 115° to 118°C. Recrystallization from acetonitrile gave white crystals having an MP of 120° to 122°C. (Fisher-Johns method); 124°C. (DSC method). Analysis — Calculated $C_9H_{10}N_4O_5S$ (percent): C, 37.76; H, 3.46. Found (percent): C, 37.60; H, 3.73.

Using the above procedure but replacing ethylene glycol by 1,3-propylene glycol, 1,4-butanediol, 1,3-pentanediol, 2,3-hexanediol, 1,5-heptane-diol, and 2,2-dimethyl-1,6-hexanediol, 2,5-diethyl-1,6-hexanediol there are obtained respectively: 3-hydroxypropyl, 4-hydroxybutyl, 3-hydroxy-pentyl, 3-hydroxy-2-methylpentyl, 5-hydroxyheptyl, 6-hydroxy-2,2-di-methylhexyl, and 6-hydroxy-2,5-diethylhexyl 4-azidosulfonylcarbanilate.

Example 1:   A mixture of 1.7 g. of a poly(maleic anhydride co-methyl vinyl ether) (having an average molecular weight of 250,000: Gantrez AN-119), 1.57 g. of Palacet Scarlet B, and 1.43 g. of 2-hydroxyethyl 4-azidosulfonylcarbanilate was dissolved in 30 ml. of anhydrous pyridine. The resulting solution was heated at 80°C. for 12 hours. The product so

obtained was evaporated to dryness and the residue was dissolved in a mixture of equal parts of acetone and water. The solution was acidified by the
addition of 100 ml. of 5 N hydrochloric acid. The solid which separated
was isolated by filtration, dried, and dissolved in 50 ml. of a mixture of
acetone and methanol. There was thus obtained a solution containing a
photosensitive modified poly(maleic anhydride co–methyl vinyl ether) in
which 1 in 2 of the recurring units in the chain are moieties having the
formula:

$$-CH-CH-\underset{\underset{OCH_3}{|}}{CH}-CH_2-$$
$$\underset{COOP}{|}\quad\underset{COOQ}{|}$$

where one of P and Q is hydrogen and the other is

$$-CH_2-CH_2-O-\overset{\overset{O}{\|}}{C}-NH-\left\langle\!\!\!\!=\!\!\!\!\right\rangle-SO_2N_3$$

and 1 in 2 of the recurring units are moieties of the formula:

$$-CH-CH-\underset{\underset{OCH_3}{|}}{CH}-CH_2-$$
$$\underset{COOA}{|}\quad\underset{COOB}{|}$$

where one of A and B is hydrogen and the other is

$$-CH_2-CH_2-\underset{\underset{C_2H_5}{|}}{N}-\left\langle\!\!\!\!=\!\!\!\!\right\rangle-N=N-\left\langle\!\!\!\!=\!\!\!\!\right\rangle-NO_2$$

A film was cast by spraying a small sample of the above polymer solution
on a polyethylene sheet using a whirler. The plate was irradiated by exposure for 5 minutes to a mercury arc lamp (Hanovia type SH). When the irradiation was complete, the film was found to be bonded to the polyethylene
substrate and was highly resistant to removal by various abrasive forces and
was insoluble in acetone and other polar solvents.

Example 2: Using the procedure described in Example 1 but replacing the
poly(maleic anhydride co–methyl vinyl ether) there was used by a poly (maleic anhydride co–styrene) having an average molecular weight of 100,000,
there was obtained the corresponding modified poly(maleic anhydride co-
styrene) in which approximately 1 in 2 of the recurring units in the chain
were moieties having the formula shown on the following page.

$$C_6H_5$$
$$-CH-\!\!-\!\!-CH-\overset{|}{C}H-CH_2-$$
$$\underset{COOP}{\quad}\quad\underset{COOQ}{\quad}$$

where one of P and Q is hydrogen and the other is

$$-CH_2-CH_2-O-\overset{O}{\overset{\|}{C}}-NH-\!\!\left\langle\;\;\right\rangle\!\!-SO_2N_3$$

and 1 in 2 of the recurring units were moieties of the formula:

$$C_6H_5$$
$$-CH-\!\!-\!\!-CH-\overset{|}{C}H-CH_2-$$
$$\underset{COOA}{\quad}\quad\underset{COOB}{\quad}$$

where one of A and B is hydrogen and the other is

$$-CH_2-CH_2-\underset{C_2H_5}{\overset{|}{N}}-\!\!\left\langle\;\;\right\rangle\!\!-N\!\!=\!\!N-\!\!\left\langle\;\;\right\rangle\!\!-NO_2$$

Example 3: A film was cast from the solution of light sensitive modified copolymer of Example 1, on polyethylene foil as substrate. On top of the film was placed a master representing a negative of a pattern of dots to be reproduced. The film was exposed for 2 minutes to the light emitted by a Hanovia type SH mercury arc, the plane of exposure being at a distance of 15 cm. from the lamp. The exposed film was developed by immersion with agitation for 1 minute in a mixture of acetone and N,N-dimethylformamide (100 to 1 by volume). The image so obtained exhibited a high degree of resolution.

Polymeric Anthraquinone Derivatives

A process described by F.F. Rogers, Jr.; U.S. Patent 3,591,661; July 6, 1971; assigned to E.I. du Pont de Nemours and Company provides light sensitizing copolymers of alpha olefin having from 2 to 4 carbon atoms and about 0.01 to 10 mol percent of monomer having the structural formula:

where R is an alkylene group having 1 to 4 carbon atoms and R₁ is selected
from hydrogen and methyl. These copolymers, in addition to providing desirable formed structures by themselves, are particularly useful in the cross-
linking of polyethylene and polypropylene, and there is accordingly also
provided a cross-linked polymeric composition comprising a blend of the
above copolymer and an alpha olefin selected from polyethylene and polypropylene where the copolymer comprises at least about 0.1% of the blend.

The polymeric photosensitizers can be prepared by reaction of an anthraquinone derivative containing an active hydrogen atom with an olefin-carboxylic acid halide copolymer following the general procedures described
in U.S. Patent 3,441,545. Particularly suitable active hydrogen groups
which will react with the acid halide groups of the olefin-carboxylic acid
halide copolymers include alcoholic hydroxyl groups, amino groups containing at least one hydrogen bonded to nitrogen, enolizable carbon hydrogen
groups, mercapto groups, thiocarboxylic acid groups and phosphino groups.
Preferred derivatives are those with an alcoholic hydroxyl.

Acid copolymers which can be used for conversion to acid halide copolymers
for making the desired polymers include copolymers of ethylene with acrylic
acid, methacrylic acid, itaconic acid or maleic acid and combinations of
these acids. Of those, copolymers of ethylene with acrylic and methacrylic
acids are particularly preferred.

In the following examples, which illustrate the process, parts and percentages are by weight. In these examples, Melt Index is determined according
to ASTM-D-1238 and dynamic zero strength temperature is determined
according to ASTM-D-1430.

Example 1:   Part A — In the preparation of 2-bromomethylanthraquinone,
125 parts of 2-methylanthraquinone, 100 parts of N-bromosuccinimide,
125 parts of benzoyl peroxide and 3,685 parts of reagent grade carbon
tetrachloride are placed into a reaction vessel filled with a mechanical
stirrer and distillation head. The reaction mixture is brought to reflux
with stirring and about 475 parts of solvent is distilled from the mixture.

The mixture is further heated at reflux for 40 hours after which it is cooled
and the product which separates is filtered, dried, extracted with water at
70°C. and again dried. This material is recrystallized from acetic acid to
yield 51 parts of the desired product with MP of 198° to 201°C. This is
used in the preparation of 2-acetoxymethylanthraquinone described below.

Part B — In the preparation of 2-acetoxymethylanthraquinone, a mixture of
50 parts of 2-bromomethylanthraquinone, 87 parts of sodium acetate in
790 parts of acetic acid is heated at reflux for 12 hours, cooled and the

solid material which separates is washed with warm water to yield 31 parts of product. It is identified as the compound of Part B melting in the range of 144° to 149°C.

Part C — In the preparation of 2-hydroxymethylanthraquinone, a mixture of 5 parts of 2-acetoxymethylanthraquinone and 120 parts of ethanol is heated in a steam bath to partial solution followed by addition of 40 parts of concentrated hydrochloric acid and further heating for about 15 minutes to effect complete solution. After an added 30 minute period of refluxing, the solution is cooled, and the solid product which forms is filtered and dried. There is obtained 3.15 parts of product having an MP of 190° to 190.5°C., which is identified as 2-hydroxymethylanthraquinone.

Part D — In the synthesis of the copolymer, twenty-five parts of an ethylene-methacrylic acid copolymer having a melt index of 4.9 and containing 3% methacrylic acid is reacted with excess thionyl chloride in Perclene solvent at 80°C. for 3 hours. The resulting ethylene methacrylyl chloride, together with a molar equivalent of 2-hydroxymethylanthraquinone, is dissolved in excess tetrachloroethylene.

The solution is heated at reflux for 70 hours. Excess isopropanol is added to to quench any unreacted acid chloride, after which the reaction mixture is heated at reflux for an additional 18 hours. The product, isolated by precipitation with additional isopropanol, is a creamy white powder which is identified as 2-methacryloxymethylanthraquinone/ethylene copolymer.

Example 2:   The copolymer prepared in Example 1 is pressed into a self-supporting film and subjected to irradiation by a 1,000-watt lamp for a period of 30 seconds. The irradiated film exhibits increased resistance to grease and oil as well as increased dimensional stability, indicating that cross-linking has taken place.

Example 3:   A film of a blend of the copolymer prepared in Example 1 and polypropylene is prepared by repeated pressing of the sample at 190°C. The amount of anthraquinone moiety in the finished film comprises 0.6% by weight. The film sample is exposed to a medium pressure mercury ultraviolet lamp for 20 seconds. The zero strength temperature of the irradiated film is 226°C. compared to 186°C. for the unexposed sample.

Example 4:   Two polyethylene film samples are prepared. Sample A contains the ethylene/methacryloxymethylanthraquinone copolymer of this process and Sample B, a control sample, contains an equal percentage of the ethylene/acryloxybenzophenone photosensitizing copolymer described in U.S. Patent 3,214,492. The two samples are exposed simultaneously to the radiation from a medium pressure mercury lamp.

After 2 seconds exposure of the Sample A, a zero strength temperature of 173°C. is measured. After 4 seconds exposure of the Sample B, a zero strength temperature of 140°C. is measured. After exposure for approximately 10 seconds, both Sample A and Sample B exhibit a zero strength temperature of about 230°C. Thus, in the early stages of the irradiation, the photosensitizer of the process appears to promote cross-linking about twice as rapidly as that in Sample B.

Polysulfonates

J.A. Arcesi and F.J. Rauner; U.S. Patent 3,640,722; February 8, 1972; assigned to Eastman Kodak Company describe light-sensitive polysulfonates and the use of such polymers in the preparation of photographic images. It is known in the photographic art to reproduce images by processes which involve imagewise exposure of a coating of a radiation-sensitive material, the solubility of which is differentially unmodified by the action of radiation, and subsequent treatment of the coating with a solvent or solvent system which preferentially removes portions of the coating in accordance with its exposure to light.

Such processes have been employed to prepare lithographic printing plates, stencils, photoresists, and similar photomechanical images. Among the radiation-sensitive materials which have been used in such processes, are light-sensitive polymers which are insolubilized or hardened on exposure to light.

Typical of these light-sensitive polymers are the cinnamic acid esters of hydroxy polymers such as are described in Minsk et al. U.S. Patent 2,690,966, Minsk U.S. Patent 2,725,372, and Robertson et al., U.S. Patent 2,732,301; cinnamylidene malonate polyesters such as are described in Michiels et al. U.S. Patent 2,956,878, and Clement et al. U.S. Patent 3,173,787 and polyester and polycarbonates containing the styryl ketone group such as are described in Borden et al. U.S. Patent 3,453,237.

Since many of these light-sensitive polymers are not alkali resistant, but are hydrolyzed upon contact with strong alkali, it is necessary that resists prepared from these polymers be used only with acidic etchants. In those instances where it is desired that an alkaline etchant be used, other photosensitive compositions must be used, such as those based on azide sensitized rubbers described in Hepher et al. U.S. Patent 2,852,379 and Sagura et al. U.S. Patent 2,940,853 and similar compositions in which a nonlight-sensitive component, such as the rubber is sensitized to light by a light-sensitive monomer, such as the azide. Polysulfonates are known to be alkali resistant and it would be desirable if a light-sensitive polysulfonate were prepared for use in the preparation of alkali resistant photoresists.

Among the polyesters named in the abovementioned Borden et al. U.S. Patent 3,453,237, is one prepared from a sulfur-containing acid such as thionyl chloride. However, preparation of such polymers is uneconomic since thionyl chloride tends to hydrolyze rapidly and hence large excesses of the thionyl chloride reactant must be used.

The light-sensitive polysulfonates of this process contain recurring units derived from an aromatic disulfonyl chloride and recurring units derived from a bisphenol containing the styryl ketone group

$$\left(\text{⬡}-\overset{}{\underset{|}{C}}=\overset{}{\underset{|}{C}}-\overset{O}{\overset{\|}{C}}-\right)$$

These polymers exhibit good light sensitivity and upon exposure to actinic radiation cross-link and harden to give alkali-resistant materials which are useful as photoresists. The polymers can be prepared by reacting an aromatic disulfonyl chloride and a bisphenol reactant, preferably in stoichiometric amounts, under basic conditions in the presence of a suitable catalyst.

Representative aromatic disulfonyl chlorides which are useful in preparing light-sensitive polysulfonates of this process include: benzenedisulfonyl chlorides such as 1,3-benzenedisulfonyl chloride, 1,4-benzenedisulfonyl chloride, 1-chloro-2,4-benzenedisulfonyl chloride, 1-bromo-3,5-benzene-disulfonyl chloride and 1-nitro-3,5-benzenedisulfonyl chloride.

A particularly useful class of aromatic disulfonyl chlorides can be represented by the structural formula:

$$Cl-\overset{O}{\underset{O}{\overset{\|}{\underset{\|}{S}}}} \quad \overset{O}{\underset{O}{\overset{\|}{\underset{\|}{S}}}}-Cl$$

$$R_1 \qquad R_1$$

where each R_1 is independently a hydrogen atom or a lower alkyl group typically having one to four carbon atoms, e.g., methyl, ethyl, propyl, etc.

Light-sensitive bisphenol reactants which can be used to prepare light-sensitive polysulfonates include dihydroxy chalcones and dihydroxy dibenzal ketones such as: 4,4'-dihydroxychalcone, 2,6-bis(3-hydroxybenzal)cyclohexanone, 2,6-bis(4-hydroxybenzal)cyclohexanone, di-m-hydroxybenzal-acetone, divanillalacetone and divanillalcyclohexanone. A particularly useful class of light-sensitive bisphenol reactants have the structural formula as shown on the following page.

where R_2 and R_3 are each hydrogen atoms or lower alkoxy groups typically having one to four carbon atoms, e.g., methoxy, ethoxy, etc., and R_4 and R_5 are each a hydrogen atom or together represent the hydrocarbon radical necessary to complete a saturated ring of five to six carbon atoms. The following examples illustrate the process.

Example 1: In a 1,000-ml. resin kettle equipped with a dropping funnel and a metal stirring rod are placed 11.4 g. (0.05 mol) of bisphenol A, 17.6 g. (0.05 mol) of divanillalcyclopentanone and 200 ml. of 1N sodium hydroxide (0.2 mol). The solution is stirred for 25 minutes and 0.55 g. (1% weight based on the weight of reactants) of benzyltriethyl ammonium chloride is added. The solution is stirred an additional 5 minutes. To this stirred solution is added dropwise over a 10 minute period, a solution of 27.5 g. (0.1 mol) of 1,3-benzenedisulfonyl chloride in 200 ml. of methylene chloride.

The reaction mixture is stirred vigorously for 1 hour and 20 minutes. At the endpoint of the reaction, a yellow color develops and the stirring is terminated. The polymer is isolated by dripping the mixture into hot water with vigorous stirring. The polymer is filtered, washed on the filter with water, then with methanol, and dried in a vacuum oven at 50°C. The polymer is isolated in better than 90% yield (47 grams). The inherent viscosity of 60:40 mixture of cyclohexanone and 4-butyrolactone (0.25 g. polymer per 100 cc solution at 25°C.) is 0.24.

Example 2: In a 1,000 ml. resin kettle equipped with a dropping funnel, a condenser and an efficient stirrer are placed 9.12 g. (0.04 mol) of bis-phenol A, 13.04 g. (0.04 mol) of divanillalacetone and 160 ml. of 1N sodium hydroxide (0.16 mol). The solution is stirred for 15 minutes and 0.44 g. of benzyltriethylammonium chloride is added (1% by weight based on the weight of the reactants). The solution is stirred for an additional 5 minutes.

To this stirred solution, at room temperature, is added dropwise over a 10 minute period 22.0 g. (0.08 mol) of 1,3-benzenedisulfonyl chloride dissolved in 160 ml. of methylene chloride. After 2 hours, the wine-red color characteristic of the disodium salt of the divanillalacetone disappears and the reaction mixture turns yellow in color. The stirring is terminated after 2 hours. The polymer is isolated by dipping the mixture into hot water with vigorous stirring. The resulting mixture is filtered and the polymer washed on the filter with water, then with methanol, and then dried in a vacuum oven at 40°C. for 24 hours. The yellow polymer is isolated in better than 90% yield. The inherent viscosity, measured as in Example 1, is 0.68.

Examples 3 to 17:   Additional polymers are prepared by the procedure described in Example 2 with the exceptions that the reaction times are 3.5 hours, a 5 to 10% excess of sodium hydroxide is used, and the addition of the disulfonyl chlorides is rapid (1 to 2 minutes). The reactants used to prepare these additional polymers are given in Table 1. The bisphenol re-reactants employed are identified as follows: BPA=Bisphenol A, HDVA=1, 5-bis(4-hydroxy-3-methoxyphenyl)pentan-3-one (hydrogenated divanillalacetone), DVA=divanillalacetone and DVCP=divanillalcyclopentanone.

TABLE 1

Example	Bisphenol (nonlight-sensitive)	Mole percent	Bisphenol (light-sensitive)	Mole percent	Aryl disulfonyl chloride	Mole percent
3	BPA	25	DVA	25	1,3-benzenedisulfonyl chloride	50
4	BPA	25	DVA	25	1-methyl-2,4-benzenedisulfonyl chloride	50
5	BPA	25	DVA	25	....do....	50
6	BPA	25	DVA	25	....do....	50
7	BPA	25	DVA	25	1,3-dimethyl-4,6-benzenedisulfonyl chloride	50
8	HDVA	25	DVA	25	1,3-benzenedisulfonyl chloride	50
9	HDVA	25	DVA	25	1-methyl-2,4-benzenedisulfonyl chloride	50
10	HDVA	25	DVA	25	1,3-dimethyl-4,6-benzenedisulfonyl chloride	50
11	BPA	25	DVCP	25	1,3-benzenedisulfonyl chloride	50
12	BPA	25	DVCP	25	1-methyl-2,4-benzenedisulfonyl chloride	50
13	BPA	25	DVCP	25	....do....	50
14	BPA	25	DVCP	25	1,3-dimethyl-4,6-benzenedisulfonyl chloride	50
15	HDVA	25	DVCP	25	1,3-benzenedisulfonyl chloride	50
16	HDVA	25	DVCP	25	1-methyl-2,4-benzenedisulfonyl chloride	50
17	HDVA	25	DVCP	25	1,3-dimethyl-4,6-benzenedisulfonyl chloride	50

Example 18:   Physical properties of the polymers prepared in Examples 1 to 17 are determined and are given in Table 2. All inherent viscosities (η inherent) are determined at 25°C. in a 1:1 by volume solvent mixture of phenol and chlorobenzene using 0.25 g. of polymer per 100 cc of solution, except the copolymers of Examples 1, 2 and 4, for which a 60:40 by volume solvent mixture of cyclohexanone and 4-butyrolactone is used in the same polymer concentration.

The sensitivity values are determined by the procedure of L.M. Minsk et al. "Photosensitive Polymers, I & II" Journal of Applied Polymer Science; Vol. II, No. 6, pp. 302-311 (1959). This is a measure of the relative speed of the polymer, when exposed to ultraviolet or visible light, compared with the speed of unsensitized polyvinyl cinnamate as a standard. Glass transition temperature (Tg) is the temperature at which an amorphous polymer changes from a brittle, glass-like to a rubbery, fluid state. The solvent compositions employed in preparing the 10% polymer solutions are identified as follows: (1) 1,2-dichloroethane:cyclohexanone, 80:20% by volume

(2) 1,1,2-trichloroethane:cyclohexanone, 80:20% by volume (3) 2-methoxyethyl acetate:cyclohexanone, 80:20% by volume (4) cyclohexanone: 4-butyrolactone, 80:20% by volume

Table 2

| Polymer of example | Elemental analysis | | | | | | η inherent | Sensitivity value | Tg, °C. | Solvent for at least 10 g. of polymer/100 cc. solvent |
| | Calcd. | | | Found | | | | | | |
	C	H	S	C	H	S				
1	58.6	4.1	13.0	58.6	4.1	13.0	0.24	1,100	134	4.
2	57.5	4.0	13.4	57.3	4.4	13.7	0.68	2,500	130	4.
3	57.5	4.0	13.4	56.9	3.8	13.0	0.69	2,500	....	4.
4	58.4	4.2	13.0	57.6	4.6	12.9	0.54	2,000	....	1, 2, 3 and 4.
5	58.4	4.2	13.0	57.9	4.3	13.4	1.0	2,500	....	1, 2, 3 and 4.
6	58.4	4.2	13.0	58.8	4.7	12.7	0.57	1,600	141	1, 2, 3 and 4.
7	59.2	4.5	13.0	58.8	4.8	12.6	0.59	1,100	....	1, 2, 3 and 4.
8	57.0	3.9	12.1	54.8	4.1	12.4	0.79	2,500	....	4.
9	57.4	4.1	11.8	57.0	4.4	11.9	0.47	1,000	....	1, 2, 3 and 4.
10	58.1	4.4	11.5	58.0	4.6	11.1	0.56	1,000	....	1, 2, 3 and 4.
11	58.6	4.1	13.0	58.6	4.1	13.0	0.72	7,000	147	4.
12	59.1	4.3	12.6	59.4	4.7	12.6	0.92	7,000	....	1, 2 and 4.
13	59.1	4.3	12.6	59.4	4.3	12.2	0.88	7,000	....	1, 2 and 4.
14	60.0	4.6	12.6	60.2	4.7	12.2	0.96	7,000	....	4.
15	57.4	4.2	11.8	57.8	4.6	11.4	0.78	....	....	4.
16	58.1	4.5	11.5	57.4	4.8	11.8	0.74	3,600	....	1, 2 and 4.
17	58.0	4.7	11.2	59.6	4.7	11.2	0.63	5,000	....	4.

Example 19: Photoresist on Copper — A formulation is prepared from 5 g. of the light-sensitive polysulfonate copolymer of Example 1, 30 cc of cyclohexanone, 20 cc of 4-butyrolactone. The above mixture is warmed to complete solution. The resulting solution is filtered, coated on copper, dried, exposed for 8 minutes with a carbon arc (2,000 foot candles), and developed with a mixture of cyclohexanone, propylene carbonate and methanol. The processed plate is post baked at 120°C. for 15 minutes before etching and then etched for 6 minutes at 150°C. in 42°Bé. ferric chloride. The polymer shows no signs of degradation.

Example 20: Photoresist on Aluminum — Using the formulation and procedure of Example 19, the polymer is coated on aluminum and etched in 20% sodium hydroxide at 75°C. for periods of 10 and 20 minutes. No degradation of the resist is observed.

Example 21: Lithographic Printing Plate — A formulation is prepared from 0.8 g. of the polymer of Example 1, 12 cc of cyclohexanone, 8 cc of 4-butyrolactone. The mixture is heated to aid solution, then coated on an anodized lithographic aluminum plate, exposed for about 8 minutes at 2,000 foot candles, and developed with cyclohexanone. The plate is run on a duplicator press and over 300 impressions are made.

Example 22: Photoresist on Copper — A formulation is prepared from 13.3 g. of copolymer of Example 2, 80 cc cyclohexanone, 26 cc 4-butyrolactone and 1,2-dichloroethane. The above mixture is heated to effect solution, the solution is filtered, whirl coated on a pumiced copper support, and prebaked at 80°C. for 10 minutes. The resulting photoresist element is exposed imagewise for 5 minutes to a 95-ampere carbon arc at a distance

of 4 feet and is then tray developed in a 50:50 mixture of cyclohexanone
and 1,2-dichloroethane. The polymer in the unexposed areas is removed.
The resulting element is rinsed with acetone and post baked for 10 minutes
at 160°C. The plate is then etched for 28 minutes in 42°Bé. ferric chloride
at 150°C. The polymer shows no sign of degradation.

Substitued Polybismethyloxetane

K. Azami, H. Ohotani and H. Fukutomi; U.S. Patent 3,694,383;
September 26, 1972; assigned to Dainippon Ink and Chemicals, Incorporated,
Japan describe a process for the production of a light-sensitive polymer
which comprises substituting a light-sensitive group for the halogen atom
of a polybishalomethyloxetane.

The term "light-sensitive group," as used here, is meant to be a group which
can form a reticular structure by the setting up of a chemical reaction by
irradiation with actinic rays. For example, mention can be made of the
cinnamic acid ester group

$$-O-\underset{\underset{O}{\|}}{C}-CH-CH-\hspace{-4pt}\bigcirc$$

N,N-disubstituted dithiocarbamic acid ester group

$$-S-\underset{\underset{S}{\|}}{C}-N\underset{R'}{\overset{R}{<}}$$

azido group $-N_3$, azidobenzoic acid ester group

$$-O-\underset{\underset{O}{\|}}{C}-\hspace{-4pt}\bigcirc\hspace{-4pt}N_3$$

azidocinnamic acid ester group

$$-O-\underset{\underset{O}{\|}}{C}-CH=CH-\hspace{-4pt}\bigcirc\hspace{-4pt}N_3$$

azidobenzenesulfonic acid ester group

$$-O_3S-\hspace{-4pt}\bigcirc\hspace{-4pt}N_3$$

furan acrylic acid ester group

$$-O-\underset{\underset{O}{\|}}{C}-CH=CH-\text{(furan ring)}$$

and their derivatives. Compounds containing such light sensitive groups as mentioned are used in the substitution reaction. Included as examples of such compounds are compounds such as alkali metals, e.g., sodium or potassium, salts of cinnamic acid, alkali metals of N,N-disubstituted dithiocarbamic acid, alkali metal azides, alkali metal salts of azidobenzoic acid, alkali metal salts of azidocinnamic acid, alkali metal salts of azidobenzenesulfonic acid and alkali metal salts of furan acrylic acid.

When a light-sensitive coating composition such as described is applied to a suitable support, e.g., a metallic plate, and the organic solvent is evaporated by air-drying, heating, etc., a so-called presensitized light-sensitive coating is obtained. When this light-sensitive coating is exposed to actinic rays through either a negative film or a pattern, reaction of the light-sensitive groups takes place in the exposed areas to form a reticular structure, which becomes insoluble in solvents and has a strong resistance to acids, alkalis, solvents and the like.

On the other hand, the unexposed areas can be removed from the surface of the support by means of a solvent which can dissolve this unexposed coating or an emulsion containing such a solvent. Thus, a distinct image which is in accordance with the negative film or pattern is developed. After the development, the support can, of course, be re-exposed to actinic rays to accelerate the formation of the reticular structure. The following examples are given to illustrate the process.

Example 1: A 3-necked, 100-ml. flask equipped with an agitator, a thermometer and a reflux condenser was charged with a solution of 3.1 g. of polybischloromethyloxetane in 70 ml. of N,N-dimethylformamide, after which 7.0 g. of potassium cinnamate was added and the reaction solution was heated under reflux for 15 hours. After completion of the reaction, the reaction mixture was slowly poured into a large quantity of water which was being stirred with a homomixer to precipitate the resulting light-sensitive polymer into the water.

The precipitate was separated by means of a centrifuge, washed 2 or 3 times in methanol, and dried for 24 hours at room temperature and reduced pressure obtained by means of a vacuum pump, whereupon was obtained a powdery light-sensitive polymer in an amount of 3.5 g. When the light-sensitive polymer was analyzed for its chlorine content, it was found that 25.4%

of the chlorine atoms contained in the polybischloromethyloxetane was
substituted. Further, when an infrared absorption spectrum analysis was
carried out on the light-sensitive polymer, as characteristic absorptions of
the cinnamic acid ester group, there was an absorption due to $>C=O$ at
$1,720$ cm.$^{-1}$ and an absorption due to $>C=C<$ at $1,640$ cm.$^{-1}$. And at
$1,580$ cm.$^{-1}$ there was an absorption due to the benzene nucleus. Thus,
it was confirmed that the cinnamic ester group had been introduced into
the light-sensitive polymer.

One gram of the so obtained light-sensitive polymer was dissolved along
with 0.1 g. of 5-nitroacenaphthene in 100 ml. of methyl ethyl ketone,
after which the resulting solution was charged to a 200 ml. flash and stirred
at room temperature to prepare a light-sensitive coating composition. Next,
this light-sensitive coating composition was applied by means of the dipping
technique to the surface of a ball grained aluminum plate, after which the
solvent was removed with a drier to form a light-sensitive coating.

The Photographic Step Tablet (21 steps), a product of Eastman Kodak Co.,
was placed in contact atop this light-sensitive coating and an exposure was
made with ultraviolet rays under the conditions of an intensity of 62,000
μw./cm.^{2} and a time of 30 seconds, using a 250 w. high pressure mercury
vapor lamp. This was followed by development of the unexposed areas and
dyeing of exposed areas to cause the image to stand out. As a result, it was
found that the light-sensitive coating had an excellent sensitivity.

Example 2:   A 3-necked, 100-ml. flask equipped with a thermometer, an
agitator and a reflux condenser was charged with a solution of 3.1 g. of
polybischloromethyloxetane in 70 ml. of dimethylformamide and, after
adding 2.6 g. sodium azide, the reaction mixture was heated under reflux
for 6 hours. After completion of the reaction, the reaction mixture was
poured into methanol to precipitate a light-sensitive polymer, which was
separated by filtration to isolate a light-sensitive polymer having a rubbery
elasticity.

When this light-sensitive polymer was analyzed for its chlorine content, it
was found that 59.5% of the chlorine atoms contained in the polybischloromethyloxetane had been substituted. Further, when this light-sensitive polymer was analyzed as to infrared absorption spectrum, there was as a characteristic absorption of the azido group a pronounced absorption at $2,120$ cm.$^{-1}$
and absorptions due to $-C-N$ at $1,450$ cm.$^{-1}$ and $1,295$ cm.$^{-1}$. Thus it
was shown that the azido group had been introduced into the light sensitive
polymer.

Next, 3 g. of this light-sensitive polymer were dissolved along with 0.3 g.
of 5-nitroacenaphthene in 100 ml. of toluene, after which the resulting

solution was charged to a 150-ml. flash and stirred at room temperature to prepare a light-sensitive coating composition. The so obtained light-sensitive coating composition was then applied to a ball grained zinc plate to form a light-sensitive film thereon. This was followed by placing a gray scale atop the film in contact therewith and giving the film an exposure of ultraviolet rays under the conditions of an intensity of 62,220 uw./cm.^{2} and a time of 35 seconds. After completion of the exposure, development was carried out with a toluene-methyl ethyl ketone solvent mixture, whereupon the unexposed areas of the coating were removed. It was found that the light-sensitive coating had an excellent sensitivity.

Sensitized Polyesters

In a process described by J.W. Mench, B. Fulkerson and W.J. Dulmage; U.S. Patent 3,615,628; October 26, 1971; assigned to Eastman Kodak Company sensitized polyesters containing unsaturated alicyclic rings are coated upon a support material to provide negative-working photographic elements useful in the photomechanical arts for preparing not only lithographic and relief printing plates but also resist stencils for etching and other operations. The following examples illustrate the process.

Example 1: Polyvinyl alcohol (8.8 g.) is dissolved in dimethyl-formamide (100 ml.) with stirring and heating in an oil bath. The polymer dissolves at 140° to 145°C. At that temperature a solution of cis-4-cyclohexene-1, 2-dicarboxylic anhydride (30.4 g.) in dimethylformamide (60 ml.), which has previously been heated to about 90°C., is added and heating and stirring are continued for half an hour at a 140° to 150°C. solution temperature. The solution is then allowed to cool, and the polymer is precipitated in diethyl ether (2 l.), washed with two further portions of ether and dried under vacuum. In this way, 25 g. of soft gummy polymer is obtained which is soluble both in ethanol and in dilute ammonia.

Example 2: Utilizing the method of Example 1, polyvinyl alcohol (8.8 g.) is esterified with cis-4-cyclohexene-1,2-dicarboxylic anhydride (15.2 g.). The polymer (17 g.) is soluble both in a 50:50 v/v mixture of ethanol and water and in dilute ammonia.

Example 3: Polyvinyl alcohol (50 g.) is suspended for 16 hours at about 25°C. in a mixture of acetic acid (250 g.) and cis-4-cyclohexene-1,2-dicarboxylic anhydride (200 g.). After this time period, anhydrous potassium acetate (100 g.) is added and the temperature of the reaction mixture is raised to 90° to 95°C. A clear solution results in 20 minutes, and at the end of 1 hour, the solution is diluted with acetic acid and the polyvinyl cis-4-cyclohexene-1,2-dicarboxylate product is isolated by precipitation into a large quantity of acidified water. The resulting tacky precipitate

is then ground in a water slurry and washed with distilled water until free
of acidity. The resultant product is dried at about 25°C. under vacuum.

Example 4: Utilizing the method of Example 1, polyvinyl alcohol (8.8 g.)
is esterified with 5-norbornene-2,3-dicarboxylic anhydride (32.8 g.). The
polymer (14 g.) is soluble both in a mixture of cyclohexanone and ethanol
and in dilute ammonia.

Example 5: Utilizing the method of Example 1, polyvinyl alcohol (8.8 g.)
is esterified with 5-norbornene-2,3-dicarboxylic anhydride (16.4 g.). The
resulting polymer is soluble in a mixture of cyclohexanone and ethanol.

Example 6: 50 grams of cotton linters are heated at 90° to 95°C. for 20
minutes in a mixture of 2.5 g. of sulfoacetic acid, 12.5 g. of acetic anhydride and 250 g. of acetic acid. To the resulting slurry is added 200 g. of
cis-4-cyclohexene-1,2-dicarboxylic anhydride(tetrahydrophthalic anhydride)
and 125 g. of anhydrous potassium acetate. After approximately 6 hours
reaction at 90° to 95°C., a clear solution results. This is held for an additional 16 hours at 90° to 95°C., then is diluted with acetic acid and the
ester isolated by precipitation into water which has been acidified with
hydrochloric acid.

The product is washed with distilled water until acid free, then dried at
room temperature under vacuum. Alternatively, it is possible to dry the
product at elevated temperature without causing insolubility by addition of
sufficient stabilizer, such as hydroquinone, to the last wash water so that
approximately 0.5% is retained by the ester.

The polymers, such as those described in Examples 1 to 6, are rendered
advantageously light sensitive by the addition of sensitizers. Particular
sensitizers are employed in the remaining examples, which describe the
negative-working photographic elements. These sensitizers are referred to
by number, the subsequent references being as follows: Sensitizer 1 is
3% solution of 2,6-di(4-azido-benzylidene)-4-methylcyclohexanone in
cyclohexanone. Sensitizer 2 is 4% of 4,4'-diazidostilbene-2,2'-disulfonic
acid in 1% aqueous ammonia.

Example 7: A sheet of anodized aluminum is whirler coated at 200 rpm
with the following solution: resin (as in Example 1, 97 g./l. in ethanol),
1.0 ml.; sensitizer 1, 3.2 ml.; ethanol, 20.0 ml. The dry layer is
exposed for 1 minute through a negative line transparency to four 125 w.
high-pressure mercury vapor lamps, rich in ultraviolet light, placed 18"
from the exposure plane. Exposure causes the sensitized polymer to harden
in the areas of exposure. The exposed plate is developed by swabbing with
a 10% aqueous solution of trisodium phosphate, which causes the polymer

in the unexposed areas to dissolve away. The developed plate is then rinsed with water, swabbed with a mixture of gum arabic (2.5%) and phosphoric acid (5%) and linked to give a lithographic plate. One thousand impressions are typically obtained on a standard rotary lithographic printing machine with no loss of image quality.

Example 8: A lithographic plate is prepared by coating a sheet of anodized aluminum with a solution according to the method of Example 7 as follows: resin (as in Example 1, 62 g./l. in aqueous ammonia), 4.0 ml.; sensitizer 2, 6.2 ml. The layer is exposed as in Example 7 and developed by swabbing with an 0.5% aqueous solution of trisodium phosphate. Development causes the polymer in the unexposed areas to dissolve away. The plate produces clear ink impressions on a rotary lithographic printing machine.

Example 9: A sheet of flexible poly(ethyleneterephthalate) 0.003 inch copper laminate is whirl coated at 200 rpm with the following solution: resin [as in Example 2, 100 g./l. in alcohol and water (1:1)], 15 ml.; sensitizer 2, 5 ml. The dry layer is exposed for 2 minutes to the light source of Example 7. The exposed sheet is then developed by immersion in a 10% aqueous trisodium phosphate solution, followed by spraying with ethyl alcohol, thereby causing the polymer in the unexposed areas to be dissolved away. The developed element produces clear impressions on a standard rotary lithographic printing machine.

Example 10: A lithographic plate is prepared by swab coating a sheet of anodized aluminum with a solution having the following composition: resin (as in Example 5, 64 g./l. in cyclohexanone), 1.0 ml.; sensitizer 1, 2.0 ml.; ethanol/cyclohexanone mixture (1:1), 10.0 ml. The layer is exposed as described in Example 9 for 10 minutes and developed by swabbing with ethanol, which causes the polymer in the unexposed areas to be dissolved away. The developed plate produces high quality impressions on a standard rotary lithographic printing machine.

Unsaturated Monomers Containing Ionic Sites – Zwitterions

In a process described by L.D. Taylor; U.S. Patent 3,578,458; May 11, 1971; assigned to Polaroid Corporation unsaturated monomers containing ionic sites are photopolymerized and, further, ionically cross-linked through these sites. Examples of unsaturated monomers containing ionic sites are zwitterions.

The system which is the subject of the process differs primarily from other photo-resist systems in that the relief image comprises an ionically bonded polymeric substance. It is based principally on two facts: firstly, that a mixture of soluble ionic compounds having opposite ionic charges, each of which has more than one charged site, or a soluble compound with a multiplicity of opposite charges, will cross-link ionically, thereby forming insoluble adducts; and secondly, that certain soluble monomers can be photo-polymerized to form polyionic compounds. If these two premises are treated as a system and a soluble, mono-charged ionic monomer is photo-polymerized in the presence of a soluble polyionic compound of opposite charge, or if a soluble monomer with ionic sites of opposite charge is photo-polymerized, an insoluble complex held together by ionic bonds will be obtained.

For the purposes of this process, a polyionic compound is considered to include any polymer with a multiplicity of ionic sites. In the preferred system it has been found to be beneficial to add a photo-polymerization catalyst, for example, one which is capable of producing polymerization-enhancing free radicals easily upon exposure to actinic radiation.

Ionic monomers containing opposite charges are, for example, zwitterions, such as 4-vinylpyridine-N-butylsulfobetaine,

$$CH{=}CH_2$$

$$\overset{\oplus}{N}$$
$$(CH_2)_4SO_3^{\ominus}$$

The ionic character of the photopolymerizable monomer and resultant polymer in solution may be provided by the ionization of polymeric salts containing anionic or cationic substituents. Examples of suitable anions which may be pendent on the monomer are: sulfonate, carboxylate, etc. Examples of suitable cations which may be pendent are: quaternary nitrogen, sulfonium salts, phosphonium salts, etc. If the ionic monomer is photo-polymerized in the presence of an oppositely charged polyion to form the resist system the latter polyion must contain suitable ionic groups, as, for example, those mentioned above.

Example 1:   A solution having the following constituents was mixed: polystyrene sulfonic acid sodium salt, 0.2 g.; 2-hydroxy-3-methacrylyl-oxypropyltrimethylammonium chloride (G-Mac Methacrylate, Shell Chemical Co.), 0.5 g.; gelatin, 0.25 g.; rose bengale, 0.0002 g.; allyl thiourea, 0.0166 g.; and water, 3 cc. The above solution was coated onto a microslide and dried at room temperature. The coated slide was then exposed through a stencil utilizing a 500 w. projection bulb for 4 minutes. It was then washed with water and allowed to dry. The resulting relief image was of very good quality and exhibited excellent dyeability.

Example 2: The procedure of Example 1 was followed using the following solution, except that the exposure time was 10 minutes: 2-hydroxy-3-methacrylyloxypropyltrimethyl ammonium chloride (G-Mac Methacrylate, Shell Chemical Company), 2.5 g.; sodium polyacrylate (Alcogum An-25, Alco Chemical Corporation), 1.9 g.; methylene blue, 0.0001 g.; and water, 6.5 cc. The relief image was of good quality and was easily dyeable.

Example 3: The procedure of Example 1 was carried out using the following solution except that the plate was oven-dried before exposure: sodium polyacrylate (Alcogum Pa-15-A, Alco Chemical Corp.), 6.66 g.; 2-hydroxy-3-methacrylyloxypropyltrimethylammonium chloride (G-Mac Methacrylate, Shell Chemical Company), 2.5 g.; and water, 10 cc. The resulting relief image was found to be fair with rather good receptivity to dye.

Metal Diacrylates

J.B. Rust; U.S. Patent 3,615,627; October 26, 1971; assigned to Hughes Aircraft Company describes a method of preparing improved compositions of photopolymerizable polyvalent metal salts of acrylic and methacrylic acids. The phenomenon of forming photographic images characterized by the formation of polymer particles due to photo induced polymerization of polymerizable compounds has been known in the art for many years.

However, there has been the continuing problem of forming improved film compositions and enhancing photographic speed. Acrylic acid material has appeared most suitable. However, many of the polyvalent metal salts of acrylic and methacrylic acids have a limited solubility in water at room temperature and therefore are apt to crystallize before the desired photopolymerization can be effected.

Consequently, in high concentrations, difficulty is often encountered from precrystallization before photopolymerization occurs. Thus, the preparation of desired and suitable high-content acrylate photographic films, in the wet or cast state, without normal or condensation crystallization therein has been a problem. That is, with or without the removal of water, the highly concentrated metal acrylates have tendency to crystallize, or otherwise sometimes become unstable after preparation and standing before use.

In general, the polyvalent metal salts of acrylic and methacrylic acid and derived from metal oxides, metal hydroxides, metal carbonates, and mixtures of the same, have only a limited solubility in water, and when the solutions are concentrated beyond a certain point, crystallization occurs. It has been found that barium diacrylate possesses a fair solubility in water, for example, at room temperature, thus providing solutions of higher solids content. While such solutions permitted the development of photograph-ically more sensitive photosensitive films, the more highly concentrated solutions, for commercial production, are found to be unstable, sometimes spontaneously crystallizing after preparation and compounding.

With the improvement as provided here, more consistently stable composi-tions using barium diacrylate as the sole polyvalent metal component in highly concentrated metal acrylate–photoredox photosensitive compositions are now possible. Other of the metal acrylates are of lesser or more limited solubility and the more particular purpose here is to provide noncrystallizing film compositions in high and oversaturated concentrations, without any sign of crystallization or precipitation.

For example, lead diacrylate normally has only limited solubility in water as has also cadmium diacrylate, neodymium triacrylate and strontium diacry-late. It has been found, however, that when added to barium diacrylate solution, the mixed polyvalent metal acrylate solution which results can be made in high concentration without any sign of crystallization or precipita-tion. Furthermore, the same and other mixed polyvalent metal acrylate solutions, as described above, can be cast into films and dried to amorphous, noncrystalline, glass-like films. Aslo, when dried to a water-free condi-tion in bulk, the mixed polyvalent metal acrylate produces a glassy, amor-phous noncrystalline solid. The following examples illustrate the process.

Example 1: To about 100 ml. of acrylic acid, there was added approxi-mately 1 g. of activated carbon known to the trade as Nuchar. The carbon was thoroughly suspended in the acrylic acid and allowed to stand at room temperature with stirring for several days. The acrylic acid was then filtered to remove the carbon.

157.5 g. of barium hydroxide octahydrate was ground to a fine powder in a mortar and partially dissolved and suspended in 135 ml. of distilled water in a 500 ml. round bottom flask. 72 ml. of the filtered acrylic acid were added through a dropping funnel over a period of about 3/4 hour. After the addition of the acrylic acid, the nearly clear solution was allowed to stir for an additional 30 minutes. The slightly hazy solution was centrifuged yielding a clear, limpid solution of barium diacrylate having a concentra-tion of 37.8% by weight.

Example 2: About 100 ml. of acrylic acid were mixed with approximately 1 g. of activated carbon known to the trade as Nuchar. The mixture was stirred at room temperature for about 6 days, then the carbon removed by centrifugation. 157.5 g. of barium hydroxide octahydrate was partially dissolved and suspended in 157.5 ml. of distilled water and 72 ml. of centrifuged acrylic acid added dropwise over a period of 30 minutes. The hazy solution was stirred and heated at 60° to 80°C. for 3 hours, then divided into two equal parts. One part was filtered hot through a bed of filteraid known to the trade as Celite to yield a clear, limpid noncrystallizing solution of barium diacrylate having a concentration of about 35.7% by weight. This solution is designated as solution A.

The other part of the original solution was mixed with 4 gram of activated carbon known to the trade as Nuchar and heated with stirring for an additional thirty minutes at 60° to 80°C. The treated solution was then filtered hot through a bed of filteraid known to the trade as Celite yielding a clear, limpid noncrystallizing solution of barium diacrylate having a concentration of about 35.7 percent by weight.

Example 3:   141.5 g. of barium hydroxide octahydrate were ground in a mortar to a fine powder and mixed with 11.1 g. of lead oxide (litharge, PbO). The mixture was partially dissolved and suspended in 157.5 ml. of distilled water and 72 ml. of acrylic acid were added dropwise with stirring over a period of 30 minutes. After all the acrylic acid had been added, the solution was stirred and heated at 60° to 80°C. for a period of 2 hours. At this point, 8 g. of activated carbon known to the trade as Nuchar was added and stirring and heating continued for an additional 20 minutes.

The solution containing the activated carbon was filtered through a bed of filteraid known to the trade as Celite yielding a clear, limpid, noncrystallizing solution of lead diacrylate and barium diacrylate having a concentration of approximately 37% by weight. Of the diacrylates present, the barium diacrylate is approximately 90 mol percent and the lead diacrylate is about 10 mol percent. A portion of this solution was distilled under vacuum at about 35°C. to remove water. The resulting concentrated solution was noncrystallizing clear, and had a viscosity that was somewhat higher than the original solution. The barium diacrylate–lead diacrylate mixture had the same mol percent composition, but the final solution had a concentration of approximately 71.5% of solids by weight.

Example 4: 94.7 g. of powdered barium hydroxide octahydrate was mixed with 44.7 g. of lead oxide (litharge, PbO). The mixture was partially dissolved and suspended in 153.7 ml. of distilled water and 72 ml. of distilled acrylic acid added dropwise with stirring over a period of 30 minutes. After all the acrylic acid had been added, the solution was stirred and heated at 60° to 80°C. for 2 hours. At this point, 8 g. of activated carbon known to the trade as Nuchar was added and stirring and heating continued for an additional 20 minutes. The solution containing the activated carbon was filtered through a bed of filteraid known to the trade as Celite.

A clear, limpid solution was obtained while hot. The acrylate mixture contained 60 mol percent of barium diacrylate and 40 mol percent of lead diacrylate. When cooled to room temperature, the solution crystallized to a white pasty mass. From an inspection of the form of the crystals, it appeared that the lead diacrylate had been the major component to crystallize out of solution.

Example 5: Using the barium diacrylated-lead diacrylate solution of Example 3 consisting of 90 mol percent of barium diacrylate, and 10 mol percent of lead diacrylate and the barium diacrylate-lead diacrylate solution of Example 4 consisting of 60 mol percent of barium diacrylate and 40 mol percent of lead diacrylate, a series of solutions were prepared by mixing the two above solutions to give a new solution containing different mol percentages of barium and lead diacrylates respectively.

Each of the new solutions was stored and inspected at intervals and dried films of each of the new solutions were prepared by evaporation at room temperature. Table 1 gives the results obtained from the series of new solutions. Generally, it appears that noncrystallizing solutions can be prepared from the mixed diacrylates up to a mixture containing less than 40 mol percent of lead diacrylate. On the other hand, clear, amorphous, hard, glasslike dry films can only result from mixtures containing from 20 mol percent to 30 mol percent of lead diacrylate and 80 mol percent to 70 mol percent of barium diacrylate inclusive. These limits might be extended slightly by using an expanded series of the same and different metal acrylate mixtures.

TABLE 1

Mole percent			
Barium diacrylate	Lead diacrylate	Solution characteristics	Dry film characteristics
100	0	Non-crystalline to about 60% concentration.	Crystalline.
90	10	Non-crystalline to about 75% concentration.	Do.
85	15	....do...................	Do.
80	20	Non-crystalline at all concentrations.	Non-crystalline, glass-like.
75	25	....do...................	Do.
70	30	....do...................	Do.
65	35	Non-crystalline to about 45% concentration.	Crystalline.
60	40	Crystallizes at about 35% concentration.	Do.

Example 6: 126.2 g. of powdered barium hydroxide octahydrate was mixed with 12.9 g. of cadmium oxide powder. The mixture was partially dissolved and suspended in 146.7 ml. of distilled water and 72 ml. of acrylic acid added dropwise with stirring over a period of 30 minutes. After all the acrylic acid had been added, the solution was stirred and heated at 60° to 80°C. for 2 1/2 hours. At this point, 8 g. of activated carbon known to the trade as Nuchar was added and stirring and heating continued for an

additional 30 minutes. The solution containing the activated carbon was filtered hot through a bed of filteraid known to the trade as Celite yielding a clear, limpid, noncrystallizing solution of barium diacrylate and cadmium diacrylates having a concentration of approximately 38.4% by weight. Of the diacrylates present, the barium diacrylate is approximately 80 mol percent and the cadmium diacrylate is about 20 mol percent. The final solution had a pH of about 8.

When this solution was dried at room temperature, a clear, hard, transparent glass-like, noncrystalline film was formed. In order to determine the photographic properties of the above solution, a photoredox catalyst was prepared from 2.14 g. sodium p-toluenesulfonate dihydrate, 0.03 g. methylene blue and 100 ml. distilled water. A photosensitive composition was prepared in the dark by mixing 4 ml. barium-cadmium diacrylate solution above and 1 ml. photoredox catalyst solution above.

This composition was placed between two glass plates separated by a peripheral shim spacer 7 ml. thick to form a uniform film 0.18 mm. thick. This film was exposed to a spot of light having an intensity at the film surface of 10^{-2} watts/cm^2. The intensity of the light emerging from the back surface of the film was monitored with a photomultiplier tube and the electrical signal recorded on a strip chart recorder. From this data, the optical density of the photosensitive composition at the illuminated spot was calcualted as a function of exposure time. The results are given in Table 2.

TABLE 2

| | Induction period | \multicolumn{6}{c}{Optical density} |
		0.1	0.2	0.4	0.6	0.8	1.0
Time (seconds).........		5.2	6.1	6.8	7.7	8.6	9.5	13.0

Similar and comparable results were obtained with the photosensitive composition containing the photosensitive catalyst and fixing material in this film form.

β-Substituted α-Cyanacrylic Esters

A. Moschel, W. Luders and H. Steppan; U.S. Patent 3,699,086; October 17, 1972; assigned to Kalle AG, Germany describe a process for the manufacture of esters of β-substituted α-cyanoacrylic acids which are obtained by the conversion of polyhydric alcohols, with β-substituted α-cyanacrylic acid chlorides, e.g., benzylidene-α cyanacetic acid chloride, with the exclusion of short wave-length light. These esters are cross-linkable on exposure to radiation. The esters are used for the manufacture of photo-cross-linkable layers, coatings and shaped articles.

One method for making these photo-cross-linkable esters consists of first dissolving or swelling the polymer containing hydroxyl groups in a 10 to 40 fold quantity of pyridine, adding a strongly nucleophilic amine, e.g., 1,4-diazabicyclo-[2.2.2]-octane at a concentration 0.1 to 0.5% by weight relative to pyridine, and adding an acid chloride, e.g., benzylidene-α-cyanacetic acid chloride either in portions in the pure state or dissolved in a solvent which is inert but miscible with water and pyridine, e.g., dioxane, in a temperature range between 45° and 60°C.

After completion of the reaction, the pyridinium chloride which has separated out is filtered off and the filtrate is allowed to run into the 10 to 20 fold quantity of water while stirring, in the course of which the compound either separates out directly in a filtrable form or can be obtained by extraction of the aqueous solution or dispersion.

The reaction products thus obtained are largely purified from impurities, by-products and unreacted compounds containing hydroxyl groups by washing with water. Drying is effected under mild conditions under reduced pressure in the presence of drying agents such as phosphorus pentoxide.

The quantity ratio of the acid chloride and the polyfunctional hydroxyl compound may lie within wide limits determined by the requisite light sensitivity of the products and should be 40.0 to 98.0 mol percent of acid chloride relative to the hydroxyl groups present.

The process must be carried out with exclusion of short wavelength light, i.e., to avoid premature cross-linking, in the absence of light in the wavelength range of 2000 to 6000 A. These compounds may be used singly or mixed with inert or cross-linkable fillers such as vinyl polymers and copolymers in a dissolved form for the manufacture of photo-cross-linkable layers, coatings and shaped articles. Layers of such high molecular weight products become selectively insoluble at the exposed sites after exposure to radiation using an original, while the unexposed sites remain soluble and can easily be removed by a suitable solvent.

The reactions in the following examples were carried out in rooms lighted only by yellow light. The cross-linking was effected by exposing a 0.02 mm. thick film to the radiation of a xenon impulse lamp (5 kilowatts) at a distance of 1 m. for 5 to 15 minutes.

Example 1:   5.65 g. of a dried polyvinyl alcohol, a 4% aqueous solution of which had a viscosity of 4 cp./20°C. and a degree of saponification of 88%, were swollen overnight at 100°C. in 50 ml. of anhydrous pyridine and then diluted with 50 ml. more of pyridine. The mixture was cooled to 50°C. 0.250 g. of diazabicyclooctane dissolved in 5 ml. of pyridine were

added and 10.2 g. of benzylidene cyanacetic acid chloride were added in
portions at this temperature. After stirring for 8 hours, the viscous mass
was diluted with acetone to twice its volume, the pyridinium chloride was
filtered off, and the filtrate was run into a 10-fold quantity of water, while
stirring. The product was filtered off with suction, washed with water and
dried at 30°C. under reduced pressure over P_2O_5. 13 g. of a fibrous product were obtained which in the IR spectrum showed strong bands at 1,710
(ester), 1,600 (aromatic structure) and 2,220 cm.$^{-1}$ (nitrile) and in the
UV spectrum showed 302mu max. with an extinction of $E^{1\%}_{1cm.}$=537 (dioxane).

The product was entirely soluble in dioxane. A coating made on an electrolytically roughened aluminum sheet yielded a film which became insoluble in dioxane after an irradiation of 15 minutes.

Example 2: 1.41 g. of a dried polyvinyl alcohol, a 4% aq. solution of which
showed a viscosity of 4 cp./20°C. and a degree of saponification of 88%,
were dissolved in 50 ml. of dimethylformamide at about 100°C. 50 mg. of
diazabicyclooctane and 3.0 g. of triethylamine were added. 5.4 g. of
cinnamylidene cyanacetic acid chloride dissolved in 30 ml. of benzene
was added dropwise at 80°C. The mixture was stirred 6 more hrs. at 80°C.

The benzene was removed under reduced pressure, the product was filtered
off with suction from the triethylammoniumchloride and the filtrate was run
into 1 l. water, with agitation which continued for 1 hr. The product was
filtered off with suction, rinsed with 1 l. water and dried at 30°C. over
P_2O_5 under reduced pressure. 4.5 g. of a brown amorphous product, soluble in warm acetone and dioxane and in dimethyl formamide, was obtained.
The UV spectrum showed a 343mu max.; extinction of $E^{1\%}_{1cm.}$=1,330 (dioxane).
The product was dissolved in dioxane; a film was produced which, after
brief exposure to radiation (10 minutes), became insoluble in dioxane.

Example 3: 2.82 g. of a dried polyvinyl alcohol of which a 4% strength
by weight aqueous solution had a viscosity of 7.0 cp./20°C., and which
had a degree of saponification of 88%, were swollen in 50 ml. of pyridine
overnight at 100°C., 50 ml. of pyridine and 100 mg. of diazabicyclooctane were added thereto and, at 50° to 55°C., 10.8 g. of cinnamylidene
cyanacetic acid chloride were introduced in portions.

The mixture was stirred at this temperature for 8 hours, allowed to stand
overnight. 50 ml. of acetone were added thereto, and the resulting mixture was filtered. The filtrate was allowed to run into 3 l. of water while
stirring, the mixture was stirred for a further hour, and the product was
filtered off with suction and thoroughly rinsed with 3 l. of water. The product was made into a paste with 100 ml. of methanol and the product was
filtered off with suction. After drying under reduced pressure (2 hours at

50°C.) and over P_2O_5 (overnight at 25°C.), 7.5 g. of fibrous yellow-white product were obtained which in the UV-spectrum showed a maximum at 343 mu with an extinction of E $^{1\%}$ 1 $_{cm.}$ = 1,076 (dioxane). A film produced with this product became insoluble in dioxane after brief exposure to radiation (10 minutes).

POLYMERIC COATINGS

Dinorbornene Ladder Polymers

In a process described by H.O. Colomb, Jr., D.J. Trecker and T.K. Brotherton; U.S. Patent 3,658,669; April 25, 1972; assigned to Union Carbide Corporation dinorbornene compounds and dinorbornene polymers with other polymers such as the polyolefins, vinyl polymers, acrylic polymers, polyesters, polyamides, polyethers, polyureas, polyurethanes, natural polymers, etc., are readily cross-linked by irradiation. Many dinorbornenyl compounds have been described in the prior art.

It has been found that the dinorbornene compounds can be polymerized to form so-called "ladder" polymers. It has also been found that the dinorbornene compounds can be blended with other polymers to obtain readily cross-linkable compositions, and that the compositions can be cured by irradiation.

Cross-linking of the blends is induced by radiation. Two types of radiation are suitable, ionizing radiation, either particulate or nonparticulate, and nonionizing radiation. As a suitable source of particulate radiation, one can use any source which emits electrons or charged nuclei. Particle radiation can be generated from electron accelerators such as the Van de Graaff, resonance transformers, linear accelerators, insulating core transformers, radioactive elements such as cobalt 60, strontium 90, etc.

As a suitable source of nonparticle ionizing radiation, one can use any source which emits radiation in the range of from 10^{-3} to 2000 A., preferably from 5×10^{-3} to 1 A. Suitable sources are vacuum ultraviolet lamps, such as xenon or krypton arcs, and radioactive elements such as cesium-137, strontium-90, and cobalt-60. The nuclear reactors are also known to be a useful source of radiation. As a suitable source of nonionizing radiation, one can use any source which emits radiation of from 2000 to 8000 A., preferably from 2500 to 4500 A. Suitable sources are mercury arcs, carbon arcs, tungsten filament lamps, xenon arcs, krypton arcs, sunlamps, lasers, and the like.

Example 1:   A mixture was prepared containing 14.9 grams of

5-isocyanatomethylbicyclo[2.2.1]hept-2-ene, 100 grams of a polycaprolactone diol having an average molecular weight of about 2,000 (produced according to U.S. Patent 3,169,945) and one drop of dibutyltin dilaurate. The mixture was heated under a nitrogen atmosphere at about 90°C. for about 5 hours and then allowed to cool. The carbamate produced was a waxy solid having a reduced viscosity at 30°C. of 0.13 when measured from a 0.5% solution in chloroform. The structure of the carbamate was confirmed by infrared analysis as:

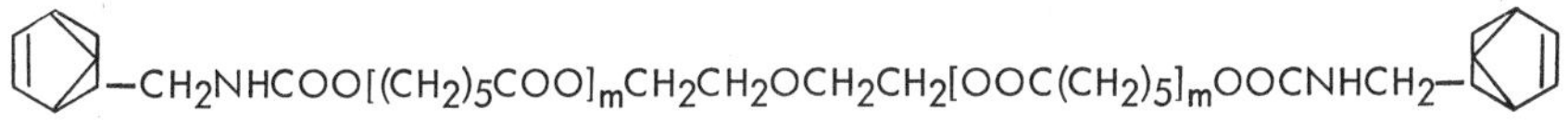

where the total number of m groups present is an average of about 16.6. Microanalytical data further confirmed that the carbamate was produced. Calculated for $C_{121.6}H_{198}O_{38.2}N_2$: C, 63.6; H, 8.7; N, 1.2. Found: C, 63.8; H, 8.7; N, 1.3.

Example 2: A mixture was prepared containing 13.2 g. of 5-isocyanatomethylbicyclo[2.2.1]hept-2-ene, 17.4 g. of polyethylene glycol having an average molecular weight of about 400 and one drop of dibutyltin dilaurate. This was stirred at room temperature for 48 hours to produce the carbamate of the formula:

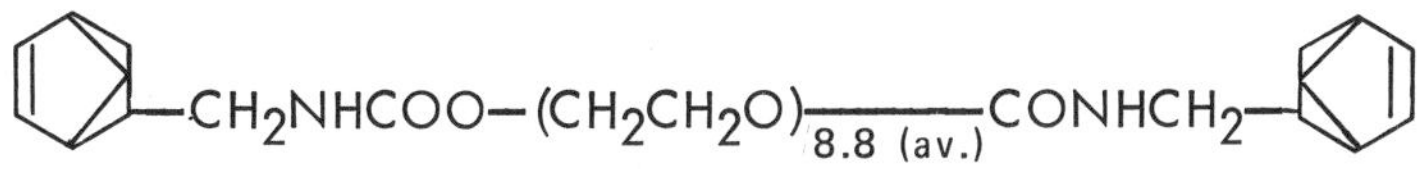

The carbamate was a viscous, clear liquid having a reduced viscosity at 30°C. of 0.05 when measured from a 0.5% solution in chloroform. Microanalytical data further confirmed that the carbamate was produced. Calculated for $C_{35.5}H_{39}O_{11.8}N_2$: C, 60.8; H, 8.4; N, 4.0. Found: C, 60.8; H, 8.6; N, 4.0. The following example illustrates the process.

Example 3: A solution of 9 g. of pentacyclo[8.2.1.1^{4,7}.0^{2,9}.0^{3,8}]-tetradeca-5,11-diene and 19 g. of acetophenone, as the photosensitizer, was placed in a Pyrex tube, degassed and sealed. The tube was irradiated with ultraviolet light using a 140 watt mercury arc at a distance of 10 cm. for a period of 20 days. The temperature ranged from 40° to 45°C. A solid white polymer was produced that was filtered, washed with acetone and dried. The polymer had a reduced viscosity of 0.01 at 30°C. when measured from a 0.1% solution in chloroform. A film was produced from the polymer whose structural formula is:

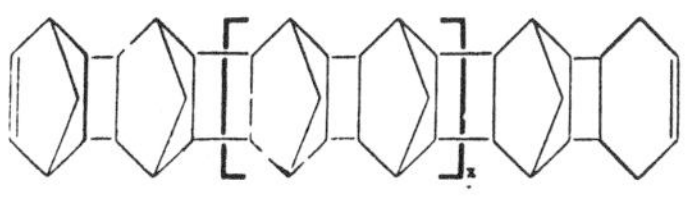

Example 4:   (a) A solution was prepared containing 20 g. of the carbamate
of Example 1 and 1 g. of acetophenone in 20 ml. of benzene.   (b) For comparison purposes, a solution of 20 g. of the polycaprolactone diol used as
the starting reactant in Example 1 and 1 g. of acetophenone in 20 ml. of
benzene was also prepared.   The two solutions, in separate sealed tubes,
were irradiated for 161 hours at about 47°C. in a manner similar to that
described in Example 3.   The contents of tube A completely gelled whereas
the contents of tube B showed no detectable change.   The tubes were opened
and distilled overnight under vacuum on a steam bath to remove the benzene
solvent.

The polymer recovered from tube A was a waxy solid that had a reduced
viscosity of 0.167 at 30°C. when measured from 0.2% solution in benzene.
A film was produced from the polymer, whose structure was confirmed by
infrared analysis as being repeating units of the formula:

$$\left[\,\text{[ring structure]}-CH_2NHCOO-\left[(CH_2)_5COO\right]_m-CH_2CH_2OCHCH_2CH_2-\left[OOC(CH_2)_5\right]_m-OOCNHCH_2-\right]_x$$

Some cross-linking had occurred and the terminal bicyclo groups had ring
unsaturation.   The contents of tube B after irradiation was unreacted polycaprolactone diol.   The reduced viscosity was 0.094 as compared to 0.097
prior to irradiation.   Viscosity was determined at 30°C. using a 0.2% solution in benzene.   This experiment shows that the dinorbornene compounds
polymerize whereas the polycaprolactone diol does not.

Example 5:   A mixture was prepared containing 29.7 g. of the carbamate
of Example 2 and 1 g. of acetophenone.   The mixture was irradiated for
269 hours at about 40°C. in a manner similar to that described in Example
3.   The reaction product was dissolved in 100 ml. of benzene and 30 ml. of
cyclohexane were added; two layers developed.   The oil layer was separated
and stripped on a rotary evaporator at 80°C. and 10.1 mm. pressure for 2
hours to remove volatiles.   The clear, viscous residue had a reduced viscosity of 0.10 when measured from a 0.5% solution in chloroform.   Infrared
analysis showed the presence of ladder polymer units of the formula:

$$\left[-CH_2NHCO_2-\left(CH_2CH_2O\right)_n-CONHCH_2-\text{[ring structure]}\right]-$$

The terminal bicycloheptene units had ring unsaturation.

Example 6: A high molecular weight polycaprolactone, having a reduced viscosity of 2.05 when measured at 25°C. using a 0.2% solution in benzene, was milled with 5% by weight of pentacyclo(8.2.1.1^{4,7}.0^{2,9}.0^{3,8})tetradeca-5,11-diene until a uniform blend was obtained. No gross changes in the mechanical properties of the polycaprolactone were noted. A plaque having a diameter of 5 inches and a thickness of 20 mils was prepared. For control purposes a similar plaque was prepared from the same polycaprolactone. The two plaques were irradiated at 40°C. with a 100 watt medium pressure ultraviolet mercury arc lamp. At intervals sections were removed to observe the effect of the irradiation treatment.

Initially, the ultraviolet irradiation caused a rapid decrease in the original molecular weight in both plaques; within 2 hours viscosity measurements indicated that the average molecular weight was only about one fourth of the original average value. However, as the irradiation continued the molecular weight of the plaque produced from the blend increased and after 20 hours of irradiation a maximum was reached.

The resulting polymer was tough and flexible. The control plaque produced from the unmodified polycaprolactone continued to degrade and after 20 hours of irradiation the polymer was weak and brittle. The results show that the presence of the dinorbornene compound reversed the effect of ultraviolet radiation on polycaprolactone from that of predominantly degradation to that of predominantly cross-linking.

Example 7: A uniform blend was prepared by milling polyethylene having a density of 0.96 g. per cc and a melt index of 0.5 decigram per minute with 5 weight percent of pentacyclo(8.2.1^{4,7}.d0^{2,9}.0^{3,8})tetradeca-5, 11-diene as the dinorbornene compound. A 5 inch plaque, about 22 mils thick, was molded from the blend (Plaque 1). A second plaque was prepared from a blend containing 10 weight percent of the same dinorbornene compound (Plaque 2). A third plaque was prepared from the unmodified polyethylene (Plaque 3) without dinorbornene compound present.

The plaques were irradiated under nitrogen by 1 mev. electrons from a Van de Graaff accelerator and the amount of irradiation necessary for effective cross-linking was determined by determination of the amount of insoluble polymer. The amount of insolubilization was determined by refluxing a weighed portion of the irradiated plaque for 4 hours in ethylbenzene under nitrogen, draining and again extracting with refluxing ethylbenzene for 1/2 hour, drying at 60°C. in a vacuum oven overnight, and reweighing to determine the amount of insoluble cross-linked polymer remaining.

It was found that the dinorbornenyl compound enhanced the cross-linking reaction to the extent that gelation occurred in the blends after irradiation

with from 1.5 to 2 megarad; whereas, gelation occurred in the unmodified
polyethylene only after irradiation with more than 3 megarads. The amount
of insolubilization achieved at various radiation dosages is tabulated below.
Eventually, at about 20 megarads, all of the plaques exhibit an insolubles
of about 80%. However, the rapid cross-linking achieved by the blend at
lower dosages was completely unobvious and unexpected.

Dose, megarads	Percent Insoluble		
	Plaque 1	Plaque 2	Plaque 3
1.5	7.5		
2.0	16	13	0
3.5	38	36	10
5.0	51	49	31
10.0	74	73	61

Maleimide Vapors

In a process described by A.N. Wright and W.F. Mathewson, Jr.; U.S.
Patent 3,619,259; November 9, 1971; assigned to General Electric
Company a thin, continuous film is formed on a substrate by the ultraviolet
surface polymerization of the vapor of an imide containing photopolymerizable organic material. Such films are useful as coatings on metallic and
nonmetallic substrates, capacitor dielectrics, cryogenic device insulation,
insulation for microelectric devices, insulation on electrically conductive
wire, and for corrosion protection.

A preferred class of imide-containing photopolymerizable organic materials
which can be employed in the process are substituted malemides. Some of
the maleimides which can be employed are included by the formula,

$$\begin{array}{c} O \\ \parallel \\ RC-C \\ \parallel \\ RC-C \\ \parallel \\ O \end{array} \!\!\!\! \diagdown \!\!\! NR'$$

where R is selected from hydrogen and alkyl radicals, and R' is selected
from hydrogen and monovalent hydrocarbon radicals. Bismaleimides also
can be used in the process which include compounds of the formula,

$$\begin{array}{c} O \qquad\qquad O \\ \parallel \qquad\qquad \parallel \\ RC-C \qquad\quad C-CR \\ \diagdown NR''N \diagup \\ RC-C \qquad\quad C-CR \\ \parallel \qquad\qquad \parallel \\ O \qquad\qquad O \end{array}$$

where R is defined above, and R" is a divalent organic radical.

FIGURE 1.1:   PHOTOPOLYMERIZED FILMS FROM SUBSTITUTED
MALEIMIDES

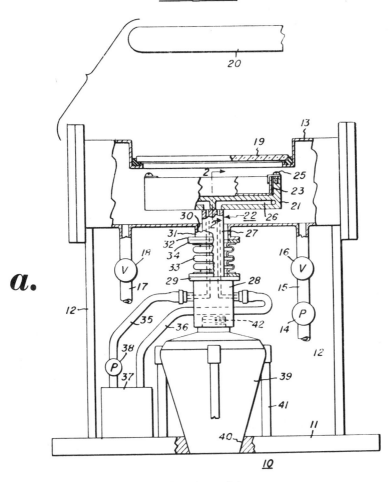

a.

Side View of Apparatus

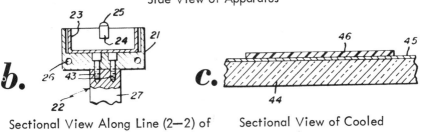

b.

Sectional View Along Line (2—2) of
Figure 1.1a

c.

Sectional View of Cooled
Substrate

(continued)

FIGURE 1.1:   (continued)

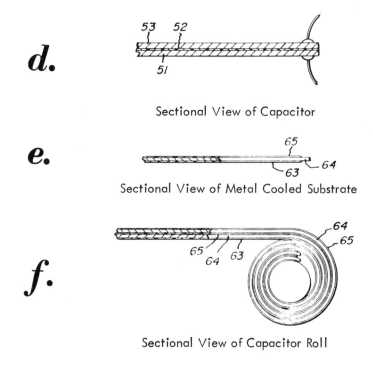

d.

53   52

51

Sectional View of Capacitor

e.

65

63   64

Sectional View of Metal Cooled Substrate

f.

64

65

65   63

64

Sectional View of Capacitor Roll

Source: A.N. Wright and W.F. Mathewson; U.S. Patent 3,619,259;
          November 9, 1971

Among the maleimides which can be employed are, e.g., maleimide,
N-methylmaleimide, N-vinylmaleimide, N-phenylmaleimide, etc., imide
derivatives of citraconic anhydride, such as methylmaleimide, methyl and
N-methylmaleimide. Among the bismaleimides there are included, e.g.,
methylene dianiline bismaleimide, and bismaleimides in which the divalent
organic radical is shown by the formulas:

where M is selected from

$$
\begin{array}{cc}
\underset{\overset{\Vert}{HC-C}}{\overset{O}{\Vert}} & \underset{\overset{\Vert}{HC-C}}{\overset{O}{\Vert}} \\
\Big\rangle N-- \text{ and } & \Big\rangle N-- \\
\underset{\overset{\Vert}{O}}{HC-C} & \underset{\overset{\Vert}{O}}{HC-C}
\end{array}
$$

The process is based on a discovery that the vapor of an imide-containing photopolymerizable organic material can be converted to a high temperature resistant supported films, coatings, or unsupported films, by surface photopolymerization with ultraviolet light. Unsupported films can be made by effecting polymerization of the vapor of an imide-containing organic material on the surface of a substrate, e.g., a metal substrate, followed by removal of the substrate by techniques such as etching.

In Figure 1.1a, an apparatus is shown generally at (10) for forming films, coatings and products having such films or coatings. A base (11) is provided on which is mounted a pair of support members (12). An enclosure (13) is positioned upon support members (12). A vacuum pump (14) is connected by a line (15) to enclosure (13) to evacuate the latter.

A control valve (16) is provided in evacuation line (15). An inlet line (17) is connected at one end to enclosure (13) and at its other end to a source of material to be supplied in gaseous form to enclosure (13). A control valve (18) is provided in line (17) to control the supply of material to enclosure (13). An ultraviolet light transmitting window (19) is shown positioned in the upper wall portion of enclosure (13).

An ultraviolet light (20), which is normally provided with a reflector is shown outside and spaced above enclosure (13) in alignment with window (19). However, light (20) can be positioned inside enclosure (13). Light (20) is supported in any suitable manner. Such a light source can provide ultraviolet light in the region of up to about 3500 A., and which is directed by its reflector through window (19) into enclosure (13). It is preferred to employ ultraviolet light in the wavelength range of between 1800 to 3000 A.

A metal hood is also positioned around the enclosure and light source. A substrate support member (21) is positioned within enclosure (13) and connected to the driven end of a driver shaft (22). A tray or container (23) is located within the upper recessed portion of member (21) to provide a container for material to be used during the operation of apparatus (10). Brackets (24) are shown at opposite ends of tray (23), which brackets are fastened by means of screws (25) to support member (21). A cooling tube (26) is imbedded in substrate support member (21) to provide cooling for the member, associated tray (23) and material placed in tray (23). Since apparatus (10) is useful for coating diamonds, borazon and other particle material, there is provided a driver shaft (22) which has an upper drive portion (27) and a lower driven portion (28). Driver portion (27) of shaft (22) has a smaller diameter than driven portion (28). Shaft (22) is shown with a flange (29) at the junction of portions (27) and (28). Driven portion (27) of shaft (22) extends through an aperture (30) in the wall of enclosure (13).

A closure (31) with an associated flange (32) extends outwardly from and surrounds aperture (30). A diaphragm (33) with a flange (34) at each end is connected by means of these flanges to associated flange (32) of closure (31) and to flange (29) on driver shaft (22). In this manner a vacuum can be maintained in enclosure (13) while shaft (22) can be vibrated. Tube (26) within substrate support member (21) continues through the interior of shaft (22) and is connected to an inlet tube (35) and an outlet tube (36).

Tubes (35) and (36) are connected to a cooling unit (37) which is shown positioned outside enclosure (13) and supported on base (31). Unit (37) consists of, e.g., a Dewar flask in which is positioned a coil connected to the ends of tubes (35) and (36), and which is filled with ice. A thermometer is positioned in the ice to record the temperature within unit (37). Other cooling units, such as a heat exchanger or a refrigeration device, can also be employed.

A circulating pump (38) is connected to inlet tube (35) to circulate a coolant through tube (35), tube (26) and outlet tube (36). A wide variety of coolants might be employed, e.g., water or ethanol. A vibrating device (39) is shown positioned in a recess (40) in base (11). A plurality of support members (41) are attached to base (11) and to device (39) to position the device. The upper end of device (39) fits into a recess (42) in the end of a driven portion (28) of shaft (22). For example, a multiimpedance driver unit might be employed for device (39).

In Figure 1.1b, there is shown a sectional view of a portion of apparatus (10) taken on line (2—2) of Figure 1.1a. In Figure 1.1b, the end of driver portion (27) of shaft (22) is shown connected to substrate support member (21) by means of threaded fasteners (43). In this manner, the drive end (27) of shaft (22) is connected to substrate support member (21) and positions this member within enclosure (13).

In Figure 1.1c, there is shown a glass substrate support (44) and a 0.25 micron thick aluminum film substrate (45) thereon. A continuous film (46) is shown adhering firmly to the upper surface of the aluminum film (45) in accordance with the method using the apparatus shown in Figure 1.1a. In Figure 1.1d, there is shown in section a capacitor which has a first electrode (51), a continuous dielectric film (52), a second electrode (53) in contact with the dielectric film (52) and electrical leads connected to each of the electrodes. Film (52) is formed on the upper surface of electrodes (51) in the apparatus shown in Figure 1.1a. Such a capacitor can also be made by employing a composite sheet having a first electrode (51), a dielectric film (52) thereon, and a second electrode (53) in contact with film (52).

The composite sheet can be cut, subsequently into a plurality of smaller sheets. Each of the smaller sheets has a pair of leads attached to its electrodes thereby forming a plurality of capacitors. In Figure 1.1e, there is shown a metal substrate (64) having a dielectric film (63) on one side and a dielectric film (65) on the opposite side. The metal substrate (64) can be flexible, such as aluminum foil, which can be treated in accordance with the process to effect the simultaneous surface polymerization of the vapor of an imide-containing photopolymerizable organic material, such as the N-phenylmaleimide.

The aluminum foil can have a thickness of about one-fourth mil, while the dielectric film on its lower and upper surface can have a thickness in the range of from 125 to 100,000 A. and preferably 1000 to 25,000 A. In Figure 1.1f, there is shown a capacitor roll which is formed by wrapping two composites of Figure 1.1e in a simultaneous manner around a mandrel.

Alternatively, a capacitor roll can be formed by winding a composite dielectric film. It is preferred to employ ultraviolet light having a wavelength in the range of 1800 to 3000 A., while up to 3500 A. can be employed. A vapor pressure for the imide-containing organic material can be in the range of from 0.1 to 4.0 mm. of mercury. It has been found further that subsequent to the formation of the above type of continuous film formed on the substrate, the substrate can be removed, for instance, by chemical etching with hydrochloric acid or hydrofluoric acid, thereby providing an unsupported body of the film. The following examples illustrate the process.

Example 1: Apparatus was set up in accordance with Figure 1.1a. A substrate support, a microscope glass slide 1" x 3", which was provided with a 0.25 micron thick aluminum film substrate thereon, was positioned on the support block in the enclosure. A stainless steel light mask 1" x 3" with a single slot was placed on the surface of the aluminum substrate.

Solid pyromellitic diimide was placed on the support block in an area to be shaded from the light source. An ultraviolet light source, in the form of an Hanovia 700 watt lamp with a reflector was positioned within the enclosure and above the upper surface of the aluminum film substrate. The window was then positioned in the upper wall of the enclosure.

The system was pumped down to a pressure of 5 microns and the control valve was closed. A metal hood was positioned around the apparatus. The pyromellitic diimide was heated by the lamp to about 205°C. to provide a vapor pressure of 100 microns. The lamp had an effective wavelength in the range from 2000 to 3500 A. During photopolymerization, which was continued for 185 minutes under the light source, the monomer pressure rose to about 1 torr.

In this operation, a bright, bluish-colored film was formed on the aluminum film substrate by ultraviolet surface photopolymerization of pyromellitic diimide in the gaseous phase. While it is not shown in the figure, a plurality of thermocouples was provided to measure the temperature of the evaporated aluminum film to provide temperature information. Cooling means for the substrate support member, which are shown in Figure 1.1a, were not employed in this example. An average temperature of 300°C. was obtained for the aluminum film.

The process was concluded by turning off the ultraviolet pump control valve, and pumping down the interior of the enclosure to a pressure of about 10 microns to remove gaseous material and any by-products therefrom. The vacuum was then broken and the window removed. The light mask was removed and the aluminum film on the glass substrate was examined. Visual examination disclosed a thin film which was continuous.

The film was measured by interferometry and showed a thickness of 1890 A. thereby providing a growth rate of 10 A. per minute. The index of refraction was about 1.5. Thus, a product was obtained from this example which comprised a glass base with an aluminum film substrate, on which a continuous, thin, imperforate electrically insulating film adhered to the upper surface of the substrate.

Example 2: In the following example, the same apparatus, monomer, and procedures are followed as in Example 1. However, a glass substrate is employed. After 185 minutes of ultraviolet surface irradiation during which the substrate temperature was controlled at a maximum temperature of 300°C., the glass film substrate was examined. A continuous film adhered to the substrate.

Example 3: In this example, the same apparatus, monomer, and procedures are followed as in Example 1. However, the lamp is positioned outside the enclosure and the substrate was evaporated aluminum. The monomer is heated by a heat gun to a temperature of about 205°C., to provide a vapor pressure of 100 microns. After 30 minutes of ultraviolet surface photopolymerization, examination of the aluminum substrate discloses a continuous film adhering to the substrate.

Example 4:   Apparatus is again set up in accordance with Figure 1.1a.
Twelve grams of 80/100 mesh diamond particles are spread on an aluminum
tray about 6" long and 1" wide. The tray is placed on the upper recessed
portion of the substrate support member. An ultraviolet light source, in the
form of an Hanovia 700 watt lamp with a reflector, is positioned within the
enclosure and in alignment with the substrate support member.

The window is then positioned in the upper wall of the enclosure. The system is pumped down to a pressure of 5 microns of mercury and the control
valve is closed. Solid pyromellitic diimide is positioned in an area which
would be shaded subsequently. A metal hood is positioned around the apparatus. The pyromellitic diimide is heated to about 205°C. by the lamp
to provide a vapor pressure of about 100 microns.

The circulating pump is started and ethanol is flowed through the substrate
support member. The vibratory device is activated thereby vibrating the
diamond particles in the tray. The cooling unit reduces the temperature of
the substrate support member, tray and diamond particles to a substrate temperature of about 220°C. resulting in a substantial rate increase in the film
formation and a shortening of the time involved.

After 30 minutes, the process is concluded by turning off the ultraviolet
light source and the vibratory device, stopping the circulating pump, removing the hood, opening the vacuum pump control valve, and pumping
down the interior of the enclosure to a pressure of about 1 micron of mercury
to remove gaseous material and any by-products therefrom. The vacuum is
then broken and the window is removed from the enclosure. The tray with
diamond particles is lifted up from the substrate support member. Visual
examination discloses an adherent film on at least a portion of the faces of
the diamond particles.

Example 5:   An abrasive grinding wheel with coated diamond particles as
described above in Example 4 is made. The wheel is 5" in diameter by
three-sixteenths inch thick. The aluminum wheel core is 4 3/4 inches in
diameter and has a 1 1/4 inch arbor hole. The abrasive rim, which is one-
eighth inch by three-sixteenths inch in cross-section, contains a 5.46 g. of
80/100 mesh diamond particles, 7.2 g. of FFF grade silicon carbide, and
3.08 g. of 5417 Bakelite phenolic resin.

The wheel core and abrasive mix are placed in a compression mold. The
loaded mold is placed on a heated platen press, pressed for 30 minutes at
about 177°C. at 20,000 lbs. of force to stops, then 4,000 lbs. with stops
removed. The wheel is then post cured for 17 hours with a maximum temperature of about 190°C. for 12 hours.

Bis-Maleimides

G.J. Smets and F.C. DeSchurijver; U.S. Patent 3,622,321; November 23, 1971; assigned to Gevaert-Agfa N.V., Belgium describe a photographic element comprising a support having a coating comprising a polymer of styrene and a bis-maleimide of the formula:

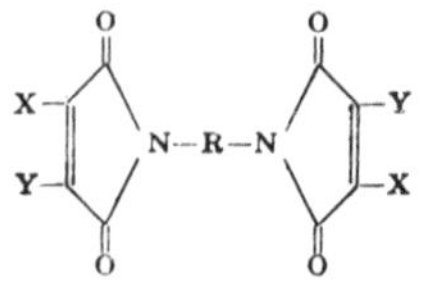

where R represents an alkylene group comprising 1 to 12 carbon atoms, a $-CH_2-Z-CH_2-$ group or an arylene group, Z being O, S or NH, and X and Y represent hydrogen, chlorine, or methyl.

It has been found that these bis-maleimides can be photocyclomerized to give intra- and extramolecular cyclomers, whereby the intramolecular cyclomers are merely formed by a ring-closure reaction in the same bis-maleimide compound, whereas extramolecular cyclomers are polymeric materials formed by a polyaddition reaction of the bis-maleimide. For facility's sake the intramolecular cyclomers will be named "cyclomers" hereinafter, whereas the extramolecular cyclomers, which actually are polymeric materials, will be named "cyclopolymers."

In the following examples exposure to ultraviolet radiation occurred in a Rayonet RS photochemical reactor equipped with a RUL 3500 A. radiation source.

Example 1: A diluted solution of hexamethylenediamine was added slowly, to avoid polymerization, to a diluted solution of maleic anhydride in ether. The product obtained was a maleamic acid, which in hot acetic anhydride (proportion by weight of maleamic acid and acetic anhydride 1 to 20) in the presence of 20% by weight of dry potassium acetate was converted into hexamethylene-bis-maleimide by heating to 90°C. Heating was continued for 5 minutes more and the mixture was then poured out on ice. The bis-maleimide precipitated and was recrystallized from acetonitrile or ethanol.

A solution of the resulting hexamethylene-bis-maleimide in acetonitrile (10^{-2} mol/l.) was exposed for 210 minutes in a photochemical reactor as described above. The reaction product was obtained in an almost quantitative yield by evaporating the solvent in vacuo after the reaction. Finally, 95% of a cyclomer corresponding to the structural formula shown on the following page.

The same result was attained when adding 5×10^{-3} mol of benzophenone as sensitizing agent to the solution. If the exposure was carried out on a 5×10^{-2} molar solution of hexamethylene-bis-maleimide with 5×10^{-3} mol/l. of benzophenone in acetonitrile, a partial polymerization took place and 15 to 20% by weight of insoluble cyclopolymer could be collected afterwards.

Example 2: Tetramethylene-bis-maleimide prepared analogously to the hexamethylene-bis-maleimide of Example 1 was dissolved and exposed as described in Example 1. 100% of the collected product was a cyclomer.

Example 3: Trimethylene-bis-maleimide was dissolved and exposed in the same conditions, as described in Example 1. 100% of cyclomer melting at 355°C. was collected.

Example 4: Dimethylene-bis-maleimide was dissolved in acetonitrile (10^{-2} mol/l.) and in the presence of 5×10^{-3} mol/l. of benzophenone exposed as described in Example 1. A mixture of a cyclomer and a high-melting cyclopolymer was collected.

Example 5: α, α'(Bis-maleimide)dimethyl ether was dissolved and exposed as described in Example 1. During exposure a crystalline product deposited. A cyclomer melting at 352°C. was collected.

Example 6: A solution of 10^{-2} mol/l. m-phenylene-bis-maleimide in acetonitrile comprising 5×10^{-3} mol/l. of benzophenone was exposed as described in Example 1. A mixture of an insoluble and a soluble cyclopolymer was collected. A polymer having a molecular weight of approximately 10,000 was collected from the soluble fraction (intrinsic viscosity in dimethyl sulfoxide: 0.2).

Example 7: 100 mg. of a copolymer of styrene and butadiene (85:15 mol percent), 20 mg. of hexamethylene-bis-maleimide and 2 mg. of Michler's ketone were dissolved in a mixture of 2 ml. of xylene and 3 ml. of methylene chloride. A 50µ thick layer was coated from this solution on a aluminum foil. The layer was dried at room temperature and subsequently exposed for 2 minutes to a test negative in a vacuum exposure unit with an ultraviolet

lamp of 300 watts at a distance of 5 cm. The unexposed portions of the
layer were then washed away with xylene. A clear hardened image, which
could be used as a resist for etching of the support or could be applied to a
negative offset printing method, was obtained.

Polyamide Resins

M. Ishii; U.S. Patent 3,684,516; August 15, 1972; assigned to Mitsui
Mining & Smelting Co., Ltd., Japan has found that photo-sensitive resin
composition comprising an alcohol soluble polyamide resin combined with
a salt of acrylic acid or methacrylic acid and acrylamide as the photo-sensitive cross-linking agent is extremely suitable for utilizing as the printing
plate. The following examples illustrate the process.

Example 1:   To 100 parts by weight of alcohol soluble nylon (Ultramid 10)
were added 15 parts by weight of zinc acrylate, 30 parts by weight of acrylamide, 2 parts by weight of benzophenone and 0.2 part by weight of hydroquinone, and the mixture thus obtained was subjected to fusion and kneading
at 160°C. for 5 minutes by means of a test roll, taken out in the form of a
sheet, folded and put in a metallic mold for glassy finishing, and subjected
to pressure molding at 160°C. for 5 minutes by means of a hot press, whereby a 1 mm. thick sheet was prepared.

Next, this sheet was bonded on a metal plate such as an aluminum plate,
thereby preparing a material for a printing plate having a metallic lining.
Subsequently, this material plate was exposed (for 4 minutes) to radiation
with an arc lamp (of 2 kva. disposed at a distance of 60 cm.) through a
negative film, and thereafter subjected to etching with methyl alcohol (or
denatured ethyl alcohol for industrial use) by employing a paddle-revolving
type etching machine (the number of rotations of paddle: 600 rpm, 900 rpm),
whereby there was obtained an etching plate as follows. Etch factor: above
30; no color loss at all. Shoulder angle: 70°; appropriate. Hardness of
exposed area (viz relief): M90. Etching speed: 0.05 mm./min.

Example 2:   The compositions prepared by severally employing calcium
acrylate, barium acrylate, lead acrylate and cadmium acrylate in lieu of
zinc acrylate employed in Example 1 were respectively molded into a sheet
in the same way as in Example 1, and subjected to exposure and etching.

The ratio of components in parts by weight was: alcohol soluble polyamide
(methoxymethylol nylon), 100; acrylate, 15; acrylamide, 20; 4,4'-di-
methoxybenzophenone, 2; p-tert-butyl catechol 0.2. Either of the compositions could produce an etching plate equal to that in Example 1.

Polyesters Prepared from Etherdiols

K. Akamatsu, T. Hagihara and T. Ishido; U.S. Patent 3,628,963; December 21, 1971; assigned to Asahi Kasei Kogyo Kabushiki Kaisha, Japan describe photosensitive compositions comprising (a) an unsaturated polyester, (b) acrylic acid and (c) a photopolymerization initiator, the unsaturated polyester being produced from an alcoholic component comprising at least one etherdiol having 1 to 4 ether-oxygen groups in the main chain and an acidic component comprising at least one unsaturated dicarboxylic acid, anhydride or dimethyl or diethyl ester thereof and having an average molecular weight of 1,500 to 50,000 and a double bond concentration of 1×10^{-2} to 2×10^{-4} mol/g.

The amount of (b) acrylic acid should be 10 to 80% by weight of the total amount of the photosensitive composition and the amount of the photopolymerization initiator should be 0.001 to 10% by weight of the total amount of the photosensitive composition. In order to increase a resolubility and a developability of the photosensitive compositions, the use of an alcoholic component having a higher molecular weight in producing an unsaturated polyester is insufficient, but it is preferable to increase a concentration of a polar group such as ether, ester or hydroxy group in the unsaturated polyester.

The use of an etherdiol having five or more ether-oxygen groups in the main chain as the alcoholic component gives an inferior water resistance and tensile strength to the photopolymerized articles. Also when an alkylenediol is used as the alcoholic component, the organic solvent resistance of the photopolymerized articles are inferior. On the other hand, an unsaturated polyester derived from an alcoholic component comprising at least one etherdiol having 1 to 4 ether-oxygen groups exhibits an excellent resolubility and developability prior to photopolymerization; water resistance and organic solvent resistance after photopolymerization. The following example illustrates the process.

Example 1:   Under an atmosphere of nitrogen gas, 0.25 mol of polyoxyethylene glycol having an average molecular weight of 200, 0.25 mol of dioxyethylene glycol, 0.23 mol of maleic acid and 0.27 mol of adipic acid were reacted at 180°C. for 6 hours to produce an unsaturated polyester having an average molecular weight of 10,000 and a double bond concentration of 1×10^{-3} mol/g.

To 100 parts of the unsaturated polyester thus obtained, there were added 0.1 part of hydroquinone, 60 parts of acrylic acid, 40 parts of styrene and 1 part of benzoin methylether and these were thoroughly mixed to produce a photosensitive composition.

To a glass cell consisting of a spacer of 10 mm. in height forming four sides of the cell, a bottom plate of a transparent glass, and a top plate of a transparent glass on which a negative film carrying a transparent image of a double eagle was tightly fixed, the photosensitive composition was placed. The negative film side of the cell was exposed for 10 minutes to the light from a 3 kw. carbon arc lamp at a distance of 75 cm. The unexposed areas were removed by washing with a 20% aqueous acetone solution to give a plastic pattern of 10 mm. in thickness precisely and clearly corresponding to the image of the negative.

Example 2: A variety of unsaturated polyesters shown in the table below were prepared in the same manner as in Example 1. To 100 parts of the unsaturated polyester, there were added 40 parts of acrylic acid, 2 parts of benzoin and 0.1 part of tert-butylcatechol and these were thoroughly mixed to produce a photosensitive composition.

Each resulting photosensitive composition was exposed for 20 minutes to 60 watt fluorescent lamps for duplicate at a distance of 5 cm. and then a tensile strength, an elongation, a water absorption were measured. Also a water dispersibility of the photosensitive composition before photopolymerization was measured. The results are shown in the table below.

Etherdiol (mole)	Unsaturated dicarboxylic acid (mole)	Double bond concentration (mole/ gram)	Average molecular weight	Tensile strength (kg./cm.²) [1]	Elongation [2]	Water absorption [3]	Water dispersibility [4]
Ethyleneglycol, 0.50.	Maleic anhydride, 0.25, sebacic acid, 0.25.	3×10^{-2}	2,300	140	B	2.0	Separated.
Trioxyethyleneglycol 0.50.	...do	2×10^{-3}	7,500	120	C	2.3	Good.
Polyoxyethyleneglycol (average molecular weight: 200), 0.50.	...do	2×10^{-3}	9,500	125	C	3.1	Do.
Polyoxyethyleneglycol (average molecular weight: 1,000), 0.50.	...do	5×10^{-4}	15,200	30	B	83	Do.
Polyoxyethyleneglycol (average molecular weight: 2,000), 0.50.	...do	2×10^{-4}	20,000	30	B	103	Do.
Propyleneglycol, 0.50.	Maleic acid, 0.25; adipic acid, 0.25.	3×10^{-2}	2,700	145	C	1.2	Separated.
Trioxypropyleneglycol, 0.50.	...do	2×10^{-3}	5,400	145	C	1.9	Good.
Polyoxypropyleneglycol (average molecular weight: 200), 0.50.	...do	2×10^{-3}	5,600	135	C	2.3	Do.
Polyoxypropyleneglycol (average molecular weight: 1,000), 0.50.	...do	5×10^{-4}	13,000	65	B	33.2	Do.
Polyoxypropyleneglycol (average molecular weight: 2,000), 0.50.	Maleic acid, 0.25; adipic acid, 0.25.	2×10^{-4}	17,000	60	B	37.2	Do.
Ethyleneglycol, 0.25, trioxypropyleneglycol, 0.25.	Maleic acid 0.25; Phthalic anhydride 0.25.	5×10^{-3}	3,700	155	C	1.9	Do.
Dioxyethyleneglycol, dioxypropyleneglycol, 0.25.	...do	2×10^{-3}	5,700	140	C	2.1	Do.

NOTE.— Conditions of measurement:
[1] Tensile strength: ASTM D638–58T.
[2] Elongation: ASTM D638–58T; A = 0–50%, B = 50–100%, C = 100–200%, D = 200–300%, E = >300%.
[3] Water absorption: ASTM D570–57T.
[4] Water dispersibility: After stirring for 10 minutes a 10% aqueous mixture solution of the photosensitive compositions by a homomixer.

Poly(2,6-Dimethyl-1,4-Diphenylene Oxide)

In a process described by F. Berardinelli and J. Gervasi; U.S. Patent 3,597,216; August 3, 1971; assigned to Celanese Corporation poly(2,6-dimethyl-1,4-phenylene oxide) is used as a photoresist in the preparation of letterpress plates requiring the use of etchants under drastic conditions, such as sulfuric acid at temperatures of 140° to 150°C. The resist is applied to the plate from a solution in a chlorinated ethylene, preferably trichloroethylene and containing a peroxide such as t-butylperbenzoate, a cross-linking agent, e.g., the diacrylate of bis-hydroxyethyl terephthalate and thioxanthen-9-one as preferred sensitizer. The resist is particularly effective as applied to plates composed of Celcon, an acetal copolymer.

Example: A composition is prepared having the following formulation:

Component	Parts by Weight
Trichloroethylene	1,329
PPO	100
t-Butylperbenzoate	35
Thioxanthen-9-one	5
BHET diacrylate	10

A solution of thioxanthen-9-one, a commercial photosensitizer, must be filtered and evaporated to dryness. The freshly recovered solid is soluble in the formulation. All of the components of the composition are dissolved in the trichloroethylene solvent. The solution may be warmed gently with solvent and PPO alone, but the other components must be added at room temperature.

Using a 3" x 3" Celcon plaque, the surface is rubbed with pumice powder until uniformly roughened, then swabbed with tetrachloroethylene. The plaque is then placed on a whirler platform at rest, about 5 ml. of the photo-resist solution poured on and the whirler started immediately. The whirler is allowed to spin for 5 seconds or less and the plaque immediately placed in an air oven at 75°C. for 2 minutes, then removed and allowed to cool at room temperature, after which the application is repeated.

A photographic negative is placed on the resist, the plaque and negative sandwiched between two pyrex plates held in place with ball bearing clips and exposed to a 400 watt mercury vapor lamp for five minutes at a distance of 2 inches. The image is developed by swirling the plaque in tetrachloro-ethylene, then drying in an oven at 100°C.

This gives a firmly adherent film approximately 0.5 ml. thick which is resistant to the etching. The etching is carried out in a bed of air fluidized finely divided silica (Cab-O-Sil) on which is deposited sulfuric acid at a 50% concentration at 140°C. The Celcon plaque is immersed in the fluidized bed and etched to a depth of 18 to 20 mils. Upon removal of the plaque from the etching bath, the PPO resist coating was unaffected, examination of the Celcon plaque surface beneath the PPO resist coating showing that it had been protected from the etching enviroment.

PRINTING INK FORMULATIONS

Epoxidized Soybean Oil Acrylates

J.F. Ackerman, J. Weisfeld, R.G. Savageau and G. Beerli; U.S. Patent 3,673,140; June 27, 1972; assigned to Inmont Corporation describe printing ink compositions preferably comprising epoxidized soybean oil acrylate or certain derivatives and a radiation sensitizer having a triplet energy between about 42 and 85 kcal/mol. The method of printing with such inks comprises exposing the inks to an amount of actinic radiation effective to polymerize the inks to a nonoffsetting state.

The process also provides alkyd-derived compositions used as vehicles for the above printing ink compositions and preferably comprising the reaction product of an alkyd formed from trimethylol propane, tall oil fatty acid, and adipic acid with the reaction product of toluene diisocyanate and the monoacrylic acid ester of ethylene glycol.

The following examples illustrate the manner of preparation and use of the vehicles and inks of this process. The epoxidized soybean oil acrylate referred to in the examples is acrylated Flexol Plasticizer EPO having about 3.6 acrylate groups per mol of product. Unless otherwise specified, the thickness of the layer of sensitized vehicle exposed was 0.1 to 0.5 mils as produced by the wedge plate technique.

Example 1: An alkyd-derived vehicle was prepared in this manner. (a) One mol each of toluene diisocyanate and the acrylic acid ester of ethylene glycol (hereafter referred to as HEA) were reacted in a flask under an oxygen blanket at a temperature in the range between 40° to 60°C. The reaction was allowed to proceed until a constant isocyanate value of about 14.5% was reached. The resulting product had a viscosity of X to Y and a Gardner color 1 to 2.

(b) One mol of trimethylol propane and 0.4 mol of tall oil fatty acid were reacted in a flask at about 220°C. The water of esterification was removed by azeotroping until a hydroxyl functionality of 2.6 or less and an acid number of about nine were reached. 0.7 mol of adipic acid was then added at a temperature of about 230°C. The resulting material had a viscosity of Z_5 at 25°C. and a Gardner color of 2 to 3.

(c) A flask provided with an oxygen blanket and heated to about 50°C. was charged with 66 parts of (b), 0.02 parts of dibutyltin dilaurate catalyst were then added to the flask, followed by 32.5 parts of (a), which were added slowly over a period of about an hour until a constant diisocyanate value (or zero) was obtained. Finally, 1.5 parts of normal propanol were added to cap any remaining free diisocyanate.

Example 2: (a) Toluene diisocyanate and HEA were reacted in the same manner as in Example 1. (b) One mol of PEP-550 and 3 mol of tall oil fatty acid were esterified by reacting them at 200° to 230°C. until an acid number of 10 or less was obtained. (c) One mol of (b) was charged to a flask having an oxygen blanket. Dibutyltin dilaurate catalyst was added at the level of 0.01% by weight of the total charge, followed by the addition of 1 mol of (a), which was added slowly over about an hour. The flask was maintained at about 50°C. The reaction was allowed to continue until a constant (or zero) isocyanate value was obtained. Finally, normal propanol was added as needed to cap any remaining free isocyanate.

PEP-550 is the propylene oxide adduct of pentaerythritol. It has a molecular weight of about 500 and is tetrafunctional in hydroxyl groups. It is a product of Wyandotte Chemicals Corporation.

Example 3: Example 2 was repeated but using 1 mol of PEP-550 to 2 mol of tall oil fatty acid in (b). Subsequently, 1 mol of (b) was reacted with 2 mol of (a). The resulting product had a viscosity of Z+.

Example 4: A printing ink was made from the following components: lithol rubine pigment, 0.155 parts; epoxidized soybean oil acrylate, 0.400 parts and Ultraflex wax, 0.045 parts. These three materials were mixed and then ground on a three-roll mill until an acceptable grind was achieved. Subsequently, 0.3 additional part of epoxidized soybean oil acrylate was added during milling. An 8 to 1 sensitizer mixture was prepared from benzophenone and Michler's ketone, respectively. 0.1 part of the sensitizer was mixed with 0.9 part of the other materials by first melting the sensitizer and then adding it to the remaining materials while stirring.

The resulting ink was tested for its roll-out properties by rolling a standard amount of it on a glass plate and then transferring ink from the glass plate to a paper stock by using a hand-held hard rubber roller. A Quickpeek color proofing kit from Thwing-Albert Company of Philadelphia was used per the directions enclosed with it. The evenness of the roll-out was excellent. A conventional wedge plate was also prepared using an NPIRI-B Wedge Plate.

Laboratory — The above wedge plate was prepared on a piece of 43-pound International Paper publication stock. The coated stock was exposed for three-eighths of a second to a beam of light from a No. 679 Hanovia mercury vapor lamp with a semielliptical reflector. The paper was exposed at about 3.82 inches from the bottom of the bulb, and the width of the reflector at its base was about 6 1/4 inches. A complete cure of the ink was obtained.

Pilot — The ink was tested in pilot operation on a web-fed four-color Webendorfer web offset press, Serial No. W-262. The press was a 4 color press with four units in line. It had an 18 inch web width. The press used was fitted with three L 5142.430 Hanovia mercury vapor lamps, each 25 inches long and 5,000 watts. Corresponding Hanovia irradiators, 43 packs and control stations were used. The lamps were mounted side by side after the fourth press station, and the dryer on the press was not used. A 43 pound machine coated publication stock substrate was printed on this press with the ink of this example at 1,000 feet per minute.

After printing, the web ran beneath the lamps at a distance of 2 3/8 inches from the bottom of the reflector. Complete cure to a nonoffsetting state was obtained at this speed. The resulting printed substrate exhibited excellent glass and color strength.

Example 5: The ink of Example 4 was prepared using 5% sensitizer rather than the 10% of Example 4. After exposure to ultraviolet light in the manner described in the Laboratory section of Example 4, a nonoffsetting cure was obtained after 1/2 second exposure and a complete cure was obtained after 0.7 seconds exposure. Again, color strength and gloss were excellent.

Pentaerythritol Triacrylate with Halogenated Initiator

R.W. Bassemir, D.J. Carlick and G.I. Nass; U.S. Patent 3,661,614; May 9, 1972; assigned to Sun Chemical Corporation describe a radiation-curable solvent-free printing ink which consists of (1) about 20 to 98 weight percent of a pentaerythritol acrylate, methacrylate, or itaconate, (2) about 2 to 80 weight percent of a halogenated aromatic, alicyclic, or aliphatic hydrocarbon photoinitiator where the halogen atoms are attached directly to the ring structure in the aromatic and alicyclic compounds and to the carbon chain in the aliphatic compounds, and (3) a colorant.

Inks and coatings made from the compositions of the process are solvent-free and, when exposed to a source of radiation, dry almost instantaneously in air at ambient temperature, thus eliminating the need for ovens as well as as avoiding the air pollution, fire hazards, odor, and so forth, that accompany the use of volatile solvents.

The inks and coatings form extremely hard and durable films on a wide variety of substrates, such as, paper; newsprint; coated paper stock; irregular, e.g., corrugated, board; metal, e.g., foils, meshes, cans, and bottle caps; woods; rubbers; polyesters, such as polyethylene terephthalate; glass; polyolefins, such as treated and untreated polyethylene and polypropylene; cellulose acetate; fabrics such as cotton, silk, and rayon. They exhibit no color change in the applied film when subjected to the required curing conditions, and they are resistant to flaking; smudging; salt spray; scuffing; rubbing; and the deteriorating effects of such substances as alcohols, oils, and fats. In addition, the compositions withstand both heat and cold, making them useful, e.g., in printing inks or coatings for containers that must be sterilized, e.g., at about 150°C. under pressure, and/or refrigerated, e.g., at less than -20°C. The following examples illustrate the process.

Example 1: A red ink was prepared from 67 parts of pentaerythritol triacrylate, 9.75 parts of Aroclor 1260, 3.25 parts of Santolite MHP, and 20 parts of lithol rubine red pigment. A glass bottle printed with this ink was exposed to a 1,200 watt Hanovia high mercury pressure lamp at a distance of 1 inch. The ink dried in less than one second and had excellent adhesion to glass as well as good grease- and rub-resistance. It withstood temperatures of 150°C. and -20°C.

Example 2: A black ink was prepared from 38.5 parts of pentaerythritol triacrylate, 38.5 parts of tripentaerythritol octoacrylate, 9.75 parts of Aroclor 1,260, 3.25 parts of Santolite MHP, and 10 parts of carbon black. A glass bottle was printed with the ink and exposed at a distance of 1 inch from a 1,200 watt Hanovia high mercury pressure lamp. The ink dried in less than one second and had excellent adhesion to glass as well as good grease- and rub-resistance. In addition, it withstood temperatures of 150°C. and -20°C.

Example 3: The procedure of Example 1 was repeated except that the amount of pentaerythritol triacrylate was 65 parts and 2 parts of Essowax 2210 (a refined paraffin wax) was included in the formulation. The results were comparable.

Example 4: The procedure of Example 1 was repeated except that a prepolymer (a mixture of dimers and trimers) of a pentaerythritol triacrylate was used instead of the pentaerythritol triacrylate monomer. The results were comparable.

Example 5: The procedure of Example 2 was repeated except that pentachlorobenzene was used instead of Aroclor 1,260. The results were comparable.

Example 6: The procedure of Example 2 was repeated except that 2-bromoethyl methyl ether was used instead of Aroclor 1,260. The results were comparable.

Example 7: The procedure of Example 2 was repeated except that chlorendic anhydride was used instead of Aroclor 1,260. The results were comparable.

Example 8:   A blue ink was prepared from 70 parts of pentaerythritol tetra-acrylate, 10 parts of pentachlorobenzene, and 20 parts of phthalocyanine blue. A glass bottle was printed with the ink and subjected to ultraviolet light as in Example 1. After an exposure of three seconds, the ink was dry and adhered well to the glass.

Example 9:   A red ink was prepared from 85 parts of a mixture of 90% of pentaerythritol triacrylate and 10% of Aroclor 4,465 and 15 parts of lithol rubine red pigment. The ink was run on a Miehle press to print coated paper. The printed paper was exposed at a distance of 1 inch from four 21 inch 2,100 watt Hanovia ultraviolet lamps. The ink dried in 0.4 second and had excellent gloss, adhesion, scratch-resistance, and rub-resistance.

Example 10:   The procedures of Example 1 through 9 were repeated except that the printed substrates were exposed by passing them on a conveyor belt beneath the beam of a Dynacote 300,000 volt linear electron accelerator at a speed of 43 feet per minute and the beam current so regulated as to produce a dose rate of 5 megarads. The results were comparable.

Example 11:   The procedures of Examples 1 through 9 were repeated except that instead of being exposed to ultraviolet light alone the printed substrates were exposed to a combination of ultraviolet light and electron beam radiation. The printed substrates were exposed by passing them on a conveyor belt beneath an ultraviolet lamp set perpendicular to the conveyor belt and rated at 100 watts per linear inch and a Dynacote 300,000 volt linear electron accelerator having a beam current so regulated as to produce a dose rate of 0.5 megarad.

Exposure to the ultraviolet light and the electron beam radiation was carried out in a variety of arrangements, e.g., ultraviolet light, then electron beam; electron beam, then ultraviolet light; ultraviolet light before and after electron beam; electron beam before and after ultraviolet radiation; and simultaneous electron beam and ultraviolet radiation. The results were comparable.

OTHER PROCESS TECHNIQUES

Sequential Irradiation

L.A. Cescon, R.L. Cohen and R. Dessauer; U.S. Patent 3,615,454; October 26, 1971; assigned to E.I. du Pont de Nemours and Company describe a multiple irradiation method which comprises providing a radiation-sensitive material comprising (1) a radiation-sensitive, multicomponent, intermolecularly reactive imageable composition whose imaging reaction is

subject to diffusion control, mixed with (2) a radiation-sensitive polymerizable composition, imaging by irradiating with imaging radiation under
imaging conditions, and deactivating the imageable composition in the unexposed areas by irradiating with the polymerizing radiation under nonimag-
ing conditions, the polymerization being effective to rigidify the material
so as to render the imaging reaction diffusion-controlled and thereby prevent
the imaging components from diffusing together and reacting.

In the imaging step it is only necessary that the imaging reaction occur
before the polymerization reaction can deactivate the system. In the deactivation (or fixing) step it is only necessary that the deactivating radiation be
applied under conditions ineffective for imaging.

Imagewise exposing the composition first to the imaging radiation then to
the deactivating radiation produces a negative image. On the other hand,
imagewise exposing the composition first to the deactivating radiation creates
a latent image, which is developed by exposing the unirradiated areas to the
imaging radiation.

Various radiation-responsive imaging systems produce a colored or other
characteristic readout product through inter-molecular reactions of two or
more reactive components. Expedients suggested for deactivating such imaging systems in the unexposed areas, thereby fixing the image, include: (1)
removing one or more of the imaging components, as by volatilization (U.S.
Patent 3,042,515) or polymerization (U.S. Patent 3,056,673); (2) volatiliz-
ing a plasticizing solvent (British Patent 1,047,569); (3) incorporating a
deactivating agent into the unexposed area to be fixed, as by spraying,
dipping or coating (British Patent 1,047,569); and (4) generating a deactivating agent in situ, including photochemically (British Patent 1,057,785).

British Patent 1,057,785 describes a sequential irradiation image-fix system
involving a photooxidant/leuco dye imaging combination that produces a
colored image at one wavelength and a second photooxidant/reductant combination that produces a deactivating agent for the imaging photooxidant at
a second wavelength.

Wainer, U.S. Patent 3,056,673, also describes a sequential irradiation
image-fix method in which organic halogen/N-vinylamine compositions
undergo two distinct photochemical reactions, both involving the vinylamine.
One at long wavelengths leads to color (subsequently developed by heating);
the other at short wavelengths polymerizes the vinylamine. Polymerizing the
amine in the uncolored area fixes the colored image by removing the vinylamine, thus making it unavailable for color formation, its polymer being
inactive in the color-forming reaction. This system describes only compositions of halogen compounds that yield free radicals at spectral ranges above

360μ and selected vinylamines such as N-vinyl carbazole. Belgian
Patent 681,944 describes photopolymerizable compositions containing a
hexaarylbiimidazole, an electron donor which may be a leuco dye, an
addition-polymerizable monomer, and optionally containing an energy transfer dye to extend the compositions' spectral sensitivity from ultraviolet to
visible light and an oxygen scavenger. In other words, it discloses that the
biimidazole/leuco dye imaging components of British Patent 1,047,569 can
photoinitiate addition-polymerizations. The energy transfer dye and the
oxygen scavenger correspond to the photoreducible dye/mild reducing agent
combination which Oster U.S. Patent 2,840,445 describes as photopolymerization.

Thus the Belgian Patent actually describes two photopolymerization initiator
systems in combination, biimidazole/leuco dye activatable in the UV and
photoreducible dye/mild reducing agent activatable in the visible. The
patent describes that images produced by imagewise exposing the disclosed
compositions to light (UV or visible) are developed by suitable means —
which exploit differences in the physical and chemical properties of the exposed and unexposed areas, such as preferentially incorporating a character-
izing dye or pigment into the underexposed areas, transferring still tacky
underexposed areas to another substrate, or washing out the underexposed
areas — to produce either images on receptors or reliefs suitable for printing.

The patent, however, does not describe or suggest multiple exposure irradiation of any of the imageable and photopolymerizable compositions such that
the color- and polymer-producing reactions can be controlled. This process
is directed to a multiple exposure irradiation method for producing fixed
images which utilizes radiation-sensitive materials containing (1) intermolecularly reactive components for producing a characteristic readout product
under one set of irradiation conditions and (2) polymerizable components
for producing rigidifying polymer under another set of irradiation conditions
effective for polymerization but ineffective for producing the readout product, and comprises the steps of producing readout product under the one set
of conditions and polymerizing without substantially producing readout product under the other set of conditions.

The polymerized matrix immobilizes the reactive components which produce
the readout product, thus preventing them from diffusing together and reacting when the composition is subsequently exposed to the readout product-
producing radiation. The composition is thus deactivated (or fixed) and the
image as represented by the actual or potential difference in readout product
concentration between the sequentially exposed areas is fixed.

The following examples illustrate the process with various radiation-sensitive
materials and describe various sets of imaging and fixing conditions.

Unless otherwise stated, deactivation (or fixing) means that the systems'
color-forming potential has been decreased to such an extent that the color
optical density resulting when the deactivated system is exposed to a contact flash from a Xenon flash lamp is substantially less than that produced
under the described imaging conditions. The Xenon flash lamp, which is
sometimes used for imaging in the following examples, is available as Hi
Co lite, and emits ultraviolet and visible light approximating sunlight at
an intensity of about 1×10^5 milliwatts/cm.^{2} for about 0.001 second flash
duration.

Example 1:   Photoimage/Thermal Fix — Polyvinyl chloride film was impregnated, by soaking with a solution containing the following ingredients:

Ingredient	Parts	Mols
2,2'-Bis(o-chlorophenyl)-4,4',5,5'-tetra-phenyl biimidazole (o-ClHABI)	0.033	5×10^{15}
Tris(p-N,N-diethylamino-o-tolyl)methane	0.050	1×10^{14}
p-Toluene sulfonic acid	0.057	3×10^{14}
Pentaerythritol triacrylate	1.98	6.6×10^{12}
Azo bis-isobutyronitrile	0.02	1.2×10^{14}
Acetone	13.0	

The film was heated at 75°C. under an IR lamp to remove the acetone. The
imaging step was carried out by irradiating one-half the film by contact-
flashing with a Xenon flash lamp to produce a deep blue color. The entire
film was then deactivated (or fixed) by heating for 45 seconds at 150°C.
Substantially no color formed. Subsequently, contact-flashing the heated
area with the Xenon lamp developed no color. The image was therefore
fixed.

To show that the image was rigidification fixed, i.e., that the imaging
components had not been destroyed but were immobilized in the polymerized
matrix produced on heating, the imaged and fixed film was swelled in an
acetone/dimethyl acetamide mixture and reirradiated with the imaging lamp.
The previously uncolored fixed area became substantially colored.

Example 2:   Two Wavelength Photoimage/Photofix/Quinone as Polymerization Initiator — An acetone solution containing the ingredients as shown on
the following page was cast as a 1 mil film on a Mylar polyester support
film and covered with 3 mil thick polypropylene film transparent to actinic
wavelengths above 300μ.

The film was imaged by covering it with a Corning 7-54 filter (transmitting
from about 240 to 420μ, maximally at about 340μ) and a stencil.

Ingredient	Parts	Mols
o-ClHABI (as in Example 1)	0.035	5.3×10^{15}
Tris(p-N,N-diethylamino-o-tolyl)methane	0.013	2.6×10^{15}
p-Toluene sulfonic acid	0.010	5.3×10^{15}
Pentaerythritol triacrylate	0.030	1×10^{12}
Pyrenequinone (approx. equimolar 1,6 and 1,8-isomers)	0.0013	5.5×10^{16}
Trimethyl nitrilotripropionate	0.30	1.1×10^{13}
Polyvinyl pyrrolidone	0.25	

Then it was contact-flashed once with the Xenon lamp described in Example 1, producing a deep blue image. The stencil was removed, the filter replaced with a Corning 0-51 filter (transmitting above 350μ and showing 60% or higher transmission above 400μ) and the film irradiated in the fixing step with a single flash of the Xenon lamp held about 6 inches from the film surface, whereupon substantially no color developed. Reexposing the thus-fixed area to the Xenon lamp's full intensity and light range, with the filter removed, again produced little or no color, demonstrating the fix.

Replacing the unsaturated triacrylate with pentaerythritol tripropionate, its saturated analog, yielded a composition which could be imaged under the above conditions but not fixed. This demonstrates that fixing depends on the polymerizable triacrylate and indicates that the quinone functions as polymerization initiator and the nitrilopropionate as a chain transfer agent in the fixing step, the quinone's concentration going too low to adequately fix either of the above compositions by the photoreduction fixing method.

Example 3:   Two Wavelength Photoimage/Photofix/Reducible Dye as Polymerization Initiator — A laminated article containing a radiation-sensitive material was prepared as in Example 2 except that Erythrosin B (20 mol/mol o-ClHABI) replaced the quinone as the visible light activated polymerization initiator. The composition was imaged by irradiating with intense ultraviolet light through a 7-54 filter, as described in Example 2, and fixed by irradiating through a Corning 3-71 filter (transmitting visible light above 460μ) with a 275 watt sunlamp (Westinghouse RS type) held 8 inches from the surface.

Examples 4 to 7:   Intensity-Controlled Photoimage/Photofix — A mixed acetone/ethanol solution containing the ingredients as shown on the following page was coated to a 1 mil dry thickness on 2 mil polypropylene, the carrier solvent being evaporated by drying for 30 minutes under an infrared lamp positioned 12 inches from the surface so as to not raise the temperature above 50°C., and the substantially dry coating covered with 5 mil Mylar polyester film to form a laminate.

Ingredient	Parts	Mols
o-ClHABI	0.059	9×10^{15}
Tris(p-N,N-diethylamino-o-tolyl)methane	0.023	4.5×10^{15}
p-Toluene sulfonic acid	0.017	9×10^{15}
Trimethyl nitrilopropionate	0.38	1.4×10^{13}
Polyvinylpyrrolidone, mol wt. 40,000	0.25	
Polyvinylpyrrolidone, mol wt. 360,000	0.25	
Pentaerythritol triacrylate	0.50	1.7×10^{13}

Fixed continuous tone nonreversal prints were made by (a) exposing the laminate through a continuous tone photographic transparency (either negative or positive) and through the polypropylene film under the fixing conditions tabulated below then (b) with the stencil removed exposing the entire surface to a single contact flash, i.e., close-to-the surface flash, of the Xenon lamp described in Example 1.

Example	Low Intensity UV Source and Exposure
4	A single flash of the Xenon flashlamp positioned 12 inches from the surface
5	5 seconds exposure to a 275-watt sunlamp 8 inches from the surface
6	10 seconds exposure to a Metalarc lamp 10 inches form the surface
7	90 seconds exposure to a fluorescent desk lamp 2 inches from the surface

Reversing the exposure order in the above examples yielded equally satisfactory fixed continuous tone watt prints.

Example 8: Intensity-Controlled Photoimage/Photofix/Projected Image — A transparent microfilm image was projected from a slide projector containing a 500 watt bulb onto the surface of a previously unexposed laminate prepared as in Examples 4 to 7 and positioned with its polypropylene film side facing the bulb at a 3 to 4 foot distance from the bulb. This exposure created a latent image in 1 to 3 minutes. One contact flash of the Xenon lamp over the entire surface developed the fixed image as a 5 to 6 times enlargement of the original.

Example 9: Photoimage/Photofix. Neutral Shade Image — A film laminate was prepared as in Examples 4 to 7 except that bis(N,N-diethylamino-o-tolyl)3,4-methylene-dioxyphenyl methane was employed as the leuco dye in place of the tris-(p-N,N-diethylamino-o-tolyl)methane. Imaging with a Xenon flash and fixing by exposure to the sun lamp of Example 5 for 10 seconds at a 10-inch distance gave a fixed grey-black image.

Examples 10 to 13:   Light Image/Heat Fix and Heat Image/Heat Fix — A
formulation containing 0.35 g. of 2,2'-(o-chlorophenyl)-4,4',5,5'-tetrakis
(m-methoxyphenyl)biimidazole, 0.228 g. of tris(p-N,N-diethylamino-o-
tolyl)methane, 0.124 g. of p-toluene sulfonic acid, 2.5 g. of pentaerythritol triacrylate, 1.1 g. of poly(methyl methacrylate) and 5 ml. of acetone
was coated to a 5 mil wet thickness on 1 mil Mylar polyester film,
warmed slightly under an IR lamp to evaporate the acetone, and laminated
with a 4 mil Mylar cover sheet.

The laminate was subjected to imaging and deactivation steps as follows:
Imaging involved either (a) heating at 130°C. for 2 minutes or (b) exposure
to a single Xenon lamp flash at 25°C. Deactivation involved (a) heating
unexposed portions of the film at 90°C. for 4 minutes or (b) heating at 65°C.
while irradiating with light at 0.1 mw./cm.^{2} for 30 seconds. The extent of
deactivation to ambient light and temperature conditions was determined by
flashing the thus treated area with the Xenon lamp at 25°C. and measuring
the extent of color formation. The results are tabulated below:

Transmission Optical Density

Treatment	Example 10	Example 11	Example 12	Example 13
(a) Light-activated imaging	1.0	1.0	-	-
(b) Heat-activated imaging	-	-	0.4	0.4
(c) Heat fixing followed by light exposure	0.1	-	0.1	-
(d) Heat and light fixing followed by light exposure	-	0.1	-	0.1

The results illustrate the versatility of the sequential exposure method in
showing that the described composition can be converted to color by heat
or light and deactivated against color formation by heat or light. The difference in optical density between the colored and deactivated areas indicates the relative effectiveness of the particular sequential operation to
produce a stable colored image.

High Intensity Continuum Light Radiation-Plasma Arc

C.L. Osborn and D.J. Trecker; U.S. Patent 3,650,669; March 21, 1972;
assigned to Union Carbide Corporation have found that the high intensity
predominantly continuum light radiation from, e.g., a swirl-flow plasma
arc radiation source causes many monomers to polymerize quite rapidly, in
some instances even in the absence of catalysts or photosensitizers. The
same effect was observed in the curing or cross-linking of polymer compositions. By comparison, low intensity ultraviolet line radiation from mercury

lamps was considerably slower and in some instances completely ineffective. Generally, high intensity predominantly continuum light radiation was effective within seconds, or at the most several minutes. Whereas, low intensity ultraviolet light radiation required appreciably longer periods of time to achieve the same effect.

The examples here show many instances in which cross-linking was achieved in seconds with high intensity predominantly continuum light radiation from a swirl-flow plasma arc radiation source versus minutes or even hours with low intensity ultraviolet line radiation from conventional mercury arc sources. The light radiation from a swirl-flow plasma arc radiation source is high intensity predominantly continuum light radiation.

The term "high intensity predominantly continuum light radiation" means continuum radiation with a source intensity of at least 350 watts per square cm. steradian when integrated throughout the entire spectral range of said continuum light radiation (about 1,000 kilowatts per square foot of source projected area) having only a minor part of the energy in peaks of bandwidths a positive amount up to 100 A., with less than about 30% of the light radiated having wavelengths shorter than 4000 A. and at least about 70% of the light energy radiated having wavelengths longer than 4000 A.

This type of high intensity continuum light radiation is illustrated by the curves shown in Figures 1.2a, 1.2b and 1.2c. These curves illustrate the high intensity predominantly continuum nature of the light radiation over the range of source intensity of from about 350 watts per square cm. steradian to about 5,000 watts per square cm. steradian. As is evident from the curves of Figures 1.2a, 1.2b and 1.2c., the light radiated is predominantly continuum light with very little light emitted as line or peak radiation (band widths less than 100 A.).

It is also evident from these figures that less than about 30% of the light radiated have wavelengths shorter than 4000 A. and that at least about 70% of the radiated light have wavelengths longer than 4000 A. This light radiation is derived from an artificial source that generates high intensity predominantly continuum light radiation with a source intensity of at least about 350 watts per square cm. steradian, when integrated throughout the entire spectral range of said continuum light radiation as abbreviated by the term: watts cm.$^{-2}$ sr.$^{-1}$; the high intensity predominantly continuum artificial light radiation has at least about 70% of the light radiated at a wavelength longer than 4000 A. and a positive amount up to about 30% of the light radiated having a wavelength shorter than 4000 A.

Generally, about 80% of the light radiated has a wavelength longer than 4000 A. and less than about 20% of the light radiated has a wavelength

shorter than 4000 A., and a source intensity that can vary from about 350
watts (about 1,000 kilowatts per square foot of source projected area) to
about 5,000 watts (about 15,000 kilowatts per square foot of source projected area) or more per square cm. steradian. A convenient source of
high intensity predominantly continuum light radiation is a swirl-flow plasma
arc light radiation apparatus. The equipment for generating high intensity
predominantly continuum light radiation by this means is known and available; many different forms are described in the literature. A highly efficient apparatus for obtaining high intensity predominantly continuum light
radiation is the swirl-flow plasma arc radiation source described in U.S.
Patent 3,364,387.

FIGURE 1.2:   PLASMA ARC IRRADIATION

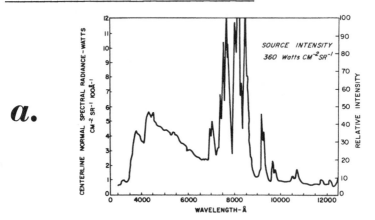

Light Radiation Curve for 18 Kilowatt Argon Source

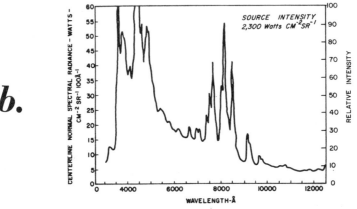

Light Radiation Curve for 60 Kilowatt Argon Source

(continued)

FIGURE 1.2:   (continued)

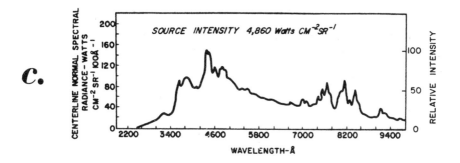

c.

Light Radiation Curve for 71 Kilowatt Argon Source

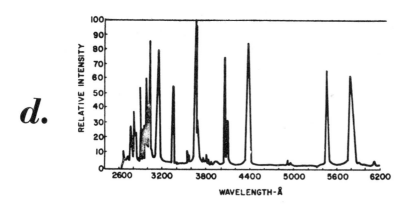

d.

Light Radiation Curve for Mercury Arc Lamp

Source: C.L. Osborn and D.J. Trecker; U.S. Patent 3,650,669; March 21, 1972

It is to be noted that in the spectra of Figures 1.2a, 1.2b and 1.2c there is a continuum of radiation throughout the entire spectral range shown. This type of continuum radiation in the ultraviolet range has not previously been obtainable from the conventional commercial mercury arcs or lamps generally available for generating ultraviolet light. The previously known means for generating ultraviolet light produced light that shows a line or peak spectrum in the ultraviolet range, as exemplified by Figure 1.2d. It is not a continuum spectrum in the ultraviolet range. In a line spectrum the major portion of useable ultraviolet light is that portion at which the

line or band in the spectrum forms a peak; in order for such energy to be
useful the material or composition that is to be treated with ultraviolet radiation must be capable of absorbing at that particular wavelength range
at which the peak appears. In the event the material or composition does
not have the ability to absorb at that particular wavelength range there is
little or no absorption or reaction. Thus, in the event the material or composition to be treated absorbs at a particular wavelength range in one of
the valleys of the spectral curve there will be little or no reaction since
there is little or no ultraviolet energy to adequately excite the system.

With a high intensity predominantly continuum radiation, as is shown by
Figures 1.2a, 1.2b and 1.2c, there is a high intensity continuum radiation
of ultraviolet energy across the entire ultraviolet wavelength range of the
spectrum shown and there is generally sufficient ultraviolet energy generated at all useful ultraviolet wavelengths to enable one to carry out reactions responsive to ultraviolet radiation without the problem of selecting
compounds that will absorb at the peak wavelength bands only.

With the high intensity continuum radiation now discovered one does not
have the problem of being unable to react materials or compositions that
absorb in the valley areas only since for all intents and purposes such valleys do not exist in high intensity continuum radiation, the high intensity
radiated light energy is essentially a continuum; it is not in peak bands.

Figure 1.2a is the light radiation curve from an 18 kilowatt argon swirl-
flow plasma arc radiation source. The measured source intensity of the
light was 360 watts per square cm. steradian; about 8% of the light had a
wavelength shorter than 4000 A. and about 92% of the light had a wavelength longer than 4000 A.

Figure 1.2b is the light radiation from a 60 kilowatt argon swirl-flow plasma
arc radiation source. The measured source intensity was about 2,300 watts
per square cm. steradian; about 10% of the light had a wavelength shorter
than 4000 A. and about 90% of the light had a wavelength longer than
4000 A.

Figure 1.2c is the light radiation from a 71 kilowatt argon swirl-flow plasma
arc radiation source. The measured source intensity was about 4,860 watts
per square cm. steradian; about 12% of the light had a wavelength shorter
than 4000 A. and about 88% of the light had a wavelength longer than
4000 A.

High intensity predominantly continuum light radiation as shown by Figures
1.2a, 1.2b and 1.2c is to be distinguished from low intensity ultraviolet
radiation generated by commercially available low, medium and high

pressure mercury arc ultraviolet lamps. These mercury arc lamps produce light emission which is primarily line or peak rather than continuum light. Figure 1.2d is a typical curve for the light radiation from a mercury arc lamp. As shown in Figure 1.2d, a major part of the light appears in bands narrower than 100 A., and much less than 70% is above 4000 A. In the following examples, which are illustrative of the process, the following test procedures were used to evaluate the products.

Tests	ASTM Numbers
Hardness (Shore A Units)	D2240-64T
Elongation, %	D638-67T
Tensile strength, psi	D638-67T
Set at break, %	D638-67T
Melt index, dgm./min.	D1248-68
Density, g./cc	D1248-68
Stiffness, psi	D638-67T
Tensile modulus, psi	D638-67T
Tensile strength, psi*	1682-59T

*On paper using a two-inch strip of paper.

Example 1: An ethylene/vinyl acetate copolymer, 300 g. having about 16.2 weight percent vinyl acetate was dissolved in 1,400 g. of toluene by heating to the reflux temperature. Methanol, 129 g., was added at 70°C., and then 23.2 g. of a 25% methanolic solution of sodium methoxide was added. The solution was refluxed for several hours, cooled, poured into methanol and filtered to recover the hydrolyzed copolymer. This was washed with methanol until neutral and dried in a vacuum oven.

The hydrolyzed copolymer, 100 g. (equivalent to 0.125 mil of vinyl alcohol), 500 ml. of benzene and 10 ml. of pyridine were charged to a 1 l. flask. 50 ml. of the solution was azeotropically distilled to remove traces of water. 20.5 g. of 5-chloroformylbicyclo[2.2.1]-hept-2-ene were added over 15 minutes to the refluxing solution which was continued for one more hour.

The solution was cooled to 40°C. and methanol was slowly added until it became cloudy; then two liters of methanol were quickly added. The ethylene/vinyl bicyclo[2.2.1]hept-2-en-5-carboxylate copolymer was filtered, washed with methanol and dried; it contained the following units in the polymer chain:

Blends of this copolymer with different additives were prepared by addition of 1 weight percent of various photosensitizers to the copolymer and milling at 120°C. until homogeneous compositions were obtained. The blends were pressed into plaques five inches in diameter and ten mils thick at mold temperatures of 120° to 150°C. One inch square samples were cut from these plaques and irradiated for various periods of time by exposure to the high intensity predominantly continuum light radiation emanating from a 50 kilowatt swirl-flow plasma arc radiation source employing argon as the gaseous medium.

The samples were irradiated at a distance of one foot from the arc. After irradiation the films were extracted with hot ethylbenzene to constant weight to determine the extent of cure that had been achieved. The cured copolymer was insoluble whereas the uncured copolymer was soluble in the hot solvent. It was found that about 90% cure was achieved in about one minute.

Samples cured using a 550 watt mercury lamp required about 20 minutes of exposure to obtain cures similar to those obtained in 20 seconds with the light radiation from the swirl-flow plasma arc radiation (SFPA) source. The results are tabulated below:

	SFPA			Mercury Lamp				
Exposure Time, minutes	1/12	1/6	1/3	1	2	5	20	60	120
Additive	Percent Insoluble Copolymer								
p-Methoxybenzophenone	77	86	88	96	48	72	91	93	92
3-Chloroxanthone	32	63	64	89	25	47	76	86	-
Benzophenone	58	66	85	89	42	61	85	87	92
None	1	8	34	54	25	32	34	62	-

Example 2: A commercially available silicone rubber gumstock having an average molecular weight of about 400,000 containing about 64.5% of a solid dimethylsiloxane (98.8)/methyl vinyl siloxane (0.2) copolymer, about 9.5% of a liquid low molecular weight ethoxy end-blocked poly(dimethylsiloxane) and 26% silica filler was blended on a roll mill with different additives until homogeneous mixtures were obtained. The homogeneous blends were cold pressed into 100 mil sheets which were irradiated at a distance of two feet with the swirl-flow plasma arc radiation source described in Example 1.

The results show the unexpected rapid curing achieved by the process even on polymeric compositions containing inert inorganic fillers; they also show that rapid, uniform curing can be obtained through thick sheets. Cross-linking was not observed on the silicone rubber when the additive was not added.

Additive	Conc. wt. percent	Exposure time, seconds	Hardness (Shore A units)	Elongation, percent	Tensile, p.s.i.	Set at break, percent
None		300	•••	•••	<100	•••
Silanic hydrogen crosslinker*	3.2	180 300	16 17	90 200	780 833	68 47
Dichlorobenzoyl peroxide**	1.6	30 60 120	40 47 52	365 700 930	610 610 500	15 6 0
2,5-dimethyl-2,5-di(t-butyl peroxy) hexane	0.64	60 120 180	26 30 33	840 1,040 1,175	1,010 900 770	17 10 6
Benzophenone	0.64	40 60 120	15 23 34	60 90 940	625 700 975	70 49 12
Silanic hydrogen crosslinker*	3.2	30	18	75	500	30
Benzophenone	0.64	60 120	20 20	200 400	725 750	27 19
Dichlorobenzoyl peroxide**	1.6	20	45	870	630	4
Benzophenone	0.64	30 60 90	50 52 48	1,215 1,250 1,140	610 620 500	1 1 1
Benzophenone	0.64	30	18	130	915	60
2,5-dimethyl-2,5-di(t-butyl peroxy) hexane	0.64	60 90 120	29 35 26	820 700 630	1,000 850 850	24 12 13

*Poly(methyl hydrogen siloxane) end-blocked with trimethylsilyl groups; viscosity 30 cps. at 25° C.
**As a 50 percent paste in a low molecular weight poly(dimethylsiloxane).
***Too low to measure.

Example 3: A 100% solids coating composition was prepared by dissolving 30 g. of poly(methyl methacrylate) having a reduced viscosity of 0.38 using a 0.5 weight percent benzene solution in a mixture of 52.5 g. of 2-butoxyethyl acrylate, 12.5 g. of neopentyl glycol diacrylate, 5 g. of 5-norbornen-2-yl-methyl 5-norbornene-2-carboxylate and 3 g. of benzil.

The coating was then applied to Bonderite No. 37 steel sheets with a wire-wound rod so as to apply a wet film of 0.5 mil thickness. The coated panel was exposed to the high intensity predominantly continuum light radiation from a 50 kilowatt argon swirl-flow plasma arc at a distance of 2 feet for a period of 5 seconds. The cured coating exhibited a Sward hardness of 22 and a reverse impact of 150 inch-pounds.

LIGHT SENSITIVE ADDITIVES

PHOTOINITIATORS

Alpha-Methylolbenzoin Ethers

H. Hoffmann, H. Hartmann, C.H. Krauch and O. Volkert; U.S. Patent 3,689,565; September 5, 1972 have found that α-methylolbenzoin ethers having the formula:

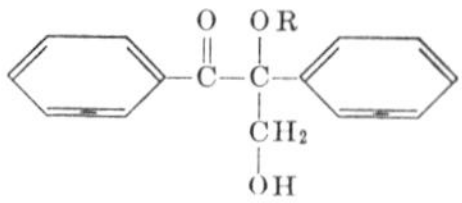

where R denotes an alkyl or alkoxyalkyl radical having one to eight carbon atoms and particularly one to four carbon atoms are particularly advantageous photoinitiators. Production is generally carried out by dissolving benzoin in an excess of a compound bearing the radical R (alcohol) and etherifying the benzoin under conventional conditions in the solution. A polar solvent which stabilizes carbanions, particularly dimethyl sulfoxide or dimethylformamide, is added and a 10% molar excess of formaldehyde, preferably in the form of paraformaldehyde in a solvent of the same type, particularly dimethylsulfoxide or dimethylformamide, is added to a solution of alkali metal hydroxide in alcohol.

The pH during the methylolation should be in the neutral or weakly alkaline range, preferably from 7 to 9. After the reaction is over, the methylolation product may be precipitated with water. A particular advantage of the α-methylolbenzoin ethers lies in their good thermal stability combined with good photoinitiating action. The improved thermal resistance may be clearly seen when comparing the ratio of the velocity constants of the photochemical decay k_p to the velocity constants of the thermal decay k_t of the individual benzoin compounds in a system containing photopolymerizable monomers, the following values being obtained at 50°C:

$$\text{C}_6\text{H}_5-\overset{\text{O}}{\underset{}{\text{C}}}-\overset{\text{OR}}{\underset{\text{R}^1}{\text{C}}}-\text{C}_6\text{H}_5$$

R	R^1	k_p/k_t
H	H	180
CH3	H	7,500
CH(CH3)2	H	7,900
H	CH3	8,300
H	C6H5	4,700
CH3	CH2OH	13,600

Improved thermal stability of photoinitiators is of very great importance for their use in practice. In the production of photopolymerizable coatings, plates, sheeting or film the substances being photopolymerized, for example in the process of dissolving or in laminating film or sheeting onto substrates or in the pressing of plates, are exposed to considerable thermal stress, often having to be kept at temperatures above 100°C. for long period. This is especially the case in the production of photopolymerizable elements from which relief printing plates are to be prepared later by exposure to light followed by washing out, especially when the elements have been prepared in a press. Even in the light-curing process itself, however, undesirable thermal stresses may occur, for example by the heat of polymerization or when using light sources characterized by poor energy conversion rates as regards the output of light.

The photopolymerizing substances should therefore (in order not to harden in procedures involving thermal stresses and not to polymerize in dark areas in optical information fixation) contain photoinitiators which (a) have substantial thermal stability and (b) decompose photochemically with a good quantum yield into radicals which are able to initiate the desired polymerization or cross-linking. The α-methylolbenzoin ethers satisfy requirements in practice to a far greater extent than the benzoin compounds previously used. The following are examples of α-methylolbenzoin ethers according to this process: the methyl, ethyl, butyl, hexyl, octyl, 2-ethylhexyl and cyclohexyl ethers and also 2-alkoxyethyl ethers such as the 2-methoxymethyl ether. It is preferred to use the methyl ether.

Benzoin-Silyl Esters

K. Fuhr, H. Vernaleken, H.-G. Heine, H. Rudolph and H. Schnell; U.S. Patent 3,636,026; January 18, 1972; assigned to Farbenfabriken Bayer AG, Germany describe α-substituted benzoin-silyl esters which are useful photosensitizers for the photopolymerization of polymerizable compounds or compound mixtures including mixtures of unsaturated polyesters and copolymerizable monomeric compounds. Typical examples of such benzoin-silyl esters are: benzoin-trimethylsilyl ester, $BP_{0.15}$ 118°C., MP 77°C.; α-methylbenzoin-trimethylsilyl ester, $BP_{0.03}$ 120°C., MP 32°C.; α-ethylbenzoin-trimethylsilyl ester, $BP_{0.04}$ 110° to 113°C., n_D^{20} 1.5377; α-allylbenzoin-trimethylsilyl ester, $BP_{0.01}$ 112° to 115°C., n_D^{20} 1.5432; α-phenylbenzoin-trimethylsilyl ester, $BP_{0.02}$ 155° to 158°C., MP 43° to 44°C.; 4,4'-dimethylbenzoin-trimethylsilyl ester, $BP_{0.05}$ 140°C., n_D^{20} 1.5412; 4,4'-dimethoxybenzoin-trimethylsilyl ester, $BP_{0.05}$ 180°C., n_D^{20} 1.5598.

The compounds can be manufactured in a manner which is in itself known by reaction of benzoins with substances which split off trialkylsilyl groups, for example trimethylsilyl chloride, hexamethyldisilazane and N-trimethylsilylacetamide, optionally in polar solvents such as pyridine, dimethylsulfoxide and dimethylformamide. Benzoin-trimethylsilyl ester can be manufactured in accordance with the following instruction: a solution of 50 g. of benzoin (0.236 mol) in 100 ml. of hexamethyldisilazane is heated for 1 hour to the boil with exclusion of moisture after the addition of 2 drops of concentrated sulfuric acid.

During this time the sump temperature is 125°C. During the course of the reaction ammonia is split off. Thereafter the excess hexamethyldisilazane is distilled off under normal pressure and the remainder is distilled off in vacuo. A double distillation of benzoin-trimethylsilyl ester follows in order to purify the crude product. The boiling point of the benzoin-trimethylsilyl ester is $BP_{0.15}$ 118°C. and the yield is 55 g., equal to 81%.

Example 1: 10 g. of extracted and freshly distilled acrylic acid methyl ester are mixed with 0.1 g. of two known sensitizers and one sensitizer according to the process. Illumination is carried out with a mercury vapor high pressure lamp (Philips HPK 125 W/L) through quartz glass in a waterbath at 24°C., at a distance of 10 cm. Here the solution of the sensitizer in the monomer is contained in a quartz glass of internal diameter 1.7 cm. under a nitrogen atmosphere. The time of illumination is 2 1/2 minutes. Immediately after the illumination the quartz glass is introduced into an acetone/solid carbon dioxide mixture in order to prevent a thermal polymerization. The solution of the polymer in the monomer and the solid polymer constituents which are present on the inside of the quartz glass on the side facing the mercury vapor high pressure lamp are introduced into a small round flask by means of small quantities of a solvent (methylene chloride). Thereafter unpolymerized monomeric constituents and the solvent are distilled off in a rotating evaporator. After drying in a vacuum drying cabinet to constant weight at 60°C., the amount of polymer is determined. Table 1 contains a summary of the amounts of poly(acrylic acid methyl ester) obtained with various sensitizers.

TABLE 1

Additives in Percent by Weight	Amount of Poly(Acrylic Acid Methyl Ester) in Percent by Weight
Benzoin, 1	16.5
Benzoin-ethyl ether, 1	24.4
Benzoin-trimethylsilyl ester, 1	28.2

If an initiator is not present, the amount of polymer is less than 0.1%.

Example 2:   An unsaturated polyester manufactured by condensation of 152 parts by weight of maleic anhydride, 141 parts by weight of phthalic anhydride and 195 parts by weight of propanediol-1,2 is mixed with 0.045 parts by weight of hydroquinone and dissolved in styrene to give a 65% by weight solution. Two parts by weight of two different known photosensitizers on the one hand and of three different photosensitizers according to the process on the other hand are added to 100 parts by weight at a time of this form in which the resin is supplied and the mixture is stored with exclusion of light at 60°C. until it gels. Table 2 contains the sensitizers employed and the values of the storage stability at 60°C.

TABLE 2

Sensitizer	Storage Stability at 60°C., days
Benzoin	<1
Benzoin-ethyl ether	<1
α-methylbenzoin-trimethylsilyl ester	>10
α-ethylbenzoin-trimethylsilyl ester	>10
α-phenylbenzoin-trimethylsilyl ester	>10

Benzoin Ethers and Organic Phosphines

A process described by B. Behrens and H. Delius; U.S. Patent 3,699,022;

October 17, 1972; assigned to Reichhold-Albert Chemie AG, Germany
relates to preparations curable by UV irradiation, such as molding, impregnating and coating compositions, of mixtures, stabilized in the usual manner, of unsaturated polyester resins and copolymerizable, monomeric compounds and photoinitiators, which optionally additionally contain polymerization initiators and/or metal accelerators and/or paraffin or wax or
wax-like substances, characterized in that they contain, as the photoinitiator, an overall combination consisting of: (a) benzoin ethers, and (b) a
subcombination of at least two different compounds of trivalent phosphorus,
consisting of (b') organic esters of phosphorus acid of general formula:

$$P{\overset{\displaystyle O-R_1}{\underset{\displaystyle O-R_3}{-O-R_2}}}$$

where R_1, R_2 and R_3 can be identical or different and represent aliphatic,
cycloaliphatic, aromatic, araliphatic or heterocyclic radicals (but one of
the radicals R_1, R_2 or R_3 must always be an aromatic radical) and (b") of
organic derivatives of phosphine of general formula:

$$P{\overset{\displaystyle R_1}{\underset{\displaystyle R_3}{-R_2}}}$$

where R_1, R_2 and R_3 can be identical or different and represent aliphatic,
cycloaliphatic, aromatic, araliphatic or heterocyclic radicals (but one of
the radicals R_1, R_2 or R_3 must always be an aromatic radical) and the percentages by weight mentioned relate to the total weight of unsaturated
polyester resin and copolymerizable monomers.

Known systems can be cured both by the action of high energy radiation
from, for example mercury vapor high pressure lamps, and by the action of
radiation of less high energy, for example under super-actinic radiation
from fluorescent lamps which possess a coating of fluorescent substances
which emit blue-violet light and rays in the longer wavelength UV, for
example in the range of 3000 to 5800 A. When using these benzoin ethers
as sensitizers, it is possible to photopolymerize mixtures of unsaturated
polyesters with unsaturated vinyl compounds, such as styrene, at room temperature in a relatively short time.

A content of benzoin ethers of about 2% by weight, relative to the mixture,
is necessary to achieve adequate activity. However, after irradiation under
mercury vapor high pressure lamps these mixtures show an intense yellow
coloration which does not fully disappear again even at the end of several

days. German Patent 1,934,637 describes unsaturated polyester preparations curable by irradiation which contain an unsaturated polyester with a hydroxyl number of 55 to 75, a copolymerizable monomer, and stabilizer combination of 200 to 800 parts per million of an alkyl phosphite and 10 to 50 parts per million of a copper salt of organic acids; in addition, 0.1 to 5% of benzoin or of a benzoin substituted in α-position by a C_1 to C_8 alkyl radical are used as a photoinitiator. Curing takes place by irradiation with light of wavelengths 2500 to 4000 A. These polyester preparations show improved storage stability in the dark, but do not show a shortened curing time relative to comparable photopolymerizable polyester preparations stabilized in other ways.

It is the object of this process to improve unsaturated polyester preparations which are curable by UV irradiation in at least four directions conjointly relative to the state of the art, namely: (1) To improve, quite considerably, the dark storage stability at room temperature and at elevated temperatures up to about 60°C.; (2) to increase the polymerization speed quite considerably, so that the UV radiation time can be correspondingly shortened; (3) to reduce the discoloration of the cured polyester preparation by reducing the sensitizer concentration used, while having a relatively short irradiation time; and (4) to obtain a nontacky surface when light-curing such air-drying polyester resins that contain proportional amounts of β,γ-unsaturated ether alcohols. These tasks are solved through benzoin ethers in certain combinations with compounds of trivalent phosphorus being contained in the unsaturated polyester preparation.

To test the storage stability in the dark of the preparations according to the process, 61 parts by weight of an unsaturated polyester resin from 1 mol of maleic anhydride, 1 mol of phthalic anhydride, and 2.18 mols of propanediol-1,2, stabilized in the customary manner with 0.015% by weight of hydroquinone, were mixed with 39 parts by weight of styrene and mixed (as in Table 1) with various overall combinations of (a) benzoin-ethyl ether, (b') triphenylphosphite and (b'') triphenylphosphine dissolved in styrene.

TABLE 1

Benzoin-ethyl ether [1]	Triphenyl-phosphite [1]	Triphenyl-phosphine [1]	Storage stability in the dark (60°) in days
0.25	0.2	0.1	24
0.5	0.2	0.1	24
0.5	0.4	0.1	24
0.5	0.6	0.2	22

[1] In parts by weight, relative to 110 parts by weight of mixture.

The following tests and examples illustrate the process.

Mixture A:　100 parts by weight of a polymerizable mixture consisting of 67 parts by weight of polyester resin (manufactured from 1 mol of maleic anhydride, 1 mol of phthalic anhydride, and 2.18 mols of propanediol-1,2) and 33 parts by weight of styrene, stabilized with 0.015% by weight of hydroquinone, and having a viscosity of approximately 1,500 cp. at 20°C. and an acid number of 28, are diluted with a further 10 parts by weight of styrene and mixed with 0.5 part by weight of a 10% strength by weight solution of paraffin (MP 52° to 53°C.) in toluene, in order to manufacture a coating composition.

Mixture B:　100 parts by weight of a polymerizable mixture consisting of 67 parts by weight of polyester resin (manufactured from 2 mols of maleic anhydride, 1 mol of phthalic anhydride and 3.08 mols of propanediol-1,2) and 33 parts by weight of styrene, stabilized with 0.014% by weight of hydroquinone, and having a viscosity of approximately 1,100 cp. at 20°C. and an acid number of 30, are diluted with a further 10 parts by weight of styrene and with 0.5 part by weight of a 10% strength by weight solution of paraffin (MP 52° to 53°C.) in toluene, in order to manufacture a coating composition.

Comparison Tests 1 through 31:　Various benzoin ethers and esters of phosphorus acid, or various benzoin ethers and triphenylphosphine, are added to Mixture A in the concentrations indicated in Table 2. The solutions thus obtained are cast in an approximately 1 mm. thick layer onto a glass plate and irradiated with the radiation from a UV lamp (El-Vak, Luminotest) at a distance of 17 cm. The gelling times are indicated in Table 2. By way of comparison, the gelling times of the mixtures free of phosphite and of the mixture free of phosphine, under the same conditions, are also given.

TABLE 2

Experiment number	Benzoin-ether added [1]	Phosphite added [1]	Triphenyl-phosphine [1]	Gelling time in seconds
1	0.5 benzoinmethyl ether			55
2	do	0.6 triphenyl phosphite		45
3	do	0.8 triphenyl phosphite		45
4	do	1.0 triphenyl phosphite		40
5	1.0 benzoinmethyl ether			35
6	do	0.6 triphenyl phosphite		32
7	do	0.8 triphenyl phosphite		30
8	do	1.0 triphenyl phosphite		30
9	0.5 benzoinethyl ether			55
10	do	0.6 triphenyl phosphite		45
11	do	0.8 triphenyl phosphite		43
12	do	1.0 triphenyl phosphite		45
13	do	5.0 triphenyl phosphite		37
14	do	10.0 triphenyl phosphite		29
15	do	20.0 triphenyl phosphite		32
16	do	0.6 trisnonylphenyl-phosphite		50
17	0.5 benzoinmethyl ether		0.2	18
18	do		0.6	15
19	do		1.0	15
20	1.0 benzoinmethyl ether		0.2	12
21	do		0.6	10
22	do		1.0	8
23	0.25 benzoinethyl ether		0.2	40
24	do		0.6	25
25	do		1.0	20
26	0.5 benzoinethyl ether		0.2	30
27	do		0.6	20
28	do		1.0	15
29	1.0 benzoinethyl ether		0.2	18
30	do		0.6	12
31	do		1.0	9

[1] In parts by weight, relative to 110 parts by weight of mixture A.

Comparison Experiments 32 through 34: Benzoin-ethyl ether and triphenyl phosphite or benzoin-ethyl ether and triphenylphosphine are added to mixture B in the concentrations indicated in Table 3. Using the same conditions as indicated in Comparison Tests 1 through 31, the gelling times indicated in Table 3 are achieved.

TABLE 3

Experiment No.	Benzoin-ethylether [1]	Triphenyl phosphite [1]	Triphenyl phosphine [1]	Gelling time in seconds
32	0.5			45
33	0.5	0.6		30
34	0.5		0.2	14

[1] In parts by weight, relative to 110 parts by weight of mixture B.

As indicated in Table 2, tests 2 to 4, 6 to 8, 10 to 16 and in Table 3, test 33, the addition of 0.6 to 20% by weight phosphites relative to 110 parts by weight of mixture A, respectively 110 parts by weight of mixture B, causes a decrease of the gelling time during UV irradiation. An addition of triphenylphosphine according to Table 2, tests 17 to 31, respectively, and Table 3, test 34, causes a substantial decrease of the gelling time during UV irradiation, but the storage stability in the dark is also rather deteriorated.

As can be seen in Table 1 the gelling time can be shortened by the use of the overall combination according to the process as well as the storage stability in the dark can be improved. The addition of the ingredients of the overall combination, namely benzoin ethers (a), phosphite compounds (b') and phosphine compounds (b"), can be ensured in any order. In the most preferred order first the phosphite compound (b') then the phosphine compound (b") and last the benzoin ether compound (a) is added. But it is substantial that the named ingredients (a), (b') and (b") are contained in the polyester preparation as a combination before starting the UV irradiation.

Examples 1 through 11: Various benzoin ethers (a), triphenyl phosphite (b') and triphenyl phosphine (b"), in the concentrations indicated in Table 4, are added to the mixture A which has been indicated. Under the same light exposure conditions as indicated in Comparison Tests 1 through 31, the gelling times indicated in Table 4 are achieved.

Examples 12 and 13: Benzoin-ethyl ether (a), triphenyl phosphite (b') and triphenylphosphine (b"), in the concentrations indicated in Table 5, are added to the Mixture B which has been indicated. Under the same light exposure conditions as indicated in Comparison Tests 1 through 21, the gelling times indicated in Table 5 are achieved. From the gelling times indicated in Table 4 and Table 5 the described synergistic effect can be derived which appears when the overall combination (a) and (b) (b') and (b") is being used. This is surprising so far as the esters of the phosphorus acid during conventional curing of unsaturated polyester resins and copolymerizable monomer compounds by peroxides are known to be strong inhibitors already in small amounts.

TABLE 4

Example No.	Sensitiser added [1]	Triphenyl phosphite added [1]	Triphenyl phosphine added [1]	Gelling time in seconds
1	0.25 benzoinemethylether	0.2	0.1	45
2	0.25 benzoinmethylether	0.2	0.2	30
3	0.5 benzoinmethylether	0.2	0.1	30
4	0.5 benzoinmethylether	0.4	0.2	10
5	0.25 benzoinethylether	0.2	0.1	30
6	0.25 benzoinethylether	0.4	0.2	25
7	0.5 benzoinethylether	0.2	0.1	16
8	do	0.4	0.1	15
9	do	0.4	0.2	10
10	do	0.6	0.2	9
11	1.0 benzoinethylether	0.4	0.4	5

[1] See footnote ([1]) at bottom of Table 2.

TABLE 5

Example No.	Benzoin-ethylether [1]	Triphenyl-phosphite added [1]	Triphenyl-phosphine added [1]	Gelling time in seconds
12	0.5	0.2	0.1	17
13	0.5	0.6	0.2	12

[1] See footnote ([1]) at bottom of Table 3.

Examples 14 through 20:   0.5 part by weight of benzoin-ethyl ether (a), 0.6 part by weight of triphenyl phosphite (b') and 0.2 part by weight of triphenylphosphine (b"), and additionally various conventional polymerization initiators or accelerators, in the concentrations indicated in Table 6, are added to the Mixture A which has been indicated. Under the same light exposure conditions as indicated for Comparison Tests 1 through 31, the gelling times indicated in Table 6 are achieved.

TABLE 6

Example No.	Additives in parts by weight, relative to 110 parts by weight of mixture A	Gelling time in seconds
14	2 methyl-ethyl-ketone peroxide (50% by weight)	35
15	1 benzoyl peroxide paste (50% by weight)	18
16	1 cobalt octoate (6% by weight of Co)	23
17	1 zirconium octoate (6% by weight of Zr)	39
18	2 azobisisobutyronitrile	16
19	1 ethyl acetoacetate	16
20	2 dimethylaniline (10% by weight solution)	31

Example 21:   1.5 parts by weight of benzoin-ethyl ether (a), 0.6 part by weight of triphenyl phosphite (b') and 0.2 part by weight of triphenylphosphine (b") are added to 90 parts by weight of the Mixture B yet without addition of paraffin which has been indicated. In order to manufacture a roller filling composition, 15 parts by weight of monostyrene, 50 parts by weight of talc, 25 parts by weight of blanc fix (barium sulfate), 80 parts by weight of ground gypsum (calcium sulfate) and 1 part by weight of high disperse silica (Aerosil 380) are worked into this preparation, which is applied to a chipboard at a coating thickness of 250μ.

After an irradiation time of 30 seconds in a UV light curing tunnel, the roller filling composition has cured and can, after cooling to room temperature, be rubbed down without clogging of the wet emery paper of grade 400 which is used. The UV light curing channel is equipped with a high pressure burner and two HTQ7 mercury vapor high pressure lamps (length of each 755 mm., diameter 12 mm., spaced at 15 cm. from one another). Cooling is by air, and the irradiation distance is 20 cm.

Example 22:   An air-drying unsaturated polyester resin is manufactured in a known manner by condensation of 306 g. of fumaric acid, 133 g. of tetrahydrophthalic anhydride, 368 g. of diglycol and 113 g. of pentaerythritoltriallyl ether in the presence of 0.3 g. of hydroquinone. The mixture of 65 parts by weight of this polyester resin and 35 parts by weight of styrene has a viscosity of approximately 700 cp. at 20°C. and an acid number of 25. To manufacture a lacquer solution, 10 parts by weight of styrene, 1.5 parts by weight of benzoin-ethyl ether (a), 0.6 part by weight of triphenyl phosphite (b') and 0.2 part by weight of triphenylphosphine (b") are added to 100 parts by weight of this mixture.

When applied to a glass plate at a film thickness of 100μ, and after a period of irradiation of 30 seconds under the mercury vapor high pressure lamps described in Example 1, and subsequent cooling to room temperature, this lacquer solution yields a nontacky, highly glossy film. The above result is surprising because an appropriate mix free of phosphite and free of phosphine has only very unsatisfactory storage stability in the dark especially at higher temperatures up to about 60°C., and it does not gel under UV irradiation conditions described in Comparison Tests 1 through 31 even after irradiation times of about 120 seconds.

Bis(p-Aminophenyl-Unsaturated) Ketone

A process described by M.D. Baum and C.P. Henry, Jr.; U.S. Patent 3,652,275; March 28, 1972; assigned to E.I. du Pont de Nemours and Company is directed to photodissociable hexaarylbiimidazoles in combination with a bis(p-aminophenyl-α,β-unsaturated) ketone sensitizer that

absorbs in the visible light wavelengths. Hexaarylbiimidazoles dissociate
upon exposure to ultraviolet light to form stable colored triarylimidazolyl
radicals useful as light screens as described in British Patent 997,396. Such
dissociation is useful in hexaarylbiimidazole-leuco dye compositions in that
the triarylimidazolyl radical, formed as described above, oxidizes the leuco
form of the dye to the colored form. Thus, colored images are obtained
making the compositions useful in imaging applications, as described in
U.S. Patent 3,445,234.

The hexaarylbiimidazoles in general absorb largely and maximally at ultraviolet wavelength below 300 mμ and to some much lesser extent at wavelengths as high as 430 mμ. Thus, while any of the compositions described
above containing the hexaarylbiimidazole are sensitive to radiation over
substantially the whole ultraviolet range, they respond most efficiently to
radiation that corresponds to or substantially overlaps the region of maximum
absorption. It is not always practical to irradiate fully into this region.
For example, in some imaging applications, it is desired to cover the photosensitive hexaarylbiimidazole-leuco dye imaging composition with a transparent film. Some film materials, such as Mylar and Cronar commercial
polyesters, otherwise suitable, are not transparent below 300 mμ, and thus
prevent such short wavelength activating radiation from reaching the biimidazole, with consequent loss in efficiency.

Further, many commercially important ultraviolet sources, such as cathode
ray tubes widely useful in imaging devices that convert electrical to light
energy and transmit such light as images to photosensitive surfaces (plates,
papers, films), emit mainly in the near ultraviolet and above, owing in part
to limitations in the available phosphors and in part to the screening by the
fiber optic face plate of radiation below 300 mμ. Thus, imaging with such
radiation sources is not entirely satisfactory as to the imaging speeds and
optical densities that the hexaarylbiimidazole-leuco dye systems can in-
herently provide. Thus, as the activating radiation contains increasing
proportions of visible components or as components closer to the ultraviolet
region are filtered out, hexaarylbiimidazole activation becomes less efficient as to the amount of energy utilized.

U.S. Patent 3,479,185, describes photopolymerizable compositions containing a monomer, a free radical producing agent such as leuco triphenyl-
amine dye, and a hexaarylbiimidazole. Moreover, these photopolymerizable compositions can optionally contain an energy-transfer dye such as
Erythrosin (C.I. Acid Red 51), Rose Bengal (C.I. Acid Red 94), Eosin Y
(C.I. Acid Red 87), or Phloxin B (C.I. Acid Red 92). These dyes extend
the sensitivity of the three-component system into the visible spectral region
and also increase the speed of polymerization. The resulting four-component system can initiate polymerization with exposure to visible light only,

is stable, and does not lose sensitivity on aging. In the four-component system, the absorption of energy by the dye induces the same reaction from the lophine dimer combination as direct irradiation of the lophine dimer in the three-component system. Other work describes photopolymerizable compositions containing hexaarylbiimidazoles and p-aminophenyl ketones such as Michler's ketone. The ketone sensitizers extend the spectral sensitivity of the biimidazoles in the visible region of the spectrum.

This process enhances the efficiency of the hexaarylbiimidazole systems described above, especially the photopolymerizable compositions, in the visible light region of absorption through the use of selected bis(p-aminophenyl-α, β-unsaturated) ketone sensitizers. The process is directed to a photoactivatible composition comprising a mixture of the following:

(a) A hexaarylbiimidazole that has its principal light absorption bands in the ultraviolet region of the electromagnetic radiation spectrum and is dissociable to triarylimidazolyl radicals on irradiation with such absorbable ultraviolet light, and

(b) a sensitizing amount of bis(p-aminophenyl-α, β-unsaturated) ketone of the formula

where R_1 is alkyl of 1 to 4 carbon atoms, or hydrogen; R_2 is alkyl of 1 to 4 carbon atoms, or hydrogen; R_3 is hydrogen, alkyl of 1 to 4 carbon atoms, chlorine or methoxy; R_4 and R_5 are each hydrogen, alkyl of 1 to 4 carbon atoms or phenyl; with the proviso that R_4 and R_5 can be taken together and are $-CH_2-CH_2-$, $-CH_2-CH_2-CH_2-$, or $-CH_2-CH_2-CH_2-CH_2-$; and n is 0 or 1; the ketone having its principal light absorption bands in the visible regions of the electromagnetic radiation spectrum, and, optionally,

(c) a leuco dye that is oxidizable to dye by triarylimidazolyl radicals.

The process is also directed to a photopolymerizable composition which comprises (a) and (b), as defined above, and additionally,

(d) an addition-polymerizable, ethylenically unsaturated monomer, and optionally,

(e) a photooxidizable amine which may be component (c) above, and
optionally,

(f) a chain transfer agent.

Thus, it has been found that an α,β-unsaturated ketone as defined above,
which absorbs light at longer wavelengths than the hexaarylbiimidazoles
can transfer such absorbed long wavelength light energy to the hexaarylbiimidazoles, i.e., the α,β-unsaturated ketone can sensitize the hexaarylbiimidazole, thus converting it to the triarylimidazolyl radical. By thus
extending the spectral sensitivity of the hexaarylbiimidazoles to wavelengths
they do not normally absorb or absorb only weakly, the α,β-unsaturated
ketone significantly enhances their utility as light screens, photooxidants
and photopolymerization initiators.

The preparation of these sensitizing ketones involves an acid or base cata-
lyzed bis-condensation reaction of 2 mols of the appropriate dialkylamino-
benzaldehyde or cinnamaldehyde with one mol of a ketone such as cyclopentanone, cyclohexanone, acetone, etc. Preferred compounds 1 and 2
are described in Chem. Zent, 1908, p. 637 to 639; the other cycloalkanone
derivatives are readily prepared by comparable syntheses. Michler's ketone
vinylogs are described in U.S. Patents 3,257,202; 3,265,497; 2,860,983
and 2,860,984. The following examples illustrate the process.

Example 1:   The ketone sensitizers were prepared by known methods. Dicondensation was easily accomplished in 2B denatured alcohol using sodium
hydroxide catalysis, generally with a short reflux period. A high-melting
product precipitated in 50 to 95% yield upon cooling; filtration followed
by a 2B alcohol wash yielded a product of sufficient purity to use directly
in further reactions. The compounds structures were verified by NMR
spectra, supplemented by elemental analyses in some instances.

Spectral properties of these compounds were quite unusual. The vinyl protons are at extremely low field, admixed with the aromatic signal in the
NMR. The IR spectra (solution or mull) showed the carbonyl absorption at
about 6.15 microns, superimposed on the aromatic and vinyl absorption.
The ultraviolet and visible absorption spectra are shown in Table 1. Double
bond configuration is not shown; a mixture is suspected. The ketone compounds have the following structure.

(1)

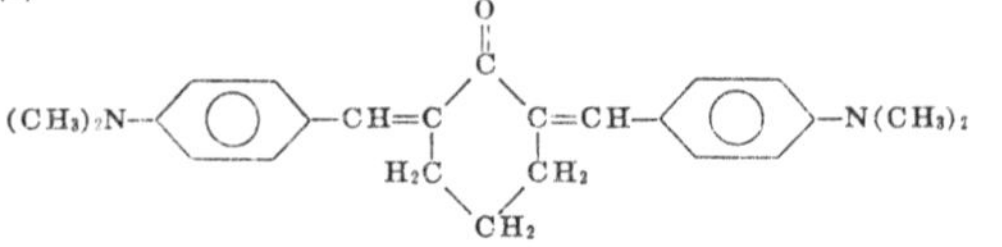

(2)

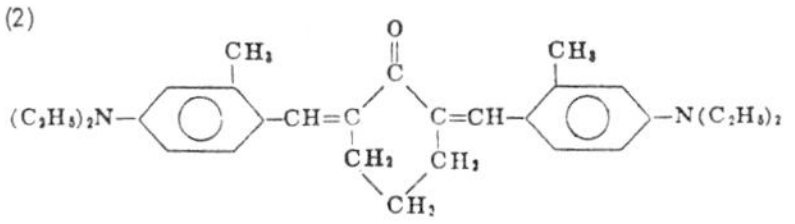

(3)

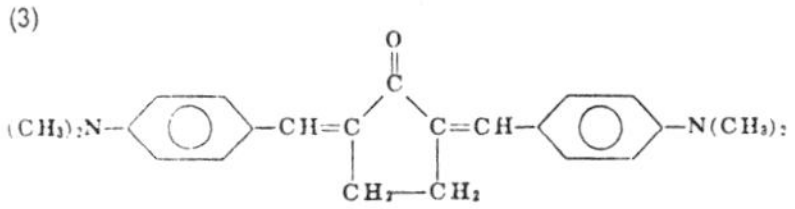

(4)

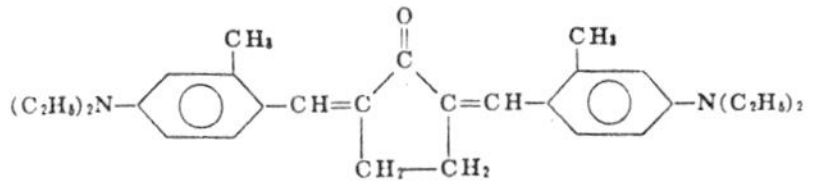

(5)

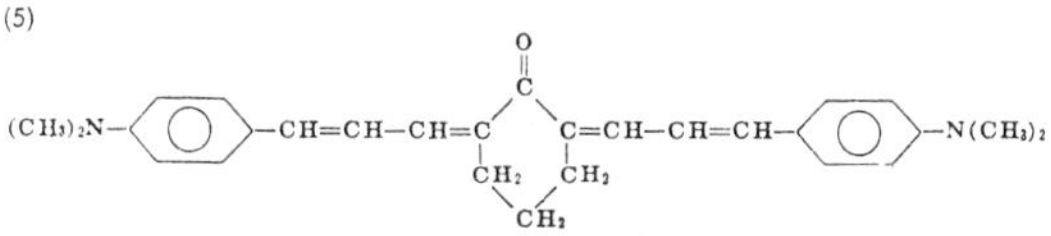

(6)

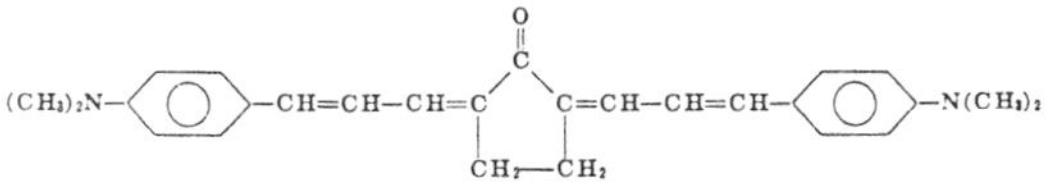

(7)

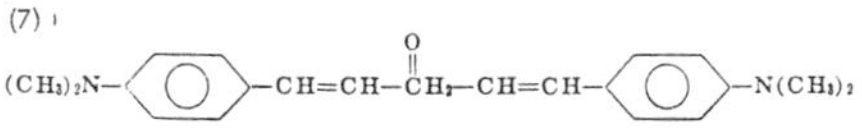

TABLE 1: ULTRAVIOLET AND VISIBLE ABSORPTION SPECTRA OF ALKANONE CONDENSATION PRODUCTS IN CHLOROFORM

Sensitizer	λmax., mμ	Extinction Coefficient
(1)	310	13,200
	434	19,500
(2)	284	22,800
	447	28,300
(3)	275	14,900
	317	7,700
	446	53,500
(4)	272	17,600
	314	4,200
	326	4,300
	430	33,600
	455	34,400
(5)	318	17,000
	480	51,100
(6)	315	18,600
	505	64,000
(7)	263	14,800
	330	6,300
	442	41,300
(8)	280	18,500
	480	64,000

Preparation of 2,6-bis(4'-dimethylaminobenzylidene)cyclohexanone (1) is
as follows. A solution of 10.0 g. (0.102 mol) of cyclohexanone, 30.4 g.
(0.204 mol) of 4-dimethylaminobenzaldehyde, and 25 ml. of 25% aqueous
sodium hydroxide in 500 ml. of 2B alcohol was refluxed 3 hours, then cooled
in an ice bath. The resultant orange-red crystals were filtered off and
washed with 2B alcohol. The melting point of the product was 248° to
251°C. and the yield was 24 g. (59%). The NMR, IR and UV analyses
agreed with the proposed product. Calculated for $C_{24}H_{28}N_2O$: C, 79.86;
H, 7.83; N, 7.77. Found: C, 78.89; H, 7.53; N, 7.92.

Preparation of 2,6-bis(2'-methyl-4'-diethylaminobenzylidene)cyclohexa-
none (2) is as follows. One hundred and six and six-tenths grams of crude,
filtered 2-methyl-4-dimethylaminobenzaldehyde was condensed with 29 g.
(0.296 mol) of cyclohexanone in the same manner as for (1) above to obtain
65 g. (51.5%) of orange crystals, MP 181° to 182.5°C. NMR, IR and UV
analyses agreed with the proposed product.

Preparation of 2,5-bis(4'-dimethylaminobenzylidene)cyclopentanone (3) is
as follows. Dicondensation of 56 g. (0.38 mol) of 4-dimethylaminobenz-
aldehyde with 16.5 g. (0.2 mol) of cyclopentanone in the same manner as
for (1) above led to 63 g. (95%) of orange crystals, MP 300° to 303°C.
NMR, IR and UV analyses agreed with the proposed product.

Preparation of 2,5-bis(2'-methyl-4'-diethylaminobenzylidene)cyclopenta-
none (4) is as follows. 204 g. of crude, filtered 2-methyl-4-diethylamino-
benzaldehyde and 35 g. (0.417 mol) of cyclopentanone were condensed in
the manner described above to obtain 161 g. (95%) of deep red crystals,
MP 167° to 170°C. NMR and IR analyses agreed with the proposed product.

Dicondensation product of 4-dimethylaminocinnamaldehyde and cyclohexanone (5) is as follows. Condensation of 5.3 g. of 4-dimethylaminocinnam-
aldehyde with 1.5 g. of cyclohexanone in the manner described for (1)
above gave 4.5 g. (70%) of deep red magenta crystals, MP 258° to 261°C.,
and after recrystallization from alcohol/chloroform, MP 260° to 263°C.,
with acceptable IR, UV and NMR characteristics.

Dicondensation product of 4-dimethylaminocinnamaldehyde with cyclopentanone (6) is as follows. Synthesis by the procedure of (5) above gave
a 93% yield of very deep maroon crystals, MP 258° to 264°C., with acceptable IR, UV and NMR spectra.

Preparation of 1,3-bis(4-dimethylaminobenzylidine)acetone (7) is as follows.
Attempts to run this reaction at reflux temperature of alcohol led to untrac-
table tars. But when 5.8 g. of acetone (0.10 mol) and 29.84 g. (0.20 mol)
of 4-dimethylaminobenzaldehyde were stirred for 7 hours at room temperature

under nitrogen, then cooled to ice temperature, filtration gave orange crystals, 6.4 g. (19%), MP 181° to 190°C. (dec.) after washing with 2B alcohol and drying.

In a like manner, the ethyl homolog of ketone (3) was obtained, MP 187° to 189.5°C., having the following structure:

(8)

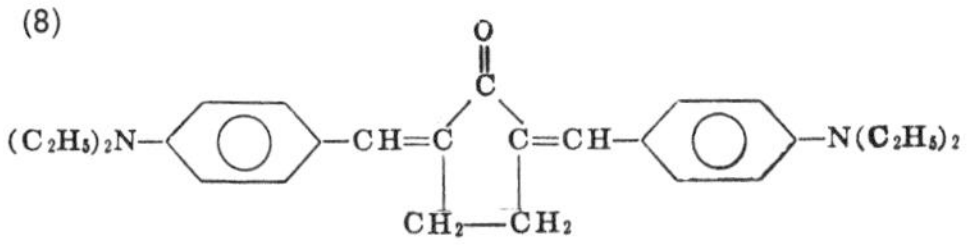

Example 2: Photopolymerization — Formulations were prepared from the following ingredients:

TABLE 2

Formulation	A	B	C	D	E	F	G	H
Acetone (ml.)	80	80	80	80	80	80	80	80
Cellulose acetate butyrate (grams) (Eastman Chemicals EAB-381-20)	10	10	10	10	10	10	10	10
Triethyleneglycol dimethacrylate (ml.)	10	10	10	10	10	10	10	10
Ketone (4) (grams)	0.050	0.050	0.050	0.050				0.050
2-Mercaptobenzoxazole (grams)		0.010		0.010	0.010	0.010	0.010	
2,2'-(o-chlorophenyl-4,4'-5,5'-tetrakis (m-methoxyphenyl) biimidazole (grams)			0.300			0.300		0.300
2,2'-bis (o-chlorophenyl)-4,4',5,5'-tetraphenylbiimidazole (grams)				0.300			0.300	

The formulations were coated on 3 mil Melinex X503 polyester film, warmed slightly to evaporate the acetone, and laminated with 0.5 mil (50S) Mylar polyester film. Polymerization was determined by dusting with pigments that do not adhere to polymerized areas of the delaminated film.

(a) Films prepared from formulations C, D, F and G were exposed to light of intensity 1.5 mw./cm.^{2} from a mercury-vapor lamp; two Corning filters 7-54 and one 0-52 filter were used to give 40 mμ bands of incident light centered near 366 μ. Under these conditions, formulation F was fully photopolymerized in 8 seconds, formulation G in 16 seconds, illustrating that hexaarylbiimidazole of F is approximately 2 times faster than that of G. Formulation C exhibited full photopolymerization in 8 seconds, D in 16 seconds. Thus, at near UV wavelengths (366 mμ), ketone (4) has no effect on the polymerization speeds which were the same as G and F.

(b) Repeating (a), but with light of intensity 10.0 mw./cm.^{2} (mercury-vapor lamp) with a wavelength range of about 40 mμ centered near 430 mμ, resulting from the use of one Corning 7-59 and one 3-74 filter, gave entirely different results. Under these conditions, formulation D exhibited a

photopolymerization rate (16 sec.) about four times that of G (64 sec.).
Similarly, the ketone sensitized formulation C is about three times faster
than the unsensitized formulation F. These data illustrate the efficiency
of ketone (4) in photopolymerization.

(c) Repeating (b), including film formulation B, but with light of intensity
25.0 mw./cm.^{2} and at wavelength greater than 430 mμ, resulting from the
use of one Corning 3-72 filter and one 1-69 filter, gave still different results. Under these conditions, formulations F and G (no sensitizer) showed
no photopolymerization even after 4-minute exposures. Formulations C
and D, on the other hand, exhibited photopolymerization rates as shown in
(a), namely 8 and 16 seconds, respectively. Under these long wavelength
irradiation conditions, there is, apparently, no absorption of the biimidazole, hence no photodissociation or photoinduced polymerization.

The presence of ketone (4), on the other hand, provides a photopolymerization rate equal to irradiation with near ultraviolet light (a). The results
obtained with formulation B (no biimidazole) are informative. Film formulation B exhibited photopolymerization, but only after about 32 seconds
exposure. Thus, ketone (4) and 2-mercaptobenzoxazole can initiate photopolymerization, but at rates 2 to 4 times slower than when a biimidazole is
present.

(d) Film formulations A, E, and H were irradiated as in (c) above. Formulations A and E showed no photopolymerization with exposures up to 4 minutes. These results indicate that the ketone sensitizer alone (formulation A)
is ineffective for initiating photopolymerization, and that the chain transfer
agent alone (formulation E) is equally ineffective. Formulation H exhibited
no photopolymerization on exposure for 2 minutes, slight polymerization
when irradiated 4 minutes. This result shows that photopolymerization can
occur in the absence of a chain transfer agent, but that photopolymerization
is markedly improved when a chain transfer agent is present.

Examples 3 through 8: Photopolymerization — Mylar (1 mil thick) polyester
film was coated to a wet thickness of 6 mil using an acetone solution of
cellulose acetate butyrate (13.2 g.), triethyleneglycol dimethacrylate
(12.5 ml.), 2,2'-bis(o-chlorophenyl)-4,4',5,5'-tetrakis-(m-methoxyphenyl)biimidazole (3.0 g.), 2-mercaptobenzoxazole (0.10 g.) and various
amounts of Michler's ketone (MK), p,p-bis(dimethylamino)benzophenone,
and/or ketone sensitizers of this process, and laminated as in Example 2.

The films were irradiated at two different wavelengths, obtained by the
use of suitable filters. Irradiation with light at about 366 mμ, incident
intensity of 1.00 mw./cm.^{2}, was obtained using two Corning 7-54 filters
and one 0-52 filter. Irradiation at 430 millimicrons, incident intensity of

10 mw./cm.^{2}, was obtained using one Corning 7-59 and one 3-74 filter. The irradiation time required to give complete photopolymerization at the two wavelengths is shown in Table 3.

TABLE 3

Example number	Additive	$I_1 = 10.0$ mw./cm.^{2} ~430 mJ (7-59, 3-74) Time (sec.)	$I_1 = 1.00$ mw./cm.^{2} ~366 mμ (7-54(2), 0-52) Time (sec.)
3	None (control)	8	2
4	0.5 g. Michler's ketone (MK)	5	1
5	1.5 g. MK	4	1
6	50 mg. ketone sensitizer (8)	2	2
7	72 mg. ketone sensitizer (8), and 100 mg. MK	2	1
8	50 mg. ketone sensitizer(2)	3	2

The examples are informative. It is obvious that when irradiating in the near UV (Example 3, ≈ 366 mμ), photopolymerization is rapid. Further, addition of Michler's ketone essentially doubles the rate of photopolymerization, while ketone sensitizers (2) and (8) afford no improvement over the control.

Irradiation in the visible (≈430 mμ), however, presents quite a different picture. Firstly, the control (no additive) is four times slower at ≈430 mμ vs. ≈366 mμ. Secondly, the addition of Michler's ketone, in relatively large amounts, increases the photopolymerization rate by about 2 times. Ketone sensitizers (8) and (2), however, are more efficient than Michler's ketone at increasing rate of photopolymerization, and required relatively small quantities of sensitizer to achieve this improvement. Clearly, the ketone sensitizers of this process are more beneficial at visible wavelengths. Indeed, with ketone sensitizer (2), the photopolymerization rate at ≈ 430 mμ is equal to unsensitized photopolymerization in the near ultraviolet, ≈ 366 mμ.

Polymeric Phenones

A process described by F. Agolini; U.S. Patent 3,641,217; February 8, 1972; assigned to E.I. du Pont de Nemours and Company provides light sensitizing polymers of a monomer having the following structural formula, where R is selected from methyl and phenyl and R$_1$ is selected from hydrogen and methyl.

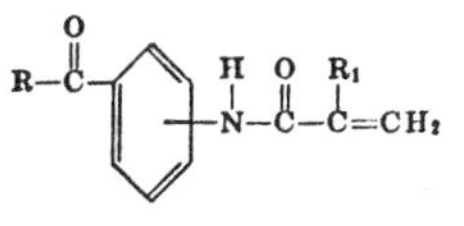

The process also involves copolymers of from 0.01 to 10 mol percent of the above monomers with alpha olefins having from 2 to 4 carbon atoms. These copolymers, in addition to providing desirable formed structures by themselves, are particularly useful in the cross-linking of polyethylene and polypropylene, and there is accordingly also provided a cross-linked polymeric composition comprising a blend of the above copolymer and an alpha olefin selected from polyethylene and polypropylene where the copolymer comprises at least about from 0.1% of the blend. The copolymers and polymer blends of the process can be formed into various shaped articles such as self-supporting films, laminates, coatings, filaments and tubing.

The shaped article is then exposed to radiation having a wavelength of 2000 to 7000 A., preferably ultraviolet radiation of 2000 to 4000 A., for a period of time sufficient to produce cross-linking, such period being at least 0.1 second under high energy xenon radiation but usually from 5 seconds to about 30 minutes under conventional radiation means, e.g., sunlamps and sunlight. After irradiation, besides exhibiting increased strength, the shaped articles display increased modulus (stiffness), improved resistance to grease and oil, increased resistance to stress-cracking and an improvement in their high temperature properties.

The shaped articles, particularly the self-supporting films, find utility in packaging applications where high oil and grease resistance is required, i.e., containers for potato chips, bacon rind, etc. The shaped articles are also useful in industrial construction; for example, as protective sheeting that is resistant to "creep." Sheets containing the copolymers that had been exposed to radiation are also useful in photoreproduction processes. In the following examples, which illustrate the process, parts and percentages are by weight. In these examples, Dynamic Zero Strength Temperature (abbreviated DZST) is determined according to ASTM-D-1430.

Example 1:   Preparation of 4-Acrylamidobenzophenone — In a vessel fitted with mechanical stirrer and thermometer there is placed 100 g. of benzanilide, 80 g. of benzoic anhydride and 800 g. of polyphosphoric acid. The reaction mixture is heated with stirring at a temperature of from 150° to 155°C. for 3 hours after which the mixture is cooled to 95°C. and then slowly added to one liter of water with stirring. The precipitate which forms is filtered and washed with further stirring in 40% sodium hydroxide solution (500 ml.) to take up any byproduct benzoic acid. The mixture is filtered and the product hydrolyzed in a mixture of 60 ml. of concentrated sulfuric acid and 75 ml. of ethanol over a period of 2 hours.

The mixture is poured into 500 ml. of water whereupon the amine sulfate is precipitated. The product is further washed with 500 ml. of ether to remove the ethyl benzoate and the remainder is then hydrolyzed in 500 ml. of 40%

sodium hydroxide solution on a steam heater for one hour. The solid material is collected, washed with water and dried to give a yield of 61 g. of product, identified as 4-aminobenzophenone, having a melting point of 118° to 120°C. To a solution of 20 g. of sublimed p-aminobenzophenone in 50 ml. of anhydrous dimethylacetamide, cooled with an ice-water bath, there is added dropwise 9.1 g. of freshly distilled acryloyl chloride over a period of 5 minutes. The reaction mixture is further stirred for approximately 25 minutes and is then precipitated by the addition of 150 ml. of water. The light yellow product is filtered, washed with water and dried, is then recrystallized from hot methanol and finally dried in an oven at 50°C. under reduced pressure.

There is obtained 17 g. of a colorless product having a melting point of 143° to 144°C. An additional 5 g. of product is obtained by adding 10 ml. of water to the mother liquor. The product is identified as 4-acrylamidobenzophenone. Elemental analysis: calculated for 4-acrylamidobenzophenone (%): C,76.47; H, 5.22; N, 5.57. Found (%): C, 76.29; 75.51; H, 5.39, 5.44; N, 5.72, 5.64. If the above procedure is repeated, using p-aminoacetophenone instead of p-aminobenzophenone, 4-acrylamidoacetophenone will be obtained as a product.

Copolymerization of 4-Acrylamidobenzophenone with Ethylene — A solution of one gram of 4-acrylamidobenzophenone dissolved in 100 ml. of anhydrous benzene is placed in a pressure reactor along with 20 ml. of catalyst solution (1.5 ml. of tertiary butyl peracetate in 150 ml. of cyclohexane). The reactor is then pressured with ethylene to 2,000 atmospheres and held at 140°C. until no more ethylene is absorbed. The reaction product is washed 3 times with 200 ml. of methanol and dried in an oven at 63°C. under reduced pressure. There is obtained 17.7 g. of copolymer having an inherent viscosity of 0.66 measured at a concentration of 0.5% in xylene at 120°C. The product contains 0.27% of nitrogen.

Irradiation of Copolymer — The prepared copolymer is milled into polyethylene (Alathon 15) to give a blend containing 0.5% phtosensitizer. The blend is pressed into film having a thickness of 1.06 ml. Samples of the film are subjected to irradiation with a General Electric UA3, 300 watt medium pressure mercury vapor lamp at a distance of 3 inches from the film sample. The effect of the irradiation on the dynamic zero strength temperatures of the various samples is then determined.

The results are shown in tabular form below. The increase in DZST is indicative of photocross-linking in the polymer film, and the increase in zero strength temperature is also correlative with improved heat sealing performances.

Exposure Time (sec.)	DZST, °C.
0	103
5	129
10	157
20	216
30	218

Example 2:   Example 1 is repeated, except that the film samples contain 0.4% of the photosensitizer and the thickness of the film is 1.03 mil. The results are shown below:

Exposure Time (sec.)	DZST, °C.
0	101
5	110
10	128
20	137
30	156

N-Alkoxy Heterocyclics for Vinyl Polymerization

In a process described by P.W. Jenkins, D.W. Heseltine, and J.D. Mee; U.S. Patent 3,699,026; October 17, 1972 energy sensitive compounds which contain a heterocyclic nitrogen atom substituted with an —OR group are photochemical initiators for the polymerization of vinyl monomers.

TABLE 1:   PHOTOCHEMICAL INITIATING COMPOUNDS

(1)	3-ethyl-1'-methoxyoxa-2'-pyridocarbocyanine perchlorate
(2)	1'-ethoxy-3-ethyoxy-2'-pyridocarbocyanine tetrafluoroborate
(3)	3'-ethyl-1-methoxy-2-pyridothiacyanine iodide
(4)	1-ethoxy-3'-ethyl-2-pyridothiacyanine tetrafluoroborate
(5)	1-benzyloxy-3'-ethyl-2-pyridothiacyanine iodide
(6)	3'-ethyl-1-methoxy-2-pyridothiacarbocyanine iodide
(7)	1-ethoxy-3'-ethyl-2-pyridothiacarbocyanine tetrafluoroborate
(8)	anhydro-3'-ethyl-1-(3-sulfopropoxy)-2-pyridothiacarbocyanine hydroxide
(9)	1-benzyloxy-3'-ethyl-2-pyridothiacarbocyanine perchlorate
(10)	3'-ethyl-1-methoxy-2-pyridothiadicarbocyanine perchlorate
(11)	1'-methoxy-1,3,3-trimethylindo-2'-pyridocarbocyanine picrate
(12)	3'-ethyl-1-methoxy-4',5'-benzo-2-pyridothiacarbocyanine perchlorate
(13)	1-ethoxy-3'-ethyl-4',5-benzo-2-pyridothiacarbocyanine tetrafluoroborate
(14)	1-ethoxy-2-methylpyridinium tetrafluoroborate

The initiators of this process can be prepared by the method described by Jenkins, Heseltine and Mee in U.S. Patent 3,615,432.

Typical applications include photosensitive emulsions, printing plates such as lithographic plates and other photocopying applications. A printing plate can be made by coating a solution of monomer which on polymerization forms a hydrophobic polymer, solvent and initiator on a support such as paper, glass, plastic, metal, etc. The coating can then be exposed in an image-wise manner to a pattern of actinic radiation. Polymerization takes place only in the exposed areas. If the combination of initiator and monomer are very active such that polymerization continues into the unexposed areas, then it may be desirable to add a polymerization retarder so as to confine the polymerization to the light struck areas.

After exposure the element can be washed with a solvent for the monomer so that the unexposed areas are washed off and only polymerized areas remain. The final image is suitable for direct printing by inking the image and contacting it with a suitable surface such as paper, either directly or through an offset roller. The affinity for the ink is also related to the degree of polymerization and thus an extremely true and accurate reproduction can be made. The photo image generally has a higher affinity for most inks than the backing. The method is suitable for line copying and also for half-tone and for continuous-tone prints.

In color photography a three plate system can be used. Each of the plates is coated with a polymerizable solution containing a different dye initiator which absorbs a different primary color. After exposure of the three elements to a colored subject, each plate is washed to remove the unpolymerized areas and dried. One plate is inked with red, a second with blue and a third with green and, in register, each of the images is transferred to a single sheet of paper to produce a color reproduction of the original.

Example 1: A solution of 0.5 g. of N-vinylcarbazole in 2 ml. of dichloromethane (dried over Linde Type 3A molecular sieves) is prepared. Nitrogen gas is bubbled through the solution until the volume is reduced to that of a solution made up to 0.5 g. of N-vinylcarbazole in 1.5 ml. dichloromethane. The solution is transferred to a test tube containing a magnetic stirring bar and 2.5 mg. of the compound to be tested. Nitrogen is bubbled for 1 min. through the solution and the test tube is stoppered. The test tube is partially immersed in an acetone bath contained in a petri dish.

Dry Ice is used as the coolant and the temperature is monitored by means of a blank solution placed alongside the test solution which contains a thermocouple. After chilling and while stirring, the test solution is irradiated with the light from a 150 watt xenon arc lamp (filtered with Corning CS 3-73 and CS 3-74 filters if the test compound absorbs at wavelengths >400 nm.; unfiltered if the test compound absorbs in the ultraviolet). A visual comparison of the change in viscosity relative to the viscosity of a blank solution is summarized for a number of the initiators in Table 2.

TABLE 2

Initiator compound[1]	Anion	Temperature, °C.	Time of- irradiation, minutes	Final viscosity
Blank		20 to (−27)	90	Free flowing.[2]
1	ClO₄	20 to 30	30	Viscous.
12	ClO₄	20 to 27	30	Do.
14	BF₄	(−15) to (−20)	5½	Solid.
4	BF₄	do	30	Viscous.
13	BF₄	do	½	Do.
7	BF₄	do	30	Do.
2	BF₄	do	10	Do.
11	Picrate	20 to 27	30	Slightly viscous.[3]

[1] See Table 1
[2] After chilling

Example 2: An aqueous solution containing 10% by weight acrylamide and 0.005% by weight of Compound 14 is irradiated with a 500 watt tungsten filament projector lamp placed 10 inches from the sample for 2 minutes. The solution turns into a viscous mass containing polyacrylamide.

Example 3: A solution containing 20% by weight of vinylcarbazole in dichloromethane and 0.1% of Compound 12 are coated on a glass plate and irradiated through a photographic negative with visible light for 15 minutes. The plate is then washed with a solvent to remove the unirradiated and unpolymerized portions. The remaining relief is then inked and applied to paper. A high quality reproduction is produced.

N-Alkoxy Heterocyclics as Cross-Linking Agents

In a process described by P.W. Jenkins, D.W. Heseltine and J.D. Mee; U.S. Patent 3,615,453; October 26, 1971; assigned to Eastman Kodak Company polymers having hardenable groups and incorporating an energy-sensitive compound containing a heterocyclic nitrogen atom substituted with an —OR group where R is alkyl, aralkyl or acyl are cross-linked by exposure, including imagewise exposure, to electromagnetic radiation.

Typical dye cross-linking compounds used in the process include those shown below.

1. 3-ethyl-1'-methoxyoxa-2'-pyridocarbocyanine perchlorate
2. 1'-ethoxy-3-ethyloxa-2'-pyridocarbocyanine tetrafluoroborate
3. 3'-ethyl-1-methoxy-2-pyridothiacyanine iodide
4. 1-ethoxy-3-ethyl-2-pyridothiacyanine tetrafluoroborate
5. 1-benzyloxy-3'-ethyl-2-pyridothiacyanine iodide
6. 2-β-anilinovinyl-1-methoxypyridinium p-toluenesulfonate
7. 1-methoxy-2-methylpyridinium p-toluenesulfonate
8. 1-methoxy-4-methylpyridinium p-toluenesulfonate
9. anhydro-2-methyl-1-(3-sulfopropoxy)pyridinium hydroxide
10. 1-ethoxy-2-methylpyridinium tetrafluoroborate
11. 1-benzyloxy-2-methylpyridinium bromide
12. 1-ethoxy-2-methylquinolinium tetrafluoroborate
13. 1,1'-tetramethylenedioxybis(2-methylpyridinium)dibromide
14. 1,1'-tetramethylenedioxybis(4-methylpyridinium)dibromide

The cross-linking compositions are useful in the reproduction of images in that their exposure to light or other forms of energy can cause insolubilization of the cross-linking composition. Thus, when a layer of one of these compositions, initially soluble, is applied to a support and exposed photographically, the exposed areas become insoluble. The process is useful for the formation of elements wholly made from these compositions. The process also makes possible the formation of coated printing films on any base by the deposition or coating by conventional means of films or coatings of the subject cross-linkable compositions. Typical bases or supports are metal sheets (e.g., copper, aluminum, zinc and magnesium), paper, glass, cellulose acetate butyrate film, cellulose acetate film, poly(vinyl acetal) film, polystyrene film, polycarbonate film, poly(ethylene terephthalate) film, polyethylene-coated paper, nylon and metal screens.

The base or support is typically coated with a solution of the polymeric material in a suitable solvent, this solution containing dissolved or homogeneously dispersed therein an energy sensitive cross-linking initiator whereupon the solvent or solvent mixture is eliminated by known means such as evaporation, leaving a coating of the photosensitive mixture upon the base or support. Thereafter, the dried photosensitive coating can be imagewise exposed to actinic light rays or other suitable energy source. When the support material carrying the photosensitive composition is light-reflecting, there can be present, e.g., superimposed on the support, a layer or stratum absorptive of actinic light in order to minimize reflectance from the combined support of incident actinic light.

If the cross-linkable composition is water-soluble, water can be used as the solvent in coating the support. If water-insoluble compositions are used, organic solvents, mixtures of organic solvents or mixtures of organic solvents and water can be used. The elements formed wholly of or coated with the photosensitive composition are useful in photography, photomechanical reproduction processes, lithography and intaglio printing. More specific examples of such uses are offset printing, silk screen printing, duplicating pads, manifold stencil sheeting coatings, lithographic plates, relief plates, and gravure plates. The term "printing plates" as used herein is inclusive of all of these.

A specific application of the process is illustrated by a typical preparation of a printing plate. In this application a plate, usually of metal, is coated with a film of the cross-linkable composition. Alternatively, any suitable support can be coated with a film of the composition. The surface of the plate is then exposed to visible light through a transparency, e.g., a positive or negative (consisting of opaque and transparent areas).

The light induces the decomposition of the cross-linking agent to a dye base and an aldehyde, and the aldehyde in turn cross-links the polymer in the areas of the surface beneath the transparent portions of the image causing these portions to become insoluble, whereas the areas beneath the opaque portions of the image retain their pre-exposure solubility. The soluble areas of the surface are then removed by a solvent, and the insoluble, raised portions of the film which remain serve as a resist image, wherein a relief plate is formed by etching the exposed base material. The plate carrying the insolubilized raised portion, i.e., the relief image, can also be inked and used as a relief printing plate directly in the customary manner.

Example 1: A composition containing gelatin and Compound 7 in an amount sufficient to produce 3.0% aldehyde by weight of the total composition is coated on a poly(ethylene terephthalate) film support at a thickness of 6μ. The element is exposed to a transparency having opaque and transparent areas with a 140 watt ultraviolet lamp placed 3 inches away from the surface of the element causing cross-linking to occur in the exposed areas. After exposure, the element is washed with water. The unexposed areas wash off leaving only an image formed by the exposed hardened areas. This procedure is repeated for Compounds 6, 8, 9, 10, 12, 13 and 14 and similar results are obtained.

Example 2: Example 1 is repeated except varying amounts of Compound 7 are employed to show the effect on cross-linking for different concentrations of cross-linking agent. The elements are swollen by immersing them in water for 3 minutes.

The extent of swelling or percent swell is an indication of the amount of cross-linking which has occurred, smaller amounts of swelling being indicative of greater amounts of cross-linking. The results are set forth in Table 1 below.

TABLE 1

Compound Number	Percent Aldehyde Relative to Dry Vehicle	Percent Swell
Blank	0	580
7	1	302
7	2	265
7	6	85

Example 3:   This example illustrates the preparation of a lithographic plate. Compound 7 is mixed with gelatin at a level calculated to produce theoretically 6% of formaldehyde with respect to dry weight of gelatin. The composition is coated on an anodized aluminum sheet at a thickness of 0.002 in. The dried coating is exposed for 15 minutes to the light from a 140 watt Hanovia ultraviolet lamp through a definition chart. The exposed plate is washed with 85°F. water for 10 minutes and dried. The sheet is placed on a lithographic printing press and a good press copy is obtained.

Example 4:   Production of Dye Mordants with Simultaneous Hardening — Compounds with ionic or cationic groups attached, when fragmented by energy, produce an aldehyde with a potential mordanting group. A gelatin silver halide emulsion coating containing Compound 9 is prepared so that 3% of the propionaldehyde-3-sulfonic acid would be expected if 100% reaction occurred. The coating is exposed through a mask with the emulsion side toward a 150 watt xenon arc lamp placed within 1 inch of the opening in the lamp housing.

The test pattern is cooled with a stream of nitrogen during a 95 minute exposure. The 1 3/4 inch square test strip is hardened in a 2% succinaldehyde prehardening bath for 1 minute and rinsed under a flow of distilled water at 26°C. for 15 minutes. The test square is then dyed with a solution of methylene blue prepared by placing 0.1 gram of methylene blue in 10 milliliters of water, slurrying for 10 minutes and filtering. A definite mordanting is noted in the exposed areas. The strip upon standing in distilled water overnight at room temperature, retains the dyeing of the exposed area.

Additional studies with the above N-heterocyclic compounds as cross-linking agents is described by P.W. Jenkins, D.W. Heseltine and J.D. Mee; U.S. Patent 3,699,025; October 17, 1972; assigned to Eastman Kodak Company.

Ethinyl Quinoles

A process described by R. Dietrich; U.S. Patent 3,615,630; October 26, 1971; assigned to Kalle AG, Germany is concerned with a light sensitive mixture containing one or more photopolymerizable compounds and, if desired, a binder and conventional sensitizers, dyestuffs and polymerization inhibitors, and with a recording material prepared by coating a support with such a mixture. The mixture consists of or contains, besides other known ethylenically unsaturated photopolymerizable compounds which may be present, one or more ethinyl quinoles of the following general formula:

where R is H, alkyl, aryl or acyl, R' is H, alkyl, aryl, or the group

with R_5 being alkyl; R_1, R_2, R_3 and R_4 are either the same or different and are alkyl or aryl, or R_1 and R_2 or R_3 and R_4 may be members of a condensed aromatic ring which may be further substituted by NH_2, NO_2, OH, OR_6 (R_6 = alkyl) or Cl, and where the alkyl groups are straight chained or branched alkyls having from one to eight carbon atoms.

A number of compounds according to the process are listed in the following table. Because of their stability and particularly favorable photopolymerization and photoinitiating properties, such quinoles are particularly suitable which are derived from anthraquinone and which contain two conjugated polymerizable groups, e.g., in the form of the eninyl group, such as in particular the compounds corresponding to formulas 1, 2 and 3 of the table. The reproduction coatings prepared with these compounds yield distinctly readable negatives in especially deep colors. But generally, all the compounds of the class of compounds identified above yield good results.

The compounds, which are known per se, are prepared by ethinylation, i.e., the addition of acetylenes or their monosubstitution products, in the form of their alkali salts, to suitable quinones. The reaction takes place in liquid ammonia or in aprotic organic solvents, e.g., dioxane, and yields the quinoles in a single step, which compounds may be further substituted on their OH group, if desired.

Compound	R	R'	R₁	R₂	R₃	R₄
1	H	$-\overset{H}{\underset{\underset{OCH_3}{\mid}}{C}}=\overset{H}{C}$	(benzene ring)		(benzene ring)	
2	H	Same as above	Same as above			—Cl
3	H	Same as above	Same as above		Cl (on ring)	
4	H	(benzene ring)	(benzene ring) NH₂		(benzene ring)	
5	H	Same as above	(benzene ring)		Same as above	
6	H	—CH=CH₂	Same as above		Same as above	
7	H	Same as above	Same as above	H	—CH₃	
8	H	Same as above	NO₂ (benzene ring) NH₂		(benzene ring)	
9	H	Same as above	HO— (benzene ring)		Same as above	
10	—CH₃	Same as above	CH₃C— (benzene ring)		Same as above	

The ethinyl quinoles combine the function of a photopolymerizable compound with that of a photoinitiator. They are superior to many known compounds in that they dissolve easily in organic solvents, do not tend to crystallize, and in some cases even have film-forming properties, so that they may be added to the layers in relatively large quantities. As can be seen from the following examples, their polymerization initiating effect is excellent, so that a liquid monomer is completely polymerized within a few seconds. Since the compounds constitute photoinitiator and monomer in one molecule, they are capable of starting their own polymerization, so that the polymers formed are not copolymers but homopolymers. This has the advantage that these compounds may be employed without further initiators or sensitizers. Since the polymers formed are colored, distinctly visible images are produced which considerably facilitates further processing.

Example 1: 5 g. of 9-(ω-methoxy-buteninyl)-anthraquinol-9, prepared by ethinylation of anthraquinone with methoxy butenin in liquid ammonia, and 1 g. of a commercial phenol resin (cresol/formaldehyde condensate) are dissolved in a mixture of 320 ml. of acetone and 180 ml. of butyl acetate, and the solution thus prepared is then homogeneously coated onto appropriate supports of mechanically roughened aluminum foil and dried.

The aluminum foils are exposed for 5 minutes under a negative original to the light of a commercial mixed-light lamp (e.g., Philips HPR 125 W; distance between the edge of the lamp and the layer: 35 cm.). A directly legible, deep orange-brown negative image is obtained. The nonimage areas may be easily removed subsequently by means of a weakly alkaline decoating solution (e.g., according to German Patent 1,193,366), so that only the thoroughly polymerized image areas are retained. After inking with greasy printing ink, long runs may be obtained on an offset printing machine from a printing foil prepared in this manner, the prints showing an exceptionally accurate screen reproduction.

Example 2: 250 mg. of 9-(ω-methoxy-buteninyl)-2-chloroanthraquinol-9, prepared by ethinylation of 2-chloroanthraquinone with methoxy butenin, are dissolved with 50 mg. of a commercially available maleic acid/styrene copolymerizate in 20 ml. of acetone (containing 20% of butyl acetate). A support consisting of an electrolytically roughened aluminum foil (e.g., Rotablatt) is uniformly coated under subdued light with this solution and then dried with warm air. The thus-sensitized foil is exposed for 4 minutes under a negative original to the light of a xenon lamp normally used in the reproduction field (e.g., Xenokop).

A sharp, deep brown image of the original on a light background is thus obtained. The nonimage areas again can be easily dissolved away with a weakly alcoholic decoating solution (a 1% solution of Na_2SiO_3 in a mixture consisting of 20 parts by volume of water, 20 parts by volume of methanol, 10 parts by volume of glycol, and 10 parts by volume of glycerol), and, after briefly wiping it with a normal hydrophilizing agent, the finished printing foil may be inked up with greasy ink. In this case also, the printing foil shows an exceptionally good reproduction even of the finest screens. Similar results are obtained when using 9-methoxy-buteninyl-1-chloroanthraquinol(9) instead of the 2-chlorisomer.

Diacyldiazomethone Groups

A process described by U.L. Laridon, G.A. Delzenne, A.L. Poot, and H.K. Peeters; U.S. Patent 3,682,642; August 8, 1972 provides photographic elements comprising a support and a light sensitive layer containing a photopolymerizable ethylenically unsaturated organic compound and a photopolymerization initiator containing at least one diacyldiazomethane group. Examples of suitable photopolymerization initiators are:

(1)   bis(phenylsulfonyl)-diazomethane
(2)   bis(4-chlorophenylsulfonyl)-diazomethane
(3)   bis(4-tolysulfonyl)-diazomethane
(4)   bis(4-nitrophenylsulfonyl)-diazomethane
(5)   bis(2-naphthylsulfonyl)-diazomethane
(6)   (phenylsulfonyl)-(phenylcarbonyl)-diazomethane
(7)   (phenylsulfonyl)-(4-methylphenylcarbonyl)-diazomethane
(8)   (phenylsulfonyl)-(4-fluorophenylcarbonyl)-diazomethane
(9)   (phenylsulfonyl)-(4-chlorophenylcarbonyl)-diazomethane
(10) (phenylsulfonyl)-(4-bromophenylcarbonyl)-diazomethane

The following examples illustrate the process.

Example 1:   60 g. of acrylamide were dissolved in 100 ml. of distilled water and 10 mg. of erythrosine were dissolved in 100 ml. of ethylene glycol monomethyl ether. In a series of glass tubes 7.5 ml. of the above-mentioned acrylamide solution and 1 ml. of the erythrosine solution were mixed. To each tube was then added 10^{-5} mol of photopolymerization initiator as indicated below, dissolved in 4 ml. of ethylene glycol monomethyl ether.

A series of comparative test solutions of the same composition, but containing no erythrosine solution was made also. The tubes containing the different solutions were exposed to an 80 watt high-pressure mercury vapor lamp placed at a distance of 10 cm. As a result thereof and depending on the nature of the particular photopolymerization initiator used as well as on the presence or absence of erythrosine as activating dyestuff, the solutions first became viscous and afterwards became solid thus representing a polymer yield of 90 to 95%. The results of the exposure test are shown in Table 1.

TABLE 1

Photopolymerization initiator		Reaction time [1]	
No.[2]	Amount present, mg.	Without erythrosine, minutes	With erythrosine, minutes
2	4	>60	6
5	4.5	38	10
7	3	120	12
8	3	>120	12
10	3.7	60	11½

[1] Reaction time at which the polymerization mixture solidifies.
[2] The numbers refer to the numbers of the particular polymerization initiators given above.

Example 2: 3 g. of acrylamide were dissolved in a mixture of 5 ml. of ethylene glycol monomethyl ether and 5 ml. of water. To this solution were added 0.5 mg. of Rose Bengal as photopolymerization initiator and 2.5 mg. of bis(phenylcarbonyl)-diazomethane. The resulting solution was poured into a chemically and heat-resistant glass tube. Exposure occurred with an 80 watt high-pressure mercury vapor lamp placed at a distance of 10 cm. After 20 minutes of exposure solid polyacrylamide was obtained.

Example 3: 60 g. of acrylamide were dissolved in 100 ml. of water. In each of 6 glass tubes 7.5 ml. of this solution were poured and 3.2 mg. of a polymerization initiator dissolved in a quantity of ethylene glycol monomethyl ether as indicated below were added hereto. To three of these glass tubes 1 ml. of a Rose Bengal solution was added.

The tubes were exposed to an 80 watt high-pressure mercury vapor lamp placed at a distance of 15 cm. Depending on the particular composition, the solutions first became viscous and solid afterwards, thus representing a polymer yield of 90 to 95%. The results of the exposure tests are given in Table 2.

TABLE 2

Photopolymerization initiator [1] Number:	Activating dyestuff,[2] ml.	Amount of ethylene glycol monomethyl ether, ml.	Solid after (min.)—
1		4	80
1	1	3	10
4		4	60
4	1	3	10
7		4	70
7	1	3	7

[1] The numbers refer to the numbers of the particular polymerization initiators given above.
[2] 10 mg. of Rose Bengal dissolved in 10 ml. ethylene glycol monomethyl ether.

Diacylhalomethane

U.L. Laridon and G.A. Delzenne; U.S. Patent 3,615,455; October 26, 1971; assigned to Gevaert-Agfa NV, Belgium describe a process for the photopolymerization of ethylenically unsaturated organic compounds, which comprises irradiating with light of wavelengths ranging from 2500 to 5000 A. a composition comprising a photopolymerizable ethylenically unsaturated organic compound and as a photopolymerization initiator a diacylhalomethane corresponding to one of the following general formulas:

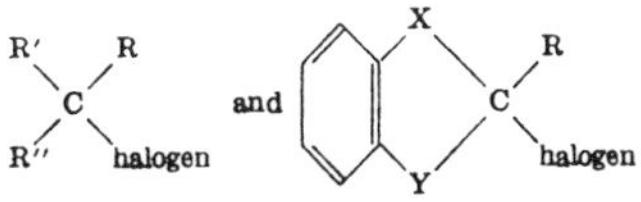

where halogen represents a halogen atom such as chlorine and bromine; R represents a hydrogen atom, a chlorine or a bromine atom, or an acetyloxy group; R' and R" (same or different) represent a benzoyl group, a nitro-benzoyl group, a dimethylamino benzoyl group, a phenylsulfonyl group, a carboxyphenylsulfonyl group, a methylphenylsulfonyl group, or a naphthoyl group; X and Y (same or different) represent a carbonyl group or a sulfonyl group. These diacylhalomethanes are obtained by halogenation of the corresponding diacylmethanes in chloroform or in acetic acid. Particularly valuable diacylhalomethanes are: 2-bromo-1,3-diphenyl-1,3-propanedione; 2,2-dibromo-1,3-diphenyl-1,3-propanedione; 2-bromo-2-hydroxy-1,3-diphenyl-1,3-propanedione, acetate; 2-bromo-2-(phenylsulfonyl)-acetophenone; and 2,2-dibromo-2-(phenylsulfonyl)-acetophenone.

Example: 1×10^{15} mol of initiator, as in the following table, was dissolved each time in 4 ml. of ethylene glycol monomethyl ether. The resulting solution was mixed with solutions of 3 g. of acrylamide in 5 ml. of water in borosilicate glass test tubes and exposed to an 80-watt high-pressure mercury vapor lamp at 10 cm. distance. The following results were attained.

		Time	
Test	Initiator	(1)	(2)
1	2,2-dibromo-1,3-diphenyl-1,3-propanedione	6	10
2	2-bromo-1,3-diphenyl-1,3-propanedione	12	19
3	2-bromo-2-hydroxy-1,3-diphenyl-1,3-propanedione, acetate	10	15
4	2-bromo-2-(phenylsulphonyl)-acetophenone	7	12
5	2,2-dibromo-2-(phenylsulphonyl)-acetophenone	7	10
6	2,2-dibromo-benzo[b]-thiophene-3 (2H)one-1,1-dioxide	10	37
7	2-bromo-2(m-carboxyphenylsulphonyl)-acetophenone	8	14
8	2,2-dibromo-1,3-indanedione	5	10
9	2-bromo-2-(p-tolylsulphonyl)-acetophenone	9	12
10	2-bromo-2-(phenylsulphonyl)-4'-nitroacetophenone	6	14
11	2-bromo-2-(phenylsulphonyl)4'-dimethylamino-acetophenone	10	>60
12	2-bromo-2-(phenylsulphonyl)-1'-acetonaphthone	4	13
13	2,2-dichloro-2-(p-tolylsulphonyl)-acetophenone	8	13

[1] In minutes needed to obtain a viscous solution.
[2] In minutes required to obtain a solid product (yield 90-95%).

Phenyl Substituted Oxido-Oxazoles

U.L. Laridon and G.A. Delzenne; U.S. Patent 3,617,279; November 2, 1971; assigned to Gevaert-Agfa NV, Belgium describe a process for the photopolymerization of ethylenically unsaturated organic compounds, which comprises irradiating with light of wavelengths ranging from 2500 to 4000 A., a composition comprising a photopolymerizable ethylenically unsaturated organic compound and as a photopolymerization initiator an oxido-oxazole corresponding to one of the general formulas:

$$
\begin{array}{c}
O \\
R_1-C \diagup \diagdown C-R_1 \\
R_7-C \underset{\parallel}{\underset{}{}} \underset{}{} N \to O
\end{array}
$$

$$
\begin{array}{c}
\Theta \qquad\qquad O \\
R_1-C \diagup \diagdown C-R_4-C \diagup \diagdown C-R_1 \\
R_1-C \underset{\parallel}{} N \to O \quad O \gets N \underset{}{} C-R_1
\end{array}
$$

where each of R_1, R_2 and R_3 (same or different) represents an alkyl group of one to three carbon atoms, or an aryl group including a substituted aryl group, and R_4 represents an alkyl group of two to three carbon atoms or an aryl group. These oxido-oxazoles may be present in the photopolymerization reaction in the form of the free base or in the form of their hydrochlorides. The latter are formed on passing hydrogen chloride through an acetic acid solution of a monoxime of an α-diketone and the corresponding aldehyde. The free oxido-oxazoles are produced from their hydrochlorides on treating them with a base. The above formulas of oxido-oxazoles are written according to the structure formulas proposed by Cornforth and Cornforth in J. Chem. Soc. (1947) 96.

Suitable oxido-oxazoles, which can be used as photopolymerization initiators according to the process are 2-phenyl-4,5-dimethyl-N-oxido-oxazole; 2-o-hydroxphenyl-4,5-dimethyl-N-oxido-oxazole; 2-p-hydroxphenyl-4,5-dimethyl-N-oxido-oxazole; 2-p-dimethylaminophenyl-4,5-dimethyl-N-oxido-oxazole; 2-m-chlorophenyl-4-phenyl-5-methyl-N-oxido-oxazole; 2,4,5-triphenyl-N-oxido-oxazole; 2-p-dimethylaminophenyl-4,5-diphenyl-N-oxido-oxazole; 4,4-bis[2-(4,5-dimethyl-N-oxido-oxazole)]-phenylene; and 4,4-bis[2-(4,5-diphenyl-N-oxido-oxazole)]-phenylene.

Example 1: Preparation of Oxido-Oxazole — In an Erlenmeyer flask were placed 360 cc of acetic acid, 54 g. of benzilmonoxime and 24 cc of distilled benzaldehyde. Through this mixture dry hydrogen chloride gas was bubbled for one hour. Already after 5 minutes a solution was obtained and 5 minutes later some product started to precipitate. Then the mixture was allowed to stay in the dark for 48 hours. The product precipitated was then filtered with suction, washed with ether, and dried under vacuum. Yield: 30 g. After recrystallization from 440 cc of benzene, 21 g. of pure 2,4,5-triphenyl-oxido-oxazole hydrochloride were obtained. MP: 172°C. The free 2,4,5-triphenyl-oxido-oxazole was obtained by dissolving 3 g. of the hydrochloride in 15 cc of methanol and on slightly alkalizing this solution by means of 25% aqueous ammonium hydroxide. The free base precipitated, was filtered with suction and finally dried under vacuum. Yield: 2.5 g. The other oxido-oxazole hydrochlorides and their corresponding free bases were prepared analogously.

Photopolymerization — To a series of solutions of 1 g. of acrylamide in 10 cc of methanol oxido-oxazole hydrochlorides were added as listed in Table 1 below. The solutions obtained were poured into test tubes of borosilicate glass and exposed by means of a mercury vapor lamp of 80 watt placed at a distance of 10 cm. The polymer formed precipitated. After 2 hours of exposure the respective yields were determined by filtration, drying and weighing of the polymer. The results are also to be found in Table 1.

TABLE 1

Initiator: hydrochloride of —	Concentration (mole/litre)	Yield (percent)
2,4,5-triphenyl-N-oxido-oxazole	4.10^{-4}	27.7
2-p-methoxyphenyl-4,5-diphenyl-N-oxido-oxazole	2.10^{-4}	42.2
2-o-hydroxyphenyl-4,5-diphenyl-N-oxido-oxazole	[1] 2.10^{-4}	59.4
2-o-hydroxyphenyl-4,5-dimethyl-N-oxido-oxazole	2.10^{-4}	32.6
2,4-diphenyl-5-methyl-N-oxido-oxazole	2.10^{-4}	38.1
2-m-chlorophenyl-4-phenyl-5-methyl-N-oxido-oxazole	2.10^{-4}	35.2

[1] In this case a mixture of cyclohexanone and acetone was used as solvent.

Example 2: 10 cc of methyl methacrylate were dissolved in 10 cc of benzene and 10^{13} mol of 2,4,5-triphenyl-N-oxido-oxazole hydrochloride was added per liter. Through this solution nitrogen was bubbled for 30 minutes and the test tube was sealed. The solution was exposed for 5 hours by means of a 300 watt mercury vapor lamp placed at 18 cm. whereupon methanol was added. The quantity of polymer obtained amounted to 10.58%. In the same circumstances a test without oxido-oxazole yielded only 1.15% of polymer.

Example 3: A series of solutions was prepared of 6 g. of acrylamide in a mixture of 10 cc of ethylene glycol monomethyl ether and 10 cc of water. To these solutions were added oxido-oxazoles as listed in Table 2 below in a concentration of 10^{13} mol/liter. Each solution obtained was poured in a test tube and exposed by means of a mercury vapor lamp of 300 watt placed at 18 cm. The polymer formed remained dissolved and was then separated after exposure by adding an excess of methanol to the polymer solution. The results obtained are also listed in Table 2.

TABLE 2

Initiator	Time of exposure (min.)	Yield (percent)
2,4,5-triphenyl-N-oxido-oxazole hydrochloride.	45	25
	60	90–95
2,4,5-triphenyl-N-oxido-oxazole	45	47
	60	90–95
2-p-dimethylaminophenyl-4,5-diphenyl-N-	15	21
oxido-oxazole hydrochloride	30	86
	45	90–95

Aromatic Nitrocompounds as Sensitizers for Diolefin Elastomers

F.L. Ramp; U.S. Patent 3,615,469; October 26, 1971; assigned to The B.F. Goodrich Company describes a method of producing etched rubber products from photosensitive diolefin polymer compounds. Improved flexible printing plates can be produced by a process that eliminates the necessity of first producing a metal plate and a master plate mold, and further provides increased accuracy to be obtained in reproducing detailed copy. Conjugated diolefin polymers are treated with aromatic nitrocompounds to make the polymers sensitive to degradation by light.

The combination of light with the aromatic nitrocompounds degrades the exposed portion of the diolefin polymers causing the polymer to become softer and more soluble in solvents. The degraded polymeric materials are removed as with solvents leaving nonexposed nondegraded areas as raised surfaces. Increased commercial versatility and utility are achieved by producing flexible printing plates.

Example 1: Using a Banbury mixer, the following components were mixed to form a uniform mixture: 100 parts synthetic cis-1,4-polyisoprene; 1 part triethanolamine; 5 parts N-octadecyl-1,3-diamino propane (Armeen TD); 20 parts silica (HiSil No. 233); and 0.25 part dicumyl peroxide (Dicup R). The uniform mixture was then calender formed into a flat sheet having a thickness of 0.125 inch and a variance of about ±0.002 inch. The mixture was heated to about 150°F. to aid forming The formed flat sheet was cured at about 350°F. for 45 minutes. About 1.5 grams of a solution (25 grams of of ethyl p-nitrobenzoate dissolved in 100 grams xylene) was brushed onto 100 square inches of the flat sheet surface and allowed to dry for about 1 hr.

at room temperature. A high contrast negative was placed on the sensitized surface. This surface was selectively exposed for about 45 minutes
to R.S. Sunlamp (275 watts) placed at a distance of about 6 inches. The
exposed areas of the flat sheet were degraded yielding about 0.025 inches
depth of relief. The degraded polymeric matter was removed with hexane
solvent in conjunction with moderate brushing action of flexible bristled
scrub brush. The etched flat sheet was air dried. Printing ink was applied
to the raised surface which reproduced detailed positive copy of printing.

Example 2:   Using a roll mill the following components were mixed to a
uniform mixture: 100 parts (wt.) cis-1,4-polyisoprene (Natsyn 400); 20
parts (wt.) cis-1,4-polybutadiene; 20 parts (wt.) silica filler; 2 parts (wt.)
triethanolamine; 0.8 part (wt.) 2,5-bis(t-butyl peroxy)-2,5-dimethylhex-
ane (Varox); and 5 parts zinc stearate. The uniform mixture was formed
into a flat sheet in the manner of Example 1 and was cured for about 45
minutes at 300°F. About 1.0 g. of the sensitizer solution of Example 1 was
roller coated to 100 square inches of the flat surface. A high contrast negative was placed on the sensitized surface which was selectively exposed as
in Example 1. Results achieved were the same as Example 1.

Example 3:   The following indicated components were mixed to a uniform
mixture which was calendered into flat sheets.

	A	B	C
Natural rubber	100	100	160
Cis-1,4-polybutadiene	10	10	10
Zinc stearate	5	5	5
Zinc oxide	1	1	1
Sulfur	2	2	2
Mercaptobenzothiazole	0.75	0.75	0.75
Tetramethylthiurammonosulfide	0.5	0.5	0.5
Diphenylguanidine	0.5	0.5	0.5
Octyl-p-nitrobenzoate		5	5
Silica			20
Relief (inches)	0.009	0.004	0.004

A. Mixture 3A was cured at 212°F. for about 15 minutes. About 12 g. of
solution (ratio of 25 g. octyl-p-nitrobenzoate in 100 g. xylene) was brush
applied to 100 square inches of ply surface. A high contrast negative was
placed upon the sheet which was then exposed to a 15 watt diazo lamp.

B. Mixture 3B was mixed with octyl-p-nitrobenzoate, and calendered
into a flat sheet. The sheet was cured at 212°F. for 15 minutes and selectively exposed through a high contrast negative for about 30 minutes to a
1,500 watt continuous spectrum lamp.

C. Mixture 3C was compounded and tested in accordance with the procedure
set forth in 3B.

Diazosulfonates and Photoreducible Dye

In a process described by S. Levinos; U.S. Patent 3,637,375; January 25, 1972 polymerization of ethylenically unsaturated vinyl compounds is effected by exposing such compounds to light in the presence of a photoinitiator comprising a combination of a light-sensitive diazosulfonate and a photoreducible dye. The further addition of amines and aldehyde/bisulfite addition products accelerates the polymerization reaction. The use of certain diazo compounds in photoinitiator compositions for photopolymer systems has been described in U.S. Patent 3,099,558 and Canadian Patent 820,828.

The activity of multicomponent photoinitiator compositions in photopolymerization systems is believed, in general, to involve a light-initiated redox reaction between the photoinitiator components with the resulting formation of free radicals. Thus, the effectiveness of a photoinitiator would appear to rely considerably upon the relative redox "potentials" of its component compounds as well as upon their respective photosensitivities. Of equal importance, however, to the practical utility of a photoinitiator composition is the degree of dark stability the components exhibit, particularly when in mutual combinations. This requirement for dark stability has heretofore seriously restricted the selection of photoinitiator components to compounds having limited relative redox potentials.

In Canadian Patent 820,828, for example, the components of a photopolymerization initiator system have been selected from diazonium salts which respectively exhibit the capability of acting as electron donors and as electron acceptors. The similarities in the constitution of such diazonium compounds as well as the inherent limitations in dark stability generally result in relatively low redox potentials and thus low rates of polymerization upon photoactivation. Similarly, in U.S. Patent 3,099,558, where diazonium salts are utilized as electron acceptors in systems with electron-donating light-absorbing dyes normally employed as sensitizers in silver halide photographic materials, redox potentials are substantially restricted as is evidenced by the inordinately long light exposures therein described.

It has been discovered that a significant increase in the photoactivity of polymerization systems can be achieved without a commensurate loss in dark stability when photoinitiator compositions are prepared by combining light-sensitive diazosulfonate compounds with photoreducible dyes. The excellent stability of diazosulfonate compounds has been found to contribue significantly to the preexposure stability of the photopolymerizable materials, yet the light sensitivity of the diazosulfonate and the apparently high redox potential in its combination with photoreducible dye result in rapid formation of free radicals which effect polymerization of the vinyl components in the system. Further, it would appear that the electron donor character of the diazosulfonate in the photoinitiator combination and the capability of the photoreducible dye to vary between the normal and leuco states tend toward the formation of a multiplicity of free radicals which result in the greatly increased yield of polymer as compared with prior systems under similar conditions of light exposure. The practical consequence of the present photoinitiator compositions is thus an increase in both dark stability and photosensitivity of the resulting photopolymerizable systems. The following examples illustrate the use of the combination of diazosulfonates and photoreducible dyes in the photopolymerization of a variety of vinyl monomers.

Example 1: A stock solution of the following composition was prepared: acrylamide, 36 g.; N,N'-methylenebisacrylamide, 2 g.; water, 24 ml.; and methyl Cellosolve acetate, 20 ml. To 20 ml. of this solution were added 100 mg. of the diazosulfonate of p-anisidine, followed by 5 drops of a 0.2% aqueous solution of methylene blue (CI 52015). A 5 ml. portion of this mixture was transferred to a 13 x 100 mm. test tube which was then exposed to the light of a 375-watt photoflood lamp positioned at a distance of 10 inches. A white opaque polymer began to form after an exposure of 42 seconds. In this as in other examples, the role of methyl Cellosolve acetate is to precipitate the photoformed polymer as a white opaque solid, thus facilitating the timing of the onset of polymerization.

Example 2: The composition used in this experiment was identical to that of Example 1 except that 100 mg. of p-diazosulfonate-1-tolyl mercapto-2,5-dimethoxybenzene was substituted for the diazosulfonate of p-anisidine. Upon exposure to light as in Example 1, polymer began to form after a period of 11 seconds. This example illustrates the difference in reactivity between diazosulfonates of different structures and particularly demonstrates the greater reactivity of diazosulfonates with substitutent groups in a position meta to the diazo group to which earlier reference has been made.

Example 3: To 5 ml. of the photopolymerizable diazosulfonate/dye sensitized composition of Example 1 were added 10 drops of a 50% aqueous solution of diethanolamine borate. Upon exposure to light as in the preceding examples, polymer began to form after a period of only 15 seconds, illustrating the accelerating effect of amine derivatives in compositions of the process.

Examples 4 through 6: Prepigmented coatings were made from the following photopolymerization compositions:

Solution 1

Gelatin	10.5 g.
Urea (gelatin softener)	0.225 g.
Water (deionized)	100 ml.

Solution 2

p-diazo-1-tolyl mercapto-2,5- dimethoxybenzene chlorozincate	1.74 g.
Water (deionized)	50 ml.

Solution 3

Phloroglucinol	0.67 g.
Ethanol	20 ml.
Water (deionized)	20 ml.

Solution 4

*p-diazosulfonate-1-tolyl mercapto- 2,5-dimethoxy benzene	4.0 g.
Water (deionized)	70 ml.
**Aerosol A-102	45 drops

> *Ground to smooth paste with
> Aerosol and water
> **Disodium ethoxylated alcohol
> half ester of sulfosuccinic acid

Solution 5

Acrylamide	4.2 g.
N, N'-methylenebisacrylamide	0.8 g.
Diethanolamine borate (50% aq. sol.)	2.0 g.
Water (deionized)	60 ml.

All solutions were added in sequence to Solution 1 with stirring to effect thorough mixing. During this mixing the combination of Solutions 2 and 3 effects an in situ formation of a black azo dye pigment. Substitution of the direct addition of colored pigment pastes or coloring dyes can also be employed to effect desired coloration of the final image. 75 ml. portions of this mixture were then dye-sensitized with the following:

(a)	Methylene blue	3.75 mg.
(b)	Thionin	3.75 mg.
(c)	Thionin	7.50 mg.
(d)	Eosin Y (CI 45380)	3.75 mg.

Coatings were made and exposed in contact with a line negative to the light of a 500 watt photoflood lamp at a distance of 12 feet, with the following times required to yield a black, polymeric relief image after swabbing with water at a temperature of 30° to 35°C.

(a) 30 seconds
(b) 30 seconds
(c) 15 seconds
(d) 60 seconds

The support employed for the coatings was a subbed, matte surface, polyester film base commonly used for drafting work. Alterations could be made
by the selective removal of image areas with a moist eraser, followed by
the pen application of India ink to effect the desired changes.

Example 7:   The following photopolymerizable composition was prepared
for phototemplate use:

Solution 1

Gelatin	5.0 g.
Urea (gelatin softener)	0.255 g.
Water (deionized)	100 ml.

Solutions 2 and 3

Same as in Examples 4 through 6

Solution 4

*p-diazosulfonate-1-tolyl mercapto- 2,5-dimethoxy benzene	4.0 g.
Polyvinyl pyrrolidone (K-30)	1.6 g.
**Arosurf 160-E 15	1.0 g.
Water (deionized)	50 ml.

*Ground to smooth paste with
other components and water
**Coco amine plus 15 mols
ethylene oxide

Solution 5

Gelatin	4.5 g.
Acrylamide	4.2 g.
N,N'-methylenebisacrylamide	0.8 g.
Methylene blue	10 mg.
Water (deionized)	80 ml.

Solutions 2, 3, 4 and 5 were added in sequence to Solution 1 with stirring
to obtain thorough mixing. This mixture was then flow coated on a 1/16"
degreased, grained aluminum sheet which had been previously brushed with
a 5% gelatin solution and allowed to dry. The light-sensitive layer was

allowed to dry under yellow safelight illumination and was then exposed for 2 minutes in contact with a line negative as in the preceding examples. After washing with water at a temperature of 30° to 35°C. to remove unexposed areas, a black positive relief image was obtained.

Dye-Sensitized Polymerization

E.J. Cerwonka; U.S. Patent 3,615,452; October 26, 1971; assigned to GAF Corporation describes a dye-sensitized photopolymerization process for vinyl monomers where the vinyl monomers are polymerized by means of a diazonium compound and a dye by a photooxidation process. The dye-sensitized photopolymerization of vinyl monomers has been the subject of several patents including U.S. Patents 2,850,455, 3,074,794, 3,097,097, and 3,145,104.

U.S. Patent 2,850,445 describes a dye-sensitized photopolymerization process where vinyl compounds are contacted with a photoreducible dye and a mild reducing agent in the presence of oxygen. U.S. Patent 3,074,794 describes a process in which a bichromate material which consists of a soluble polymer which is cross-linkable into an insoluble form by reduced bichromate and a bichromate is photopolymerized in the presence of a photoreducible dye and a reducing agent. U.S. Patent 3,097,097 also describes a process in which a polymerizable compound is photopolymerized in the presence of a photoreducible dye and a material suitable for reducing the photoexcited dye. U.S. Patent 3,145,104 describes the photopolymerization of thiol polymers in the presence of a dye-sensitizer, i.e., a photoreducible dye.

It is noted that in the above processes the photoreduction of the dye usually accompanies the initiation of the polymerization process. A mild reducing agent such as ascorbic acid, allylthiourea, or cysteine is usually incorporated in the system for this purpose. The reducing agents employed conventionally have reduction potentials of such a magnitude that the dyes are not reduced in the dark, while in the presence of actinic light the dye is reduced to its leuco form and the mild reducing agent is correspondingly oxidized. Other materials such as chelating agents including compounds such as ethylenediamine tetraacetate have also served as electron donors for the light-excited dye molecules even though this type of agent is not normally regarded as a reducing agent.

In connection with the majority of the above processes there is visible evidence of the photoreduction of the dye inasmuch as the dye which is bleached in light reverts to its original color in the dark when oxygen is available in the reaction medium. When a vinyl monomer is present in the dye-reducing agent or dye-chelating agent system, polymerization of the monomer accompanies the photoreduction of the dye. As may be seen, the dye-sensitized photopolymerization of vinyl monomers accompanied by the photoreduction of the dye employed is known in the art. As distinguished from this, the use of dye-sensitized photopolymerization processes in which the dye employed is photooxidized is relatively uncommon. One such process is described in U.S. Patent 3,147,119. The patent notes the use of zinc oxide as a mild oxidizing agent in the dye-sensitized photopolymerization of a vinyl monomer.

When the composition was exposed to actinic light, the dye was irreversibly photobleached and the vinyl monomer was polymerized. Photooxidation processes of this type have in the past been unpopular because they were believed to be too slow to be useful. Contrary to this general belief it has been found that photooxidation processes may be practically employed in photographic operations. Therefore, it is an object of this process to provide a dye-sensitized photopolymerization process where the dye employed is photooxidized.

In the dye-sensitized photopolymerization process, a diazonium salt is employed as a mild oxidizing agent. The diazonium salts have been used previously as oxidizing agents. For example, in the synthesis of aromatic hydrazine derivatives, stannous chloride or sodium sulfide are oxidized by an appropriate diazonium salt which is itself reduced in the process. Another type of reaction in which a diazonium salt is used as an oxidizing agent is in the oxidation of ethanol to acetaldehyde in an aqueous solution. Diazonium salts which are suitable include aromatic diazonium compounds with or without nuclear substituents which may include alkyl, alkoxy, acetamido, carboalkoxy, hydroxyl, aryl, halogen, sulfonate or sulfone groupings.

The most effective diazonium salts appear to be those which are the strongest oxidants. These include the unsubstituted diazonium salts or those possessing so-called negative or electron-attracting substituents such as phenyl or nitro. In this connection it should be remembered that the diazonium salt should not be so strong so as to oxidize the dye in the dark and yet be a strong enough oxidant so as to effect the polymerization when the composition is exposed to light. Suitable diazonium compounds may be those which are derived from aromatic amines such as are described in U.S. Patents 3,110,592, 2,807,545 and 2,772,972. These compounds which are all primary aromatic amines yield on diazotization the diazonium salts of interest.

The useful dyes which may be employed in connection with the photopolymerization process include those dyes which are known as desensitizing dyes in silver halide photographic emulsions, as delineated by C.E.K. Mees and T.H. James in <u>The Theory of the Photographic Process</u>, 3rd ed., 228–229, (1966). These dyes include those classes of dyes which can be reversibly photoreduced in the presence of a mild reducing agent and can also be irreversibly photooxidized or bleached in the presence of a diazonium salt. Such dye classes include the phenazines, phenthiazines, thiazolyl dyes, phthaleins, azines, oxazines, and thiazines. The following examples illustrate the process.

<u>Example 1</u>: To 5 ml. of water containing about 1 mg. methylene blue there was added in the absence of light 0.1 g. para-toluene diazonium chlorozincate. The solution was then divided into two parts. One-half of the sample was exposed to the light from a 375 watt reflector lamp at a distance of 6 inches; the other half was retained in the dark. Bleaching of the dye took place in the sample exposed to light within a few seconds. A few bubbles of gas, probably nitrogen, also appeared in the solution. The unexposed sample showed no color change after a period of several hours. The bleached sample was stored overnight in a dark room. It was noted the next day that the solution had not regained its original color.

<u>Example 2</u>: The procedure of Example 1 was repeated, substituting for the water 5 ml. of an aqueous solution of N,N'-methylene bisacrylamide, 4%. A white opaque polymer appeared in the exposed sample within a few seconds. No polymer had appeared in the unexposed sample after the lapse of several hours.

<u>Example 3</u>: The procedure of Examples 1 and 2 was repeated, substituting p-toluene diazonium chlorozincate. The same results were observed as were described for Examples 1 and 2.

<u>Example 4</u>: The procedure of Examples 1 and 2 was repeated substituting rose bengal dye for methylene blue. The same results were observed as were described for Examples 1 and 2.

<u>Example 5</u>: The following light-sensitive solution was prepared in a dark room:

Polyvinylpyrrolidone, K-90	1.000 g.
N,N'-methylenebisacrylamide (recrystallized twice from water)	0.250 g.
Methylene blue (zinc-free) (formula weight = 374)	0.037 g.
Para-toluene diazonium chlorozincate	0.200 g.
Wetsit spreading agent, 10% aqueous solution	0.08 ml.
Water, to	25.0 ml.

*alkyl sulfonate blends

The solution was poured onto a glass plate which had been appropriately subbed to receive the solution. The coated plate was whirled 5 minutes on a coating machine, and further was allowed to dry in the dark at room temperature for about an hour. A sample was cut from the plate and exposed through a Stouffer Graphic Arts step tablet to the light from a 375-watt reflector lamp for a period of 1 min. at a distance of 15". The light intensity as measured by an ultraviolet light-exposure meter reading per cm.^{2}. After the exposure the sample was washed by immersing in a tray of deionized water at room temperature for a few minutes. Further washing under a cold water tap served more completely to remove the unpolymerized areas. A polymeric image was achieved which showed seven steps of the step tablet.

Example 6: The following light sensitive solution was prepared in a dark room:

Polyvinylpyrrolidone, K-90	1.000 g.
N, N'-methylenebisacrylamide	
(recrystallized twice from water)	0.250 g.
Rose bengal (formula weight = 981)	0.098 g.
Para-toluene diazonium fluoborate	0.200 g.
Wetsit spreading agent, 10%	
aqueous solution	0.08 ml.
Water, to	25.0 ml.

The procedure for the coating of the glass plate and for the exposure of a sample thereof was the same as that described in Example 5. After the washing step had removed the unpolymerized areas, a polymeric image remained which showed nine steps of the step tablet.

The above examples are illustrative of the manner in which the photopolymerizable compositions can be used for the production of photographic polymeric relief images.

1,4-Triazines

H. Huckstadt, W. Saleck, A. Randolph and E. Ranz; U.S. Patent 3,617,280; November 2, 1971; assigned to AGFA-Gevaert AG, Germany have found that the sensitivity of silver bromide emulsions can be considerably increased without concomitant increase in fogging if the development of the exposed emulsion layers is performed in the presence of 1,4-thiazanes. Particular utility is exhibited by the following compounds:

(1)

(2)

(3)

(4)

(5)

(6)

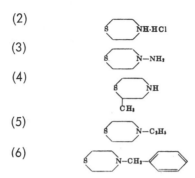

The substances for use according to the process are prepared from thiomorpholine in known manner. The preparation of thiomorpholine has been
described in J. Chem. Soc., 761-308, 1920; J. Am. Chem. Soc. 76,
1187-1188, 1954; U.S. Patent 2,761,860; and German Patent 1,104,513.
C-substituted 1,4-thiazanes are described in J. Am. Chem. Soc. 76,
2902-2906, 1954. The compounds of the process are effective for black-
white and for color photographic emulsions which contain color couplers.

Example 1:   600 mg. of saponin as wetting agent, 200 mg. of 4-hydroxy-
6-methyl-1,3,3a,7-tetraazaindene as stabilizer and 10 ml. of a 10% aqueous solution of formaldehyde are added to a silver iodobromide gelatin
emulsion which contains 60 g. of silver per liter in the form of silver halide
which has a silver iodide content of 6 mols percent. The emulsion was
divided into three equal samples and the following substances were added
per liter of emulsion to the individual samples: Sample A: comparison
sample without additive; Sample B: 300 mg. of Compound (1); Sample C:
600 mg. of Compound (3). The casting solutions are applied onto a support
of cellulose triacetate, exposed in a sensitometer customarily employed in
the art behind a gray step wedge and developed for 10 minutes in a developer of the following composition:

Sodium sulfite anhydrous	70.0 g.
Borax	7.0 g.
Hydroquinone	3.5 g.
p-Monomethyl-aminophenol	3.5 g.
Sodium citrate	7.0 g.
Potassium bromide	0.4 g.
Make up to 1 liter with water	

The results of the sensitometric test are given in the following table, where
an increase by 3° represents a sensitivity gain of one shutter stop or double
sensitivity.

TABLE 1

Sample	Sensitivity	γ	Fog
A	Blank	1.05	0.09
B	+ 1.7°	1.0	0.06
C	+ 2.5°	1.0	0.14

Example 2: A silver iodobromide gelatin emulsion containing 4.5 mol percent silver iodide which has been ripened to optimum sensitivity with gold and sulfur compounds in known manner is sensitized by the addition of 45 mg. of the following sensitizing dye per kg. of emulsion:

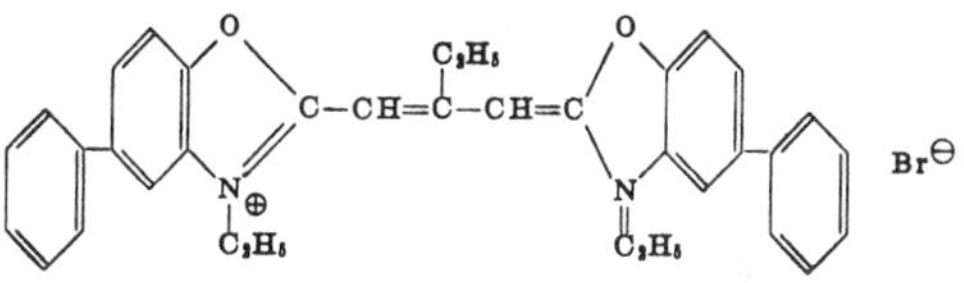

In addition, the following ingredients are added per kg. of emulsion: 1.5 g. of 1-(3'-sulfo-4'-phenoxy)-phenyl-3-heptadecyl-pyrazolone as magenta coupler; 250 mg. of 1,3,3a,7-tetraaza-4-hydroxy-6-methyl indene as stabilizer; 20 ml. of a 5% aqueous solution of saponin as wetting agent; 2.5 ml. of a 30% aqueous solution of formaldehyde as hardener. The emulsion was divided into 7 equal samples and the following substances were added per kg. of emulsion to the individual samples: Sample A: comparison sample without additive; Sample B: 1.0 g. of Compound (1); Sample C: 3.0 g. of Compound (3); Sample D: 1.0 g. of Compound (4). The samples are applied onto a support of cellulose acetate and dried. They are then exposed behind a step wedge and processed in the usual manner. Processing is carried out as follows:

Color development	7 minutes
Short stop bath	5 minutes
Washing	5 minutes
Bleaching bath	5 minutes
Washing	5 minutes
Fixing bath	5 minutes
Washing	10 minutes

The color developer has the following composition: diethyl-p-phenylene-diamine sulfate, 2.75 g.; hydroxylamine sulfate, 1.2 g.; sodium sulfite anhydrous, 2.0 g.: sodium hexametaphosphate, 2.0 g.; potassium carbonate anhydrous, 75.0 g.; potassium bromide, 2.0 g.; make up to 1 liter with water.

The other processing baths had the following composition:

Short Stop Bath

Sodium acetate	30.0 g.
Glacial acetic acid	6.0 g.
Make up to 1 liter with water	

Bleaching Bath

Potassium ferricyanide	100 g.
Potassium bromide	20 g.
Disodium phosphate	10 g.
Glacial acetic acid	4 g.
Make up to 1 liter with water	

Fixing Bath

Sodium thiosulfate	200 g.
Make up to 1 liter with water	

The color density of the magenta layers was determined with a Macbeth Quanta Log, Model TD 102 densitometer, a green color filter being interposed in the path of the beam of measuring light.

TABLE 2

Sample	Sensitivity	Gamma	Fog
A	Blank	0.61	0.16
B	+2.5°	0.63	0.18
C	+2.8°	0.65	0.16
D	+2.2°	0.59	0.16

An increase by 3° here also represents double sensitivity or a sensitivity increase of one shutter stop.

Bipyridilium Salts

A process described by T.D. Andrews, G.D. Short and I. Thomas; U.S. Patent 3,697,528; October 10, 1972; assigned to Imperial Chemical Industries Limited, England involves radiation sensitive material based on nitrogeneous dications such as bipyridyls, preferably with a water-soluble polymer as a support. Especially preferred salts for use as radiation sensitive components are salts having the following general formula.

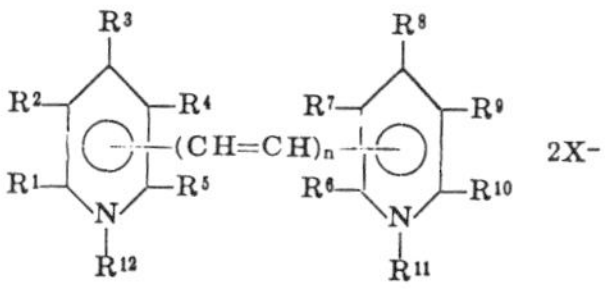

where R^{1-12} are hydrogen, halogen or organic substituents including groups between units having the above structure, which form polymeric salts; n = 0 or an integer; X^- is an anion derived from a strong acid (pKa preferably <2.5). Usually the link joins the two aromatic rings in the 4,4' or 2,2' positions, when it replaced $R^{3,8}$ or $R^{5,6}$, e.g., 2,2'-bipyridyls and 4,4'-bipyridyls. Examples of suitable salts are compounds containing the following active units (reference letter indicates appropriate cationic unit).

Formula	Name
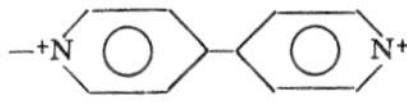	4,4'-bipyridylium (P).
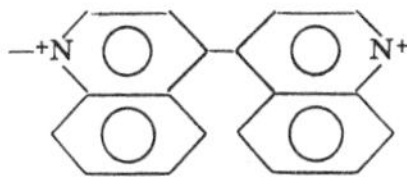	4,4'-biquinolinium (Q).
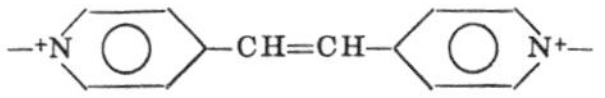	1,2-bis(4-pyridyl) ethylene (E).
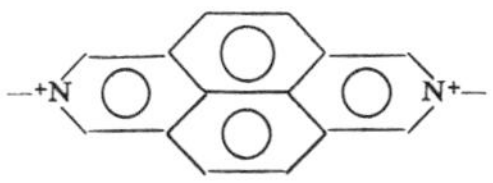	2,7-diazapyrinium (A).
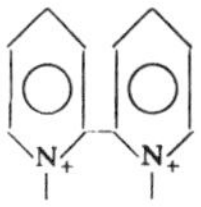	2,2'-bipyridyl (B).
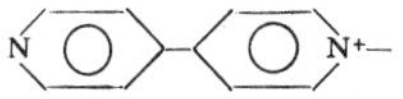	4-(4'-pyridyl) pridyinium (M).

Example 1:   15 parts of poly(vinyl alcohol) was dissolved in water made up to 100 parts at 80°C. and N,N'-dimethyl-4,4'-bipyridilium dichloride (3 parts) was dissolved after the solution had cooled to 40°C. The solution was poured onto a clean glass surface under dark-room conditions and the water was allowed to evaporate slowly at 20°C. When dry, the film was stripped from the glass and stored away from daylight.

The film was then exposed to ultraviolet light of a wavelength of 366 nm., when a blue coloration was induced. A similar effect was observed on exposure to x-rays and electrons by means of an electron beam microscope.

The image does not fade when kept in a dry atmostphere, and exhibits resolution at least as high as a conventional silver emulsion.

Example 2:   The procedure of Example 1 was repeated, substituting poly-(ammonium methacrylate) (15 parts) for the poly(vinyl alcohol). After preparing the film it was exposed to ultraviolet radiation of 366 nm. wavelength, and a similar blue coloration was noted. A sample of this film gave good light transmission in the region of 436 nm. compared with the transmission at 405 nm. The film is therefore of use in photochemical systems where it is desired to isolate the 436 nm. line of mercury lamp radiation for photochemical purposes.

Example 3:   The procedure of Example 1 was repeated substituting gelatin (15 parts) for poly(vinyl alcohol). Similar coloration was obtained on exposure to ultraviolet radiation, though the images were less stable than those obtained by methods described in Examples 1 and 2.

Example 4:   Bipyridyl (174 parts) was reacted with p-xylylene dichloride (197 parts) in methanol (1,300 parts) at 64°C. The reactants were heated under reflux conditions under an atmosphere of nitrogen in a light-proof flask while being stirred for 24 hours. The white paste obtained was diluted with methanol (2,500 parts) and ether (2,500 parts) was added to coagulate the product. This was then separated from liquors by filtration, washed with ether (3 x 500 parts) and then dried, first by suction until damp dry and then under vacuum, for 2 hours at 10 mm. Hg pressure and finally at 0.05 mm. Hg pressure for 20 hours at room temperature over phosphorus pentoxide in light-proofed apparatus. The product polymer (367 parts) is consistent with repeating units of the structure

$$-CH_2-\left\langle\bigcirc\right\rangle-CH_2-P-$$
$$2Cl-$$

3 parts of this polymer and poly(vinyl alcohol) (15 parts) were dissolved in water and a film was cast using the technique described above. On irradiation with ultraviolet light (366 nm.) a blue-purple coloration was obtained. Elemental Analysis: Calculated for $C_{18}H_{16}N_2Cl_2$: C, 65.25; H, 4.83; N, 8.46; Cl, 21.46%. Found: C, 64.57; H, 5.88; N, 8.06; Cl, 19.53%.

The solid is a pale yellow amorphous powder, soluble in water, slightly soluble in hot ethanol and methanol but insoluble in acetone, ether and nonpolar solvents. The dry solid is very hygroscopic and is light sensitive when exposed to bright sunlight, more so in the presence of an inert atmosphere.

Example 5:   The procedure of Example 4 was repeated using poly(ammonium-methacrylate) (15 parts). A similar blue-purple coloration was obtained on irradiation.

Example 6:   p-Cyanoaniline (47 parts) was refluxed with N, N'-di(2,4-dinitrophenyl)-4,4'-bipyridilium dichloride (56 parts) in ethanol/pyridine (500 parts and 200 parts respectively). The product was a compound having the structure:

$$CN-\langle \bigcirc \rangle-P-\langle \bigcirc \rangle-CN \quad 2Cl^-$$

3 parts of this material were made up into a film with poly(vinyl alcohol) using the procedure of Example 1. On exposure to ultraviolet light, a green coloration formed.

OTHER PROCESSING TECHNIQUES

Vapor Deposition on Photosensitized Substrate

K. Juna, H. Nakayama and K. Asada; U.S. Patent 3,625,744; Dec. 7, 1971; assigned to Kansai Paint Company, Limited, Japan describe a process for bringing a substance to be coated into contact with a vaporized sub-stance of an ethylenically unsaturated compound, applying actinic light rays upon the substance to be coated and thereby forming a polymer film of the compound. The following examples illustrate the process.

Examples 1 through 7:   A 1% aqueous solution of a photosensitive catalyst was applied on a polished mild steel plate (50 x 25 x 0.5 mm.) with a brush. The applied plate was allowed to dry for an hour at room tempera-ture. While the actinic light rays were being thrown on the plate from a high-pressure mercury vapor lamp of 100 w. at a distance of 50 mm., the plate was brought into contact with a vapor (180°C.) obtained by heating styrene in a 1 liter flask with a round bottom, and a polymer film was ob-tained. The thickness of the film obtained in each example was shown in Table 1 below. An infrared spectroscopic analysis showed that this polymer film was polystyrene and had a number average molecular weight of 75,000.

The tensile strength of the polymer film measured by a Tenshiron UTMU Type (manufactured by Toyo Sokki Kabushiki Kaisha, Japan, 40 mm./min.) was 6.5 kg./mm.^{2}, the breaking elongation rate was 3.2%, which was the same value as that of polystyrene coated in any conventional method. (Measurement conditions: temperature, 20°C., relative humidity, 60%.)

TABLE 1

Examples	Photosensitive catalyst	Irradia-tion time (min.)	Thick-ness of film (μ)
1	$UO_2 (NO_3)_2$	1	10
2	$UO_2 (NO_3)_2$	2	14
3	$UO_2 (NO_3)_2$ plus $FeCl_3$	2	30
4	$ZnCl_2$	2	7
5	$Ce (SO_4)_2$	2	12
6	$FeCl_3$ plus fluoresein	2	55
7	$UO_2 (NO_3)_2$ plus benzoin	2	40

Examples 8 through 16: A 1% acetone solution of a photosensitive catalyst was applied on a sheet of fine quality paper (50 x 25 mm., its tensile strength being 0.5 kg./mm.^{2}), which was allowed to dry at room temperature. The sheet of fine quality paper was then hung in a 4-neck flask of Pyrex glass in which styrene was vaporized (132°C.). A high-pressure mercury vapor lamp of 100 w. threw actinic light rays on the sheet of fine quality paper at a distance of 35 mm. for 2 minutes. Its tensile strength is improved as may be seen from Table 2.

TABLE 2

Examples	Photosensitive Catalyst	Tensile Strength (kg./mm.^{2})
8	o,o'-Dinitrodiphenyldisulfide	1.1
9	Tetramethylthiuram monosulfide	1.0
10	Azobisisobutyronitrile	0.8
11	Anthraquinone	0.8
12	Benzoin	1.1
13	Benzoyl Peroxide + $UO_2(NO_3)_2$	1.5
14	Diphenylsulfide + Thiazine	1.8
15	Azobisisobutyronitrile + Thiazine	0.8
16	$UO_2(NO_3)_2$ + Eosine	1.0

Examples 17 through 26: A 1% solution of a photosensitive catalyst was applied on a polished mild steel plate (50 x 25 x 0.5 mm.), which was allowed to dry at room temperature. The same apparatus as used in Example 8 threw actinic light rays on the mild steel plate for 2 minutes in a chamber filled with a vapor of ethylenically unsaturated compounds to obtain a film.

The thickness of each film is shown in Table 3. An infrared spectroscopic analysis showed that these films were polymers corresponding to their respective ethylenically unsaturated compounds. The number average molecular weight in Example 19 was 80,000; in Example 21, 55,000; and in Example 25, 35,000.

TABLE 3

Example	Ethylenically unsaturated compounds	Temperature of vaporized compound (° C.)	Photosensitive catalyst	Solvent	Thickness of film (μ)
17	Methyl methacrylate	90	AADC	Acetone	22
18	do	92	O U$_2$(NO$_3$)$_2$	Methanol	18
19	do	92	Riboflavin plus AADC	Acetone	36
20	do	91	Fluorescein plus FeCl$_3$	Ethanol plus methanol	12
21	Acrylonitrile	74	ZnCl$_2$	Water	7
22	do	72	UO$_2$(NO$_3$)$_2$ plus FeCl$_3$	do	3
23	do	74	UO$_2$(NO$_3$)$_2$ plus ZnCl$_2$	do	7
24	Vinyl acetate	70	UO$_2$(NO$_3$)$_2$	do	5
25	do	73	UO$_2$(NO$_3$)$_2$ plus FeCl$_3$	do	13
26	do	73	UO$_2$(NO$_3$)$_2$ plus ZnCl$_2$	do	7

NOTE.—AADC = α-aminoanthraquinone diazonium chloride·ZnCl$_2$.

Microcapsules Containing Silver Halide Grains

A process described by A.M. Gerber and V.K. Walworth; U.S. Patent 3,694,253; September 26, 1972; assigned to Polaroid Corporation is directed to a method for preparing microcapsules comprising a nucleus of silver halide surrounded by a continuous wall of a synthetic polymer. The method comprises forming a relatively thin layer of reactants including monomer and silver halide grains and exposing the thin layer to polymerizing radiation whereby the monomers preferentially polymerize around the silver halide.

The silver halides which are particularly useful in the process comprise silver chloride, silver chlorobromide, silver bromide, silver iodobromide, silver iodochlorobromide and combinations thereof, which are conventionally employed in photosensitive elements. The silver halides employed in the process may be obtained in the form of conventional silver halide emulsions or they may be composed of silver halides in the form of aqueous suspensions and precipitating the silver halides in the absence of a conventional binder material. Thus, in short, silver grains from any suitable source prepared in any conventional method may be employed without regard to size or shape.

The reactants employed in forming the capsules are deposited in a relatively thin layer and then subjected to polymerizing radiation. The thickness of the layer may vary over a relatively wide range. The minimum thickness of the layer should be slightly in excess of the desired diameter of the

capsules to be formed. The upper limit of thickness on the layer of reactants is determined by the transmission density of the layer with respect to
the ability of the incident radiation to penetrate the layer and inititate
polymerization. In a preferred form, the thickness of the layer is less than
2 times the proposed thickness of the capsules. In a particularly preferred
form, the thickness of the layer may range from 0.2 to 100 microns.

Figure 2.1 shows a cross-sectional elevational view of the process. The
reactants (11) including monomer and silver halide are premixed and disposed
in hopper (10) from which a uniform thin layer (12) is deposited on moving
belt (13) which carries the layer around transparent drum (14) having disposed therein radiation source (15) of a photopolymerizing wavelength.
Layer (16) after passing around drum (14) contains silver halide encapsulated
in a continuous polymeric wall layer. Station (17) is an optional treatment
which includes oxidizing means for the silver halide in the event that the
photopolymerizing radiation fogged the silver halide nucleus, i.e., spraying potassium ferricyanide or other suitable oxidizing agent onto the capsules. Doctor blade (18) removes capsules from belt (13) where they are
collected in hopper (19). Belt (13) as shown is an endless belt which is
passed around idler wheel (20) and is then in position to receive more reactants.

FIGURE 2.1:   METHOD FOR PREPARING MICROCAPSULES CON-
TAINING SILVER HALIDES

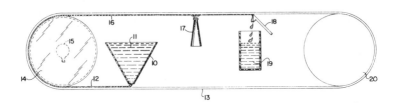

Source: A.M. Gerber and V.K. Walworth; U.S. Patent 3,694,253;
September 26, 1972

A washing step is also employed to remove any excess starting materials and
oxidizing agent. The polymerizing radiation is selected for its wavelength
and is adapted to photopolymerize the given monomer. Generally, for
vinyl compounds, any wavelength between about 250 and 700 millimicrons
may be employed. The reactants are preferably disposed in a suitable solvent for ease of distribution in the thin layer. The preferred solvent is
water.

Example: A mixture comprising 9 g. of diacetone acrylamide, 1 g. of acrylamide, 90 cc of a photosensitive silver iodobromide emulsion (18.4% silver, 3.5% gelatin) having a grain size about 0.8 micron, and 2.0 ml. of an amphoteric surfactant of the formula

$$C_8H_{17}-\overset{\overset{\displaystyle N}{\underset{\displaystyle \parallel}{}}}{C}\underset{\overset{\displaystyle |}{\underset{\displaystyle OH}{}}}{}-N\overset{CH_2}{\underset{CH_3}{}}-CH_2CH_2OCH_2COONa \atop CH_2COONa$$

was disposed on a surface in a thickness of about 1.0 micron. The emulsion was pretreated by suspending in 50 g. of emulsion 5 g. of a strong base anion exchange resin (as the hydroxide) (Dowex 21K) in a silk bag in the emulsion for 1 1/2 hours. At the end of that time the pBr was >7. The layer was exposed to a light source which emitted light of visible wavelength for about 30 seconds. Upon removal of the unreacted materials by washing with water, capsules 1.6μ in diameter comprising silver bromide, completely surrounded by a continuous polymeric wall, were observed under the optical microscope.

Figuring of Optical Elements

E. Berman, G.L. McLeod, C.H.C. Pian, S.H. Stein and J.F. Pian; U.S. Patent 3,658,528; April 25, 1972; assigned to Itek Corporation describe a method for correcting irregularities and properly contouring optical components by coating the surface to be contoured with a layer of photopolymerizable or photodepolymerizable material and using light to selectively operate on the layer to achieve the proper contour. In the case when a polymerized coating is used, the layer is deposited on the optical surface and light or similar radiant energy is selectively directed onto the outer surface of the polymer, to photodepolymerize the outer portions in accordance with the desired figuring.

The resulting depolymerization products are then removed by dissolving with a suitable solvent or are distilled away in vacuo, leaving a polymerized coating in the contour desired. In the alternate case, a layer of polymerizable material is deposited on the surface to be figured and a selective pattern of light or similar radiation is directed through the optical component, causing polymerization at the component-layer interface. The unaffected components of the layer are dissolved away with a suitable solvent, leaving the required polymerized surface on the component. The selective radiation is produced in response to digitally encoded signals obtained from a comparison of the desired contour with a contour map of the actual surface of the optical component.

Figure 2.2a shows an automated system capable of performing the process. The operation of this system will first be described before a detailed discussion of the actual polishing process is presented. An appropriately

FIGURE 2.2:   PHOTOCHEMICAL FIGURING OF OPTICAL ELEMENTS

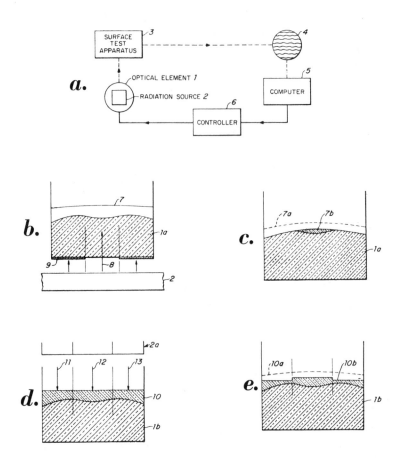

(a)   Figuring System
(b)(c)   Cross-Sectional Views of Coated Optical Components — Photo-polymerization
(d)(e)   Cross-Sectional Views of Coated Optical Components — Photo-depolymerization

Source: E. Berman, G.L. McLeod, C.H.C. Pian, S.H. Stein and
           J.F. Pian; U.S. Patent 3,658,528; April 25, 1972

coated optical blank (1), which is about to undergo the automatic figuring process, is shown positioned beneath a lamp (2) or other means for emitting suitable light or electromagnetic radiation to be used in the process. Before the photopolymerization and photodepolymerization process can begin, however, an indication of the actual contour of the blank (1) must be obtained which can be compared with the desired contour, so that the areas on the surface of the blank which require correction may be determined. Accordingly, a surface test apparatus (3) is provided for measuring the surface. While the desired measurement of the surface can be performed by a mechanical apparatus, such as by dial indicators or a transversing probe, the preferred method comprises the use of an interferometer.

An interference picture (4) of the surface is obtained which indicates the surface asymmetries and irregularities on the blank (1). This interference picture (4) is then digitally encoded in a suitable computer (5) containing information as to the desired contour of the optical component (1). The computer (5) accordingly provides information as to the differences between the existing and the desired relative elevations occuring over various areas of the blank's surface. This error information is fed to a suitable controller apparatus (6). The controller apparatus (6) converts the digital information into appropriate signals which control the operation of the means (2) for irradiating the coated optical component (1).

The irradiating means (2) may be in the form of a lamp or a bank of lamps or a flying spot capable of emitting light or other electromagnetic radiation in a particular spectral range. The signals from the controller (6) selectively operate the irradiating means (2) to accomplish the desired degree of photopolymerization or photodepolymerization, as will be more fully explained. First, with regard to the photopolymerization process, Figure 2.2b shows a portion of the surface of an optical component (1a) which has been coated with a thin layer of polymerizable material (7). After the surface has been coated, light (8) in a suitable spectral range or, more particularly, radiant energy of a proper wavelength is selectively directed through the optical component (1a) and causes polymerization at the layer-component interface.

The portions of the layer (7) which are irradiated by the light become polymerized. When the proper amount of polymerization is achieved, the unaffected residue (7a) may be removed by washing with a suitable solvent leaving the polymerized portion (7b) on the surface, forming the desired contour as shown in Figure 2.2c. Selective irradiation of the component's surface may be accomplished in a number of ways. For example, a fine beam of light, whose wavelength and intensity produce sufficient energy to chemically alter the polymer coating may be directed onto particular areas of the component's surface which require correction. The light beam may be controlled in the manner of a flying spot scanner (e.g., D'Arsonval Galvanometer Movement) so as to sweep the surface at a given rate. The beam may be intensity modulated by the controller apparatus (6) so that the sweep pattern would cause the areas requiring greater correction to be more intensely irradiated than those areas requiring less correction. In the alternative the beam sweep velocity may be modulated to alter the amount of radiation applied to given incremental areas of the element. Another alternate method using a single lamp may comprise the utilization of suitable masking on the undersurface of the component as shown in Figure 2.2b.

A mask (9) may be developed from a comparison of the interference picture (4) of the surface contour and a desired contour, such that when placed on the undersurface of the component, it will control the amount of light that reaches the various areas of the polymer coating. The use of such masking would of course obviate the need for the electronic computing apparatus (5) and (6). A further irradiation arrangement comprising the use of a bank or matrix of individually controlled lamps would also be suitable. In this arrangement, signals from the controller apparatus (6) activate the individual lamps in appropriate patterns. The lamps would be selectively energized in accordance with the location and time of exposure required to produce the desired amount of cross-linking in selected areas to correct the errors on the surface of the component.

The photodepolymerization process is illustrated in Figure 2.2d and Figure 2.2e. Figure 2.2d shows a portion of the surface of an optical component (1b) with a thin coating of a suitable polymer (10) which is appropriately photodegradable. As in the photopolymerization process, radiant energy of proper wavelength is directed onto the coated component (1b), but in this instance, the radiation strikes the polymer (10) at the air-polymer interface and radiation of three different wavelengths (11), (12) and (13) from a bank of lamps (2a) is illustrated. The outer layers of the polymer are depolymerized by the radiation in proportion to the energy falling thereon leaving an outer coating (10a) of depolymerization products.

When the desired polymer configuration is achieved, irradiation is stopped and the outer coating (10a) is removed by suitable solvent washing, thus leaving the remaining polymer (10b) in the desired contour on the optical component surface as shown in Figure 2.2e. The photosensitive materials must be susceptible to chemical alteration which will enable separation of the original material from the altered material. Such separation can be simply based on differential solvent solubility, or alternatively on the basis of volatility of one of the forms of the material. The preferred materials are those which enable simple solvent separation of the altered and original material since the use of complicated vacuum equipment is avoided. The separation of the materials should lead to removal of one with the other remaining on the surface of the component to obtain the desired result. The material which remains on the component surface should be substantially permanent, i.e., relatively stable and adherent to the component surface, especially in extremely thin thickness, and should preferably have high optical quality. The photosensitive material for use in the process can be one of two general classes. The first class embraces those substances which under the influence of light degrade, i.e., are photodegraded, while the second class include those materials which under the influence of light, undergo condensation, addition or cross-linking type reactions. The following examples illustrate the process.

Example 1: A layer of poly(methylmethacrylate) on a component is exposed to ultraviolet radiation at 130°C. through a mask prepared on the basis of the contour of the component. The exposure is carried out under high vacuum (10^{-6} mm. Hg). This coating is subsequently aluminized after normal glow discharge treatment and the Scotch tape test of the resulting mirror shows excellent adherence of the aluminum to the polymer layer.

Example 2: A layer of polyvinylbenzalacetophenone on a component is selectively exposed through the component to ultraviolet radiation using the apparatus of Figure 2.2a after which the unaffected coating is removed by solvent washing. The component is then aluminized as in Example 1. It has been found that the above-described processes may be used for the final figuring of optical component surfaces to accomplish one-quarter wavelength uniformity or better. In addition, polymer coatings achieved by both methods will satisfactorily accept an aluminized coating as required on certain reflective components.

Phosphor Screens for Color Television

S. Levinos; U.S. Patent 3,585,034; June 15, 1971; assigned to GAF Corporation describes photosensitive compositions for color television picture tube manufacture comprising an ethylenically unsaturated monomer material and a catalyst system comprising (a) a stabilized diazonium salt which functions as an electron-acceptor and (b) a photosensitive cocatalyst selected from the group consisting of (1) a complex ferric salt containing an electron-donating group and (2) a mixture comprising a simple ferric salt and an organic polycarboxylic acid capable of reducing the simple ferric salt to ferrous when subjected to actinic radiation.

The photosensitive compositions are uniquely and beneficially adapted to color television picture tube manufacturing operations based upon photographic reproduction methods. Basically, the steps involved comprise first coating the inside surface of the television picture tube viewing panel by flowing, spraying or whirling with the photosensitive composition containing

the polymerizable monomer, catalyst components and one or more optional
ingredients; as will be understood, the monomer-catalyst coating composition may be initially provided with the phosphor material or alternatively,
the phosphor may be applied by dusting or other suitable method to the
monomer-catalyst coated layer while the latter remains in a tacky condition. The coating is exposed through a shadow mask whereby to form a
latent image of the dark pattern, i.e., in the form of hardened, polymerized areas in the monomer-catalyst layer.

Alternatively, the phosphor material may be applied subsequent to the exposure operation, this procedure obviating any requirement for drying the
photosensitive layer prior to exposure. The tacky condition of the photosensitive coating, of course, facilitates adhesion of the phosphor particles.
Unexposed areas are removed by washing with deionized water. This sequence of steps is repeated for each of the remaining color aspects. A
typical composition found to be eminently suitable for use comprises:

> Polyvinyl pyrrolidone K-30: 2.5 g.
> Polyvinyl pyrrolidone K-60: (45% sol.): 16.67 g.
> Ferric ammonium oxalate (36% aq. sol.): 2 ml.
> N, N'-methylenebisacrylamide: 5.0 g.
> Ethyl alcohol (95%): 140 ml.
> Hydrochloric acid 1:1: 2.0 g.
> Citric acid: 3.0 g.
> p-Toluenediazonium chlorozincate: 5.0 g.
> Water: 185 ml.

Processing of the above composition in the manner described provides a
tricolor phosphor pattern, such procedure being implemented for each of
the red, blue and green phosphor patterns, having exceptional luminosity
and brilliance. Moreover, none of the objectionable features character-
izing the use of dichromatepolyvinyl alcohol systems is in any way evident.
The addition of zinc chloride to the above composition serves to significantly enhance the polymerization reaction rate. In order to capitalize
upon the accelerating effects of the zinc chloride, a holding period of
approximately 10 minutes is advisable prior to the wash-out step.

This allows sufficient time for the zinc chloride to exert its full catalytic
effect. Similar improvement in phosphor luminosity is obtained when the
ferric ammonium oxalate in the above-delineated formulation is substituted
in equivalent amounts by ferric ammonium citrate; similar improvement
likewise attends the replacement of p-toluenediazonium chlorozincate with
p-toluenediazonium fluosilicate, benzene diazonium fluoborate and 2,4,6-
trimethylbenzenediazonium hydrogen sulfate respectively.

POLYENE-POLYTHIOL COMPOSITIONS

CURABLE COMPOSITIONS

Synthesis of Polyene-Polythiol Liquid Components

C.L. Kehr and W.R. Wszolek; U.S. Patent 3,697,395; October 10, 1972; assigned to W.R. Grace & Co. describe photocurable liquid polymer composition which includes a liquid polyene component having a molecule containing at least two unsaturated carbon-to-carbon bonds disposed at terminal positions on a main chain of the molecule, a polythiol component having a molecule containing a multiplicity of pendant or terminally positioned —SH functional groups per average molecule, and a photocuring rate accelerator. The photocurable liquid polymer composition upon curing in the presence of actinic light forms odorless, solid, elastomeric or resinous products which may serve as sealants, coatings, adhesives and molded articles.

The use of polymeric liquid polythiol polymers which are cured to solid elastomeric products by oxidative coupling of the thiol (—SH) groups to disulfides (—S—S— groups) are known in the sealants, coatings and adhesives field. Oxidizing agents such as PbO_2 are commonly used to effect this curing reaction. These mercapto-containing compounds, however, both before and after curing with PbO_2-type curing system yield elastomeric compositions with an offensive odor which limits their usefulness generally to outdoor service. Thus, oxidatively-cured mercapto polymer systems have found restricted commercial acceptance due to their offensive odors.

A limitation of commercial liquid polymeric sealants and coatings is found in one-package systems. All the compounding ingredients, including the curing agents, are blended and charged into a tightly sealed container until used. In these commercial sealants (polysulfides, polydisulfides, polymercaptans, polyurethanes and polysilicones), the curing reaction of one-package systems is initiated by moisture from the air. Curing of such adhesives, coatings and sealants is variable, difficult to predict and control. Polyurethanes have a further disadvantage. In the curing reaction a volatile gas (CO_2) is liberated and this evolved gas tends to cause unsightly and property-weakening voids in the final product.

This process provides a photocurable liquid composition containing particular polyenes which are curable by polythiols to solid resins or elastomers. For example, when urethane-containing polyenes are compounded with polythiols, the prepared composition may be stored safely for long periods of time in the absence of actinic light. Upon exposure to actinic light such as ultraviolet light, the prepared system may be cured rapidly and controllably to a polythioether-polyurethane product which is low in cost and equal or better in reaction rate in polymer formation when compared with compositions derived from conventional technology.

Generally stated, the process provides a photocurable composition which comprises a particular polyene component, a polythiol component, and a photocuring rate accelerator.

General representative formulas for the polyenes of the process may be prepared as exemplified below:

Poly(Alkylene-Ether) Polyol Reacted with Unsaturated Monoisocyanates Forming Polyurethane Polyenes and Related Polymers

Difunctional

Trifunctional

Tetrafunctional

Tri-to-hexafunctional

$$CH_2O\text{---}\left(C_2H_4O\right)_x\left(C_3H_6O\right)_y\left(C_4H_8O\right)_z\overset{O}{\overset{\|}{C}}\text{---}\overset{R_7}{\overset{|}{N}}\text{---}\left(CH_2\right)_n\text{---}CH\text{=}CH_2$$

$$\left[R_1\text{---}\overset{|}{\underset{|}{C}}\text{---}O\text{---}\left(C_2H_4O\right)_x\left(C_3H_6O\right)_y\left(C_4H_8O\right)_z\overset{O}{\overset{\|}{C}}\text{---}\overset{R_7}{\overset{|}{N}}\text{---}\left(CH_2\right)_n\text{---}CH\text{=}CH_2\right]_p$$

$$CH_2O\text{---}\left(C_2H_4O\right)_x\left(C_3H_6O\right)_y\left(C_4H_8O\right)_z\text{---}\overset{}{\underset{\overset{\|}{O}}{C}}\text{---}\overset{}{\underset{R_7}{N}}\text{---}\left(CH_2\right)_n\text{---}CH\text{=}CH_2$$

Interconnected-modified difunctional

$$CH_2\text{=}CH\text{---}\left(CH_2\right)_n\overset{}{\underset{R_7}{N}}\text{---}\overset{O}{\overset{\|}{C}}\text{---}\left(OC_2H_4\right)_x\left(OC_3H_6\right)_y\left(OC_4H_8\right)_z\text{---}O\overset{O}{\overset{\|}{C}}\text{---}NH\text{---}\underset{}{\bigcirc}^{CH_3}\text{---}NH\text{---}\overset{O}{\overset{\|}{C}}\text{---}\Big]_p$$

$$\text{---}\left(OC_2H_4\right)_x\left(OC_3H_6\right)_y\left(OC_4H_8\right)_z\text{---}O\text{---}\overset{O}{\overset{\|}{C}}\text{---}\overset{}{\underset{R_7}{N}}\text{---}\left(CH_2\right)_n\text{---}CH\text{=}CH_2$$

Interconnected-modified tetrafunctional

$$CH_2\text{=}CH\text{---}\left(CH_2\right)_n\overset{R_7}{\underset{}{N}}\text{---}\overset{O}{\overset{\|}{C}}\text{---}O\text{---}\left(H_8C_4O\right)_z\left(H_6C_3O\right)_y\left(H_4C_2O\right)_x \quad \left(OC_2H_4\right)_x\left(OC_3H_6\right)_y\left(OC_4H_8\right)_z\text{---}O\text{---}\overset{O}{\overset{\|}{C}}\text{---}\overset{R_7}{\underset{}{N}}\text{---}\left(CH_2\right)_n\text{---}CH\text{=}CH_2$$

$$N\text{---}\left(CH_2\right)_q\text{---}N$$

$$\left(CH_2\right)_n\overset{}{\underset{R_7}{N}}\text{---}\overset{}{\underset{\overset{\|}{O}}{C}}\text{---}O\text{---}\left(H_8C_4O\right)_z\left(H_6C_3O\right)_y\left(H_4C_2O\right)_x \quad \left(OC_2H_4\right)_x\left(OC_3H_6\right)_y\left(OC_4H_8\right)_z\text{---}O\text{---}C\text{---}\overset{}{\underset{R_7}{N}}\text{---}\left(CH_2\right)_n\text{---}CH\text{=}CH_2$$

Poly(Alkylene–Ester) Polyol Reacted with Unsaturated Monoisocyanates Forming Polyurethane Polyenes and Related Polymers

Difunctional

$$CH_2\text{=}CH\text{---}\left(CH_2\right)_n\overset{}{\underset{R_7}{N}}\text{---}\overset{O}{\overset{\|}{C}}\text{---}\left[O\text{---}CH_2CH_2\text{---}O\text{---}\overset{O}{\overset{\|}{C}}\text{---}\left(CH_2\right)_4\overset{O}{\overset{\|}{C}}\text{---}\right]_n O\text{---}CH_2CH_2\text{---}O\text{---}\overset{O}{\overset{\|}{C}}\text{---}\overset{}{\underset{R_7}{N}}\text{---}\left(CH_2\right)_n\text{---}CH\text{=}CH_2$$

Interconnected-modified difunctional

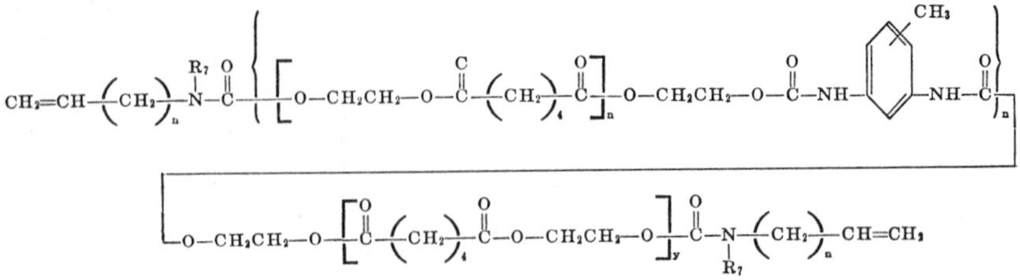

Poly(Alkylene-Ester) Polyol Reacted with Polyisocyanate and Unsaturated Monoalcohol Forming Polyurethane Polyenes and Related Polymers

Difunctional

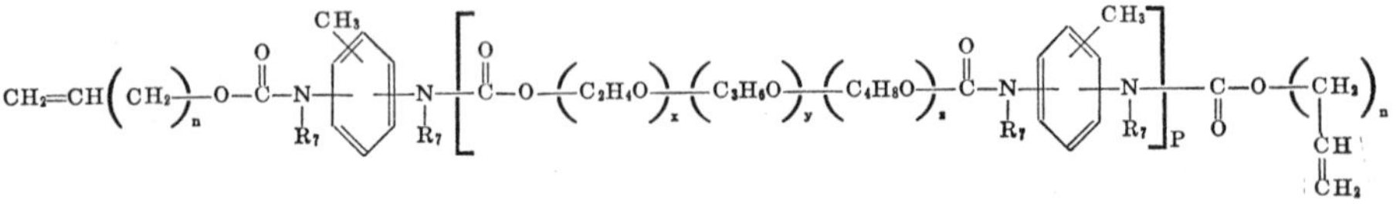

Trifunctional

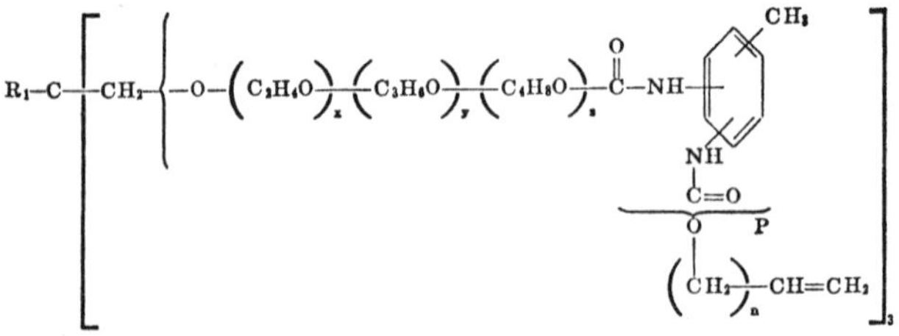

Tetrafunctional

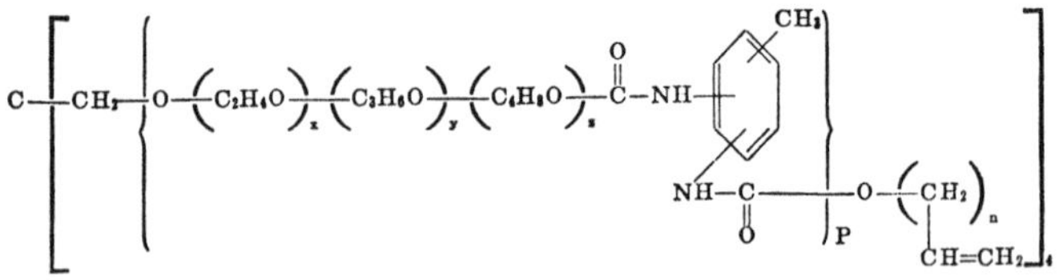

In the formulas, the sum of $x + y + z$ in each chain segment is at least 1; P is an integer of 1 or more; q is at least 2; n is at least 1; R_1 is selected from the group consisting of hydrogen, phenyl, benzyl, alkyl, cycloalkyl, and substituted phenyl; and R_7 is a member of the group consisting of $CH_2{=}CH{-}(CH_2{)_n}$, hydrogen, phenyl, cycloalkyl, and alkyl.

The class of polyenes derived from carbon to carbon unsaturated monoisocyanates may be characterized by extreme ease and versatility of manufacture when the liquid functionality desired is greater than about three. For example, consider an attempted synthesis of a polyhexene starting with an −OH terminated polyalkylene ether hexol such as "Niax" Hexol LS-490 (Union Carbide Corp.) having a molecular weight of approximately 700, and a viscosity of 18,720 cp. at 20° C. An attempt to terminate this polymer with ene groups by reacting one mole of hexol with 6 moles of tolylene diisocyanate (mixed-2,4-, -2-6-isomer product) and 6 moles of allyl alcohol proceeded nicely but resulted in a prematurely chain extended and crosslinked solid product rather than an intended liquid polyhexene.

Using the monoisocyanate route, however, this premature chain extension may be avoided and the desired polyurethane-containing liquid polyhexene may be easily prepared by a simple, one-step reaction of one mole of hexol with 6 moles of allyl isocyanate. This latter polyhexene has the added advantage of being cured to a non-yellowing polythioether polyurethane product. Similarly, the unsaturated monoisocyanate technique may be used to prepare liquid polyenes from other analagous highly functional polyols such as cellulose, polyvinyl alcohol, partially hydrolized polyvinyl acetate, and the like, and highly functional polyamines such as tetraethylene pentamine, and the like.

A general method of forming one type of polyene containing urethane groups is to react a polyol of the general formula $R_{11}{-}(OH)$, where R_{11} is a polyvalent organic moiety free of reactive carbon-to-carbon unsaturation and n is at least 2; with a polyisocyanate of the general formula $R_{12}{-}(NCO)_n$, where R_{12} is a polyvalent organic moiety free from reactive carbon-to-carbon unsaturation and n is at least 2 and a member of the group consisting of an ene-ol, yne-ol, ene-amine and yne-amine. The reaction is carried out in an inert moisture-free atmosphere (nitrogen blanket) at atmospheric pressure at a temperature in the range from 0° to about 120°C. for a period of about 5 minutes to about 25 hours.

In the case where an ene-ol or yne-ol is employed, the reaction is preferably a one step reaction wherein all the reactants are charged together. In the case where an ene-amine or yne-amine is used, the reaction is preferably a two step reaction wherein the polyol and the polyisocyanate

are reacted together and thereafter preferably at room temperature, the
ene-amine or yne-amine is added to the NCO terminated polymer formed.
The group consisting of ene-ol, yne-ol, ene-amine and yne-amine are
usually added to the reaction in an amount such that there is one carbon-
to-carbon unsaturation in the group member per hydroxy group in the polyol
and the polyol and group member are added in combination in a stoichiometric amount necessary to react with the isocyanate groups in the polyisocyanate.

A second general method of forming a polyene containing urethane groups
(or urea groups) is to react a polyol (or polyamine) with an ene-isocyanate
or an yne-isocyanate to form the corresponding polyene. The general procedure and stoichiometry of this synthesis route is similar to that described
for polyisocyanates in the preceding. In this instance, a polyol reacts with
an ene-isocyanate to form the corresponding polyene. It is found, however,
that products derived from this route, when cured in the presence of an active light source and a polythiol, may form relatively weak solid polythioether products.

To obtain stronger cured products, it is desirable to provide polar functional
groupings within the main chain backbone of the polymeric polyene. These
polar functional groupings serve as connecting linkages between multiple
repeating units in the main chain series, and serve as internal strength-
reinforcing agents by virtue of their ability to create strong interchain
attraction forces between molecules of polymer in the final cured composition.

Polythiol as used here refers to simple or complex organic compounds having
multiplicity of pendant or terminally positioned -SH functional groups per
average molecule.

On the average the polythiol must contain 2 or more -SH groups/molecule
and have a viscosity range of essentially 0 to 20 million centipoises at
70°C. as measured by a Brookfield Viscometer either alone or when in the
presence of an inert solvent, aqueous dispersion or plasticizer. Operable
polythiols in the process usually have molecular weights in the range about
50 to about 20,000, and preferably from about 100 to about 10,000.

The polythiols may be exemplified by the general formula $R_8(SH)_n$, where
n is at least 2 and R_8 is a polyvalent organic moiety free from reactive carbon-to-carbon unsaturation. Thus R_8 may contain cyclic groupings and
hetero atoms such as N, P or O and primarily contains carbon-to-carbon,
carbon-hydrogen, carbon-oxygen, or silicon-oxygen containing chain linkages free of any reactive carbon-to-carbon unsaturation.

One class of polythiols to be used with polyenes to obtain essentially odorless polythioether products are esters of thiol-containing acids of the formula $HS-R_9-COOH$, where R_9 is an organic moiety containing no reactive carbon-to-carbon unsaturation with polyhydroxy compounds of structure $R_{10}(OH)_n$, where R_{10} is an organic moiety containing no reactive carbon-to-carbon unsaturation, and n is 2 or greater. These components will react under suitable conditions to give a polythiol having the general structure $R_{10}[OC(O)-R_9-SH]_n$, where R_9 and R_{10} are organic moieties containing no reactive carbon-to-carbon unsaturation, and n is 2 or greater.

Prior to curing, the curable liquid polymer may be formulated for use as 100% solids, or disposed in organic solvents, or as dispersions or emulsions in aqueous media.

Although the mechanism of the curing reaction is not completely understood, it appears most likely that the curing reaction may be initiated by almost any actinic light source which dissociates or abstracts a hydrogen atom from an SH group, or accomplishes the equivalent thereof. Generally the rate of the curing reaction may be increased by increasing the temperature of the composition at the time of initiation of cure. In many applications, however, the curing is accomplished conveniently and economically by operating room temperature conditions. Thus for use in elastomeric sealants, it is possible merely to photoexpose the polyene, polythiol, photocuring rate accelerator admixture to ambient conditions and obtain a photocured solid elastomeric or resinous product.

A class of actinic light useful here is ultraviolet light and other forms of actinic radiation which are normally found in radiation emitted from the sun or from artificial sources such as Type RS Sunlamps, carbon arc lamps, xenon arc lamps, mercury vapor lamps, tungsten halide lamps and the like. Ultraviolet radiation may be used most efficiently of the photocurable polyene-polythiol composition contains a suitable photocuring rate accelerator. Curing periods may be adjusted to be very short and hence commercially economical by proper choice of ultraviolet source, photocuring rate accelerator and concentration thereof, temperature and molecular weight, and reactive group functionality of the polyene and polythiol. Curing periods of less than about 1 second duration are possible, especially in thin film applications such as desired for example in coatings and adhesives.

Conventional curing inhibitors or retarders which may be used in order to stabilize the components or curable compositions so as to prevent premature onset of curing may include hydroquinone; p-tert-butyl catechol; 2,6-di-tert-butyl-p-methylphenol; phenothiazine; N-phenyl-2-naphthylamine; inert gas atmospheres such as helium, argon, nitrogen and carbon dioxide; vacuum; and the like.

Specifically useful are chemical photocuring rate accelerators such as benzophenone, acetophenone, acenapthene-quinone, o-methoxy benzophenone,
thioxanthen-9-one, xanthen-9-one, 7-H-benz[d,e]anthracen-7-one, dibenzosuberone, 1-naphthaldehyde, 4,4'-bis(dimethylamino)benzophenone
and fluorene-9-one. The following examples illustrate the formation of
polyene prepolymer.

Example 1: 458 g. (0.23 mol) of commercially available liquid polymeric diisocyanate (Adiprene L-100) was charged to a dry resin kettle
maintained under a nitrogen atmosphere and equipped with a condenser,
stirrer, thermometer, and gas inlet and outlet. 37.8 g. (0.65 mol) of
allyl alcohol was charged to the kettle and the reaction was continued for
17 hours with stirring at 100°C. Thereafter the nitrogen atmosphere was
removed and the kettle was evacuated 8 hours at 100°C. 50 cc. dry benzene was added to the kettle and the reaction product was azeotroped
with benzene to remove the unreacted alcohol. This allyl terminated
liquid prepolymer had a molecular weight of approximately 2,100 and
will be referred to as Prepolymer A.

Example 2: 400 g. (0.2 mol) of Adiprene L-100 was charged to a dry
resin kettle maintained under nitrogen and equipped with a condenser,
stirrer, thermometer and gas inlet and outlet. 25.2 g. (0.43 mol) of
propargyl alcohol (HC≡C—CH_2OH) was added to the kettle and the reaction was continued with stirring for 18 hours at 160°C. Thereafter the
nitrogen atmosphere was removed and the kettle was evacuated 16 hours
at 100°C. followed by azeotropic distillations with 50 cc. water and then
50 cc. benzene to remove any excess propargyl alcohol. This HC≡C—
terminated liquid prepolymer had a viscosity of 27,500 centipoises at
70°C. and a molecular weight of 2,100 and will be referred to as Prepolymer B.

Example 3: 1 mol of commercially available poly(ethylene ether) glycol
having a molecular weight of 1,450 and a specific gravity of 1.21 was
charged to a resin kettle maintained under nitrogen and equipped with a
condenser, stirrer, thermometer and a gas inlet and outlet. 2.9 g. dibutyl tin dilaurate as a catalyst was charged to the kettle along with 2
mols tolylene-2,4-diisocyanate and 2 mols of allyl alcohol. The reaction was continued with stirring at 60°C. for 2 hours. Thereafter a
vacuum of 1 mm. was applied for 2 hours at 60°C. to remove the excess
alcohol. This CH_2=CH— terminated prepolymer had a molecular weight
of approximately 1,950 and will be referred to as Prepolymer C.

Example 4: 1 mol of a commercially available poly(propylene ether)
glycol having a molecular weight of about 1,958 and a hydroxyl number
of 57.6 was charged to a resin kettle equipped with a condenser, stirrer,

thermometer and a gas inlet and outlet. 4 g. of dibutyl tin dilaurate as a catalyst was added to the kettle along with 348 g. (2.0 mols of toly-lene-2,4-diisocyanate and 116 g. (2 mols) of allyl alcohol. The reaction was carried out for 20 minutes at room temperature under nitrogen. Excess alcohol was stripped from the reaction kettle by vacuum over a one hour period. The thus formed $CH_2=CH-$ terminated liquid prepolymer had a molecular weight of approximately 2,400 and will be referred to as Pre-polymer D.

Example 5: 750 g. of a N-containing tetrol (hydroxyl functionality = 4) (Tetronic Polyol 904) having a MW of 7,500 was placed in a reaction vessel heated at 110°C. The flask was maintained under vacuum for 1 hour. Then, under an atmosphere of nitrogen, 0.1 cc. dibutyl tin di-laurate was added and the flask was cooled to 50°C. Now 18.3 g. allyl isocyanate was added slowly, maintaining the temperature at about 95°C. for about 1 hour after the addition was completed. The thus formed poly-meric polyene (i.e., Prepolymer E) had a theoretical allyl functionality of 2.2, a theoretical hydroxyl functionality of 1.8, and a calculated molecular weight of about 7,683. The following examples illustrate the photocuring process.

Example 6: 0.01 mol of the allyl-terminated liquid prepolymer A was charged to a 2 oz. glass bottle along with a stoichiometric amount to re-act with the allyl groups, i.e. 0.0066 mols, of trimethylolpropane tris (β-mercaptopropionate) having a molecular weight of 398, and about 2% total weight of benzopehenone. The liquid reactants were stirred together and heated for 1/2 hour at 140°C. Thereafter the reactants were photo-exposed to type RS Sunlamp under ambient conditions of room temperature and pressure. After exposure, the liquid reactants became solid; self-supporting, cured, odorless, elastomeric polythioether product resulted.

Example 7: Example 6 was repeated except that 0.0033 mol of trimethylol-propane tris (β-mercaptopropionate) having a molecular weight of 398 was used. The mixture, upon photocuring formed a solid, odorless, self-support-ing, cured elastomeric polythioether polymer.

Example 8: 0.005 mol of the allyl-terminated liquid Prepolymer E was charged to a 2 oz. glass jar along with a stoichiometric amount of a poly-thiol to react with the allyl groups in Prepolymer E, 0.0036 mol of tri-methylolpropane tris(β-mercaptopropionate), and 2% total weight of ben-zophenone. The liquid reactants were stirred together briefly at room tem-perature and allowed to stand under ambient conditions. After exposure to ultraviolet light, a solid, odorless, self-supposting cured elastomeric polythioether polymer resulted.

Example 9:   1,510 g. of a commercially available polyoxypropylene glycol solid (Pluracol P 2010) was charged to a resin kettle maintained under a nitrogen atmosphere and equipped with a condenser, stirrer, thermometer and gas inlet and outlet. The reactant was degassed at room temperature for 3 hours.

265.4 g. of an 80 to 20% isomer mixture of tolylene-2,4-diisocyanate and tolylene-2,6-diisocyanate (Mondur TD 80) was charged to the kettle and heated for 2 hours at 120°C. with stirring under nitrogen. Then, 116.9 g. (2 mols) of allyl alcohol was added to the kettle and the mixture was re-fluxed for 16 hours at 120°C. Excess allyl alcohol was stripped by vacuum at 115°C. for 23 hours. The thus-formed $CH_2=CH-$ terminated polyene prepolymer had a molecular weight of approximately 2,460 to 2,500, and a viscosity of 16,000 cp. as measured on a Brookfield Viscometer at 30°C.

0.005 mol of the thus-formed polyene prepolymer were charged to a 2 oz. glass jar along with a stoichiometric amount of a polymeric dithiol. 0.5 g. of acetophenone (a UV photoinitiator) was charged to the glass jar and the mixture was immediately stirred. Thereafter the mixture was placed out-doors for UV curing. In 24 hours a solid, self-supporting, odorless, cured elastic polymer resulted.

Example 10:   3 g. of a linear saturated hydrocarbon backbone ethylene-propylene-nonconjugated diene terpolymer (Nordel) which had been visbroken until it had a reduced specific viscosity of 0.99 and contained 0.4 vinyl, 6.4 trans and 0.4 vinylidene unsaturated groups per 1,000 carbon atoms, was dissolved in 100 ml. of benzene in a glass jar. A 50% excess over the stoichiometric amount, i.e. 0.0006 mol (0.3 g.) of pentaerythritol tetrakis (β-mercaptopropionate) was added to the jar in addition to 0.5 g. acetophenone. The glass jar was placed in the sun-light outdoors under atmospheric conditions.

After 24 hours the benzene had substantially evaporated leaving a gelatinous polymeric precipitate. Acetone was added to precipitate more polymer. The polymer was filtered off, washed with acetone and dried in a vacuum oven at 60°C. 2.3 g. of the above polythioether polymer product was ex-tracted with benzene along with a "control" sample of the starting visbroken Nordel material. The "control" sample showed a gel content (benzene in-soluble) of 3.4% whereas the cured (crosslinked) solid polythioether pro-duct had a gel content of 82.8%.

Additional studies with photocurable polyene-polythiol polymer composi-tions are described by C.L. Kehr and W.R. Wszolek; U.S. Patents 3,697,397 and 3,697,402; October 10, 1972; U.S. Patent 3,700,574; October 24, 1972; all assigned to W.R. Grace & Co.

Polythiol-Phosphonitrile Polymers

In a process described by V.S. Frank, E.E. Stahly, R.G. Rice; U.S. Patent 3,676,311; July 11, 1972; assigned to W.R. Grace & Co. burn-resistant polymers are prepared by photocuring a mixture of polythiol and a phosphonitrilic polymer containing at least two reactive unsaturated carbon-to-carbon (ene) groups. The photocuring is achieved by exposing the mixture to ultraviolet light. The phosphonitrilic polymer containing at least two reactive ene groups is prepared by reacting a cyclic or linear phosphonitrilic chloride polymer, e.g., $(PNCl_2)_4$, with moieties containing reactive ene groups, e.g., allyl.

The ingredients necessary to prepare a photocurable composition which is burn-resistant are 2 to 98 parts by weight of a polythiol containing at least two thiol groups per molecule; and 98 to 2 parts by weight of a phosphonitrilic polymer containing at least two groups having reactive unsaturated carbon-to-carbon bonds.

One class of polythiols operable with polyenes in the process to obtain essentially odorless photocured polythioether compositions are esters of thiol-containing acids of the general formula: $HS-R_a-COOH$ where R_a is an organic moiety containing no "reactive" carbon-to-carbon unsaturation with polyhydroxy compounds of the general structure: $R_b-(OH)_n$ where R_b is an organic moiety containing no "reactive" carbon-to-carbon unsaturation and n is 2 or greater. These components will react under suitable conditions to give a polythiol having the general structure $R_b-[OC(O)-R_a-SH]_n$, where R_a and R_b are organic moieties containing no "reactive" carbon-to-carbon unsaturation and n is 2 or greater.

The phosphonitrilic polymers containing at least two reactive ene groups are prepared from phosphonitrilic halide polymers or polyphosphonitrilic halides. The chloride forms are the preferred polymers, as are the cyclic forms.

The conversion of the polyphosphonitrilic halides to the phosphonitrilic polymers containing at least two reactive ene groups is normally done in a solvent or reaction medium. If oxygen is present, organic solvents containing hydrogen should be avoided as the polyphosphonitrilic chloride tends to react with the organic solvents. The rate of reaction is fairly slow. No such reaction appears to occur when solvents devoid of active hydrogen in the molecule are used.

The preferred method of curing the mixture of polythiol and phosphonitrilic polymer is to expose the mixture to actinic radiation containing substantial amounts of ultraviolet radiation. Useful ultraviolet (U.V.) radiation has a wavelength in the range of about 2000 to about 4000 Angstrom units.

Sunlight contains U.V. light but the length of exposure to obtain a photo-cure may be relatively long. Various light sources can be used to obtain U.V. light. Such sources include carbon arcs, mercury arcs, fluorescent lamps with special ultraviolet light emitting phosphors, xenon arcs, argon glow lamps, and photographic flood lamps. Of these, the mercury vapor arcs, particularly the sun-lamp type, and the xenon arcs are very useful.

The sun-lamp mercury vapor arcs are at a distance of 7 to 10 inches from the photocurable material, whereas the xenon arc is placed at a distance of 24 to 40 inches from the photocurable material. With a more uniform extended source of low intrinsic brilliance, such as, a group of contiguous fluorescent lamps with special phosphors, the photocurable material can be exposed within an inch of the lamps.

The photocuring is not a true polymerization, but the terms "photocuring" and "polymerization" are used here to cover what actually occurs. The photocuring reaction depends upon a condensation reaction of thiol groups with reactive ene groups, namely:

$$R-S-H + \underset{/}{\overset{\backslash}{C}}=\underset{\backslash}{\overset{/}{C}} \xrightarrow{\text{U.V. light}} R-S-\underset{/}{\overset{\backslash}{C}}-\underset{\backslash}{\overset{/}{C}}-H$$

The photocuring reaction is initiated by ultraviolet light.

The speed of the photocuring reaction can be increased from days to within about 1 second to about minutes by the use of a chemical photocuring rate accelerator in an amount ranging from 0.0005% by weight to about 2.50% by weight of the polythiol and the phosphonitrilic polymer in the photo-curable composition. The following examples illustrate the process.

Example 1: 3.5 g. of $(PNCl_2)_3$, 3.5 grams of allyl alcohol and 40 g. of pyridine were placed in a reaction vessel and stirred by means of a mag-netic stirrer. The solution temperature rose instantly to 60°C. The solu-tion was heated to boiling (above 110°C.) for 15 minutes. After heating, the solution separated into two layers. The solution was left standing for about 24 hours. The solution was heated, with stirring to 110°C. (reflux). The solution was cooled and diluted with 40 g. acetonitrile. The solution was left standing for another 24 hours and the solution was filtered and the precipitate was air dried.

The dried solid weighed 4 grams. The filtrate was evaporated and about two grams of an opaque liquid polymer were recovered. 0.7 g. of the opaque liquid polymer were admixed with 0.8 g. of Q-43 and 0.003 g. of dibenzosuberone. "Q-43" is the commercial designation of the tetra-ester of pentaerythritol and beta-mercaptopropionic acid. The mixture was photocured by exposure for 30 minutes to U.V. light (Black-Ray No. 22) and a burn-resistant plastic resulted.

Example 2: Example 1 was repeated, except that the solid polymer (2 g.) was admixed with 1.1 g. of Q-43 and was exposed to the same U.V. light source for 30 minutes. The resultant cured plastic was burn-resistant by ASTM D-635.

Example 3: Example 1 was repeated, except that the dibenzosuberone was left out of the formulation. The ultraviolet exposure took 8 hours to effect a photocure.

Examples 4 to 12: Example 1 was repeated nine times, except that the 3.5 g. $(PNCl_2)_3$ was replaced with 3.5 g. of $(PNCl_2)_4$, 3.5 g. of linear hexamer of $(PNCl_2)$, 3.5 g. of linear undecamer of $(PNCl_2)$, 2.5 g. of $(PNF_2)_3$, 6.2 g. of $(PNBr_2)_4$, 3.95 g. of $P_3N_3BrCl_5$, 2.75 g. of $P_4N_4Cl_2F_6$, 4.8 g. of $Cl_3P(NPCl_2)NPCl_4$, and 11.4 g. of $Cl(PNCl_2)_8PCl_4$, respectively. Burn-resistant plastics were obtained.

Example 13: Example 1 was repeated, except that 30 ml. of xylene were added to the photocurable mixture to reduce the viscosity of the mixture. A burn-resistant plastic was obtained.

Example 14: Example 1 was repeated, except that the U.V. source was a 275 watt RS Westinghouse sunlamp held at a distance of 12 inches for 15 minutes. A burn-resistant plastic was obtained.

Poly(Allyl Urethanes)

In a process described by D.W. Larsen; U.S. Patent 3,645,982; February 29, 1972; assigned to W.R. Grace & Co. an allyl urethane is formed by reacting a solution of an allyl halide, an alkali metal cyanate and a polyol. The allyl urethanes are useful in the preparation of photocurable liquid polymers, that are photocured by the free radical addition of thiol groups of polythiols to the allyl double bonds.

The crucial ingredients in the photocurable composition are about 2 to about 98 parts by weight of an allyl urethane having an allyl functionality of at least two, about 98 to 2 parts by weight of a polythiol, and about 0.005 to about 50 parts by weight (based on 100 parts by weight of the above mentioned two ingredients) of a photocuring rate accelerator. The following examples illustrate the process.

Example 1: A slurry was prepared by mixing 65 g. of sodium cyanate (1 equivalent of OCN^-), 100 g. of polyol A (0.88 equivalents of OH^-) and

100 g. of dimethylformamide (BP = 153°C.). Polyol A is NIAX Polyol
LS-490, a rigid foam polyether. The slurry was refluxed and 70 g. of allyl
chloride (0.92 mole) were added slowly to the slurry so that the temperature
was not depressed below 120°C. The slurry was then filtered and the remaining dimethylformamide was washed out of the filtrant (filtered product)
with water. The product was allyl urethane.

Example 2: Five grams of the allyl urethane were mixed with one gram of
pentaerythritol tetrakis (mercaptopropionate) and one drop (0.05 ml) of
melted benzophenone. Pentaerythritol tetrakis (ß-mercaptopropionate)
is available as Mercaptate Q-43 Ester. The mixture was formed into a
thin layer and cured to a non-tacky rubbery solid in five minutes by exposure to ultraviolet light (Westinghouse 275 Watt Sunlamp).

Example 3: Examples 1 and 2 were repeated, except that 81 g. of potassium
cyanate were used in place of the sodium cyanate. A non-tacky rubbery
solid was obtained after five minutes exposure to U.V. light.

Examples 4 to 7: Example 2 was repeated four times, except that the pentaerythritol tetrakis (ß-mercaptopropionate) was replaced with trimethylolpropane tris (ß-mercaptopropionate) (1 gram), trimethylolpropane tris (thioglycolate), (1 gram), pentaerythritol tetrakis (thioglycolate), (1 gram),
polypropylene ether triol tris (ß-mercaptopropionate) (1 gram), respectively.
A non-tacky rubbery solid was obtained in each instance.

Example 8: Example 7 was repeated, except that half of the pentaerythritol
tetrakis (ß- mercaptopropionate) was replaced with 5 g. of ethylene glycol
bis (ß-mercaptopropionate). A non-tacky rubbery solid was obtained after
5 minutes exposure to U.V. light.

Examples 9 to 14: Example 2 was repeated 6 times, except that the benzophenone was replaced with cyclohexanone (1 drop), acetone (1 drop), methyl
ethyl ketone (1 drop), dibenzosuberone (1 drop), a blend of acetone (0.05 g.),
and p-diacetylbenzene (0.05 g.), and 3-acetylphenanthrene (0.05 g.) respectively. A nontacky rubbery substance was obtained in each instance.

Addition of Acrylic Acid to Improve Adhesion

E.W. Lard, U.S. Patent 3,662,022; May 9, 1972; assigned to W.R. Grace
& Co. describes a curable liquid composition consisting of: (a) a polyene
having the formula

Polyene (1)

where A is

$$\left[-O-\bigcirc-\underset{\underset{CH_3}{|}}{\overset{\overset{CH_3}{|}}{C}}-\bigcirc-O-CH_2-\underset{\underset{}{}}{\overset{\overset{OH}{|}}{CH}}-CH_2- \right]$$

and n is essentially 0 or greater (e.g., 0 to 20); and (b) a polythiol, the equivalent ratio of polyene to polythiol being about 1:0.5 to 1.5; and (c) a member selected from a group consisting of a monomeric acid having the formula $H_2C=C(R)COOH$, where R is H, CH_3 or C_2H_5 or a homopolymer or copolymer of such monomeric acid having an average molecular weight of about 140 to 1000, the group member constituting about 2 to 10% by weight of the composition.

The above composition can contain an effective amount of a photocuring rate accelerator.

Example 1: Preparation of a Polyene (1) — 5 mols of Epon 828 (average molecular weight is about 390) and twelve moles of diallylamine were mixed under an atmosphere of nitrogen and maintained at about 80 to 90°C. (under the atmosphere of nitrogen) for about 2 to 3 hours. Then unreacted diallylamine was distilled off under reduced pressure (ca. 1 to 10 mm. of mercury absolute) and the residue (substantially pure Polyene (1)) was recovered.

Example 2: Preparation of Control — A composition was prepared by mixing the Polyene (1), of Example 1 and pentaerythritol tetra-beta-mercaptopropionate in a mol ratio of 1:1 (since each is tetrafunctional, the polyene being a tetraene and the pentaerythritol tetra-beta-mercaptopropionate being a tetrathiol, this is also an equivalent ratio of 1:1) to form a first mixture. The first mixture was mixed with about 0.1% by weight of dibenzosuberone to form a second mixture.

Polymerization of Control — A portion of the second mixture was applied as a coating to smooth glass surface; the thickness of the coating was adjusted so that on curing the coating would constitute a film of solid polymer having a thickness of about 2 mm. The thus formed coating was cured by exposure for about 5 minutes to actinic light from a Westinghouse sunlamp positioned about a foot from the coating.

Tests on Control — Two replications of the above described polymerization of the control were run to yield a total of three smooth glass substrates with a coating of the cured polymer film, each film having a thickness of about 2 mm.

Two of the films were stripped from the glass substrates; one was tested for hardness using ASTM Designation D 2240 and the other for tensile modulus using ASTM Designation D 412. The third film which had been left on the glass substrate (surface) on which it was formed was tested to determine its (the film's) adhesion to the glass surface using ASTM Designation D 1000.

The results of these tests are:

Hardness ("D")	47
Tensile Modulus, psi	10,500
Adhesion, pounds per inch	0.2

Example 3:   Preparation, Polymerization, and Testing Composition —
The general procedure set forth in Example 2 for Preparation of Control was repeated (using 5 replications or runs with each run being made in triplicate) but the general procedure was modified by mixing a quantity (the quantity varying from run to run as indicated in the following table) of acrylic acid with the second mixture to form, in each instance, a third mixture.

These third mixtures were applied to glass substrates. The thickness of each coating was adjusted to yield a film which after curing would have a thickness of about 2 mm., and the coatings were cured and tested according to the general procedures used in Example 2. The results of these tests are presented in the table below:

TABLE 1

Run number:	Percent acrylic acid	Hardness ("D")	Tensile modulus, p.s.i.	Adhesion, pounds per inch
1	0.25	50	39,000	5.0
2	0.5	60	52,700	7.2
3	2.5	65	162,000	7.4
4	5	68	146,000	7.3
5	10	68	153,000	7.3

The general procedure of Example 3 was repeated; however, in each run irradiation with a high energy electron beam from a Van de Graaff electron accelerator (rather then exposure to ultraviolet light) was used to cure the compositions. The total radiation dose was 0.8 megarad. The results obtained are in the table below:

TABLE 2

Run number:	Percent acrylic acid	Hardness ("D")	Tensile modulus, p.s.i.	Adhesion, pounds per inch
1	0.25	48	38,000	4.8
2	0.5	61	52,500	7.1
3	2.5	64	162,500	7.3
4	5	69	150,500	7.4
5	10	68	158,000	7.2

Iron Compounds and Oxime Esters as Accelerators

A process described by J.L. Guthrie; U.S. Patent 3,640,923; Feb. 8, 1972; assigned to W.R. Grace & Co. relates to accelerated curing under ambient conditions of a liquid composition comprising a polyene containing at least two reactive unsaturated carbon to carbon bonds per molecule and a polythiol containing at least two thiol groups per molecule, the total combined functionality of (a) the reactive unsaturated carbon to carbon bonds per molecule in the polyene and (b) the thiol groups per molecule in the polythiol being greater than four, in the presence of a curing rate accelerator comprising a catalytic amount of iron and its compounds. The addition of a minor amount of an oxime ester to the system allows one to cure in an inert atmosphere. The polythioethers formed can be used as adhesives.

Iron compounds which are operable in the process as curing rate accelerators are many and varied and include both organic and inorganic compounds. Thus, for example, operable compounds include iron, iron salts such as sulfate, nitrate, ferricyanide, ferrocyanide, chloride and ammonium sulfate. Organic iron salts are also operable and include oxalate, stearate, naphthenate, citrate and iron chelate compounds such as acetylacetonate, benzoylacetophenonate and ferrocene. Examples of operable oximes esters include dimethylglyoxime dibenzoate, quinone dioxime dimethoxybenzoate, quinone dioxime dichlorobenzoate, diphenylglyoxime dibenzoate, glyoxime dibenzoate, quinone dioxime diacetate, terephthalaldehyde dioxime dibenzoate, dimethylglyoxime diacetate, dimethylglyoxime distearate, quinone dioxime dibenzoate and dimethylglyoxime monoacetate.

Example 1: Preparation of Polyene — To a 2 liter flask equipped with stirrer, thermometer and gas inlet and outlet was charged 450 g. (0.45 mol) of poly(tetramethylene ether) glycol, having a hydroxyl number of 112 and a molecular weight of 1,000, along with 900 g. (0.45 mol) of (polytetramethylene) ether glycol having a hydroxyl number of 56 and a molecular weight of 2,000, both commercially available. The flask was heated to 110°C. under vacuum and nitrogen and maintained for 1 hour. The flask was then cooled to approximately 70°C. whereat 0.1 g. of dibutyl tin dilaurate was added to the flask. A mixture of 78 g. (0.45 mol) of tolylene diisocyanate and 78 g. (0.92 mol) of allyl isocyanate was added to the flask dropwise with continuous stirring. The reaction was maintained at 70°C. for 1 hour after addition of all the reactants. The thus formed allyl terminated prepolymer will be referred to as Prepolymer A.

Example 2: Polyene/Polythiol Curing — 30 g. of Prepolymer A from Example 1 were mixed in an aluminum weighing dish with 5 g. of plasticizer commercially available under the trade name Benzoflex 988, 0.15 g.

of cyclohexanone oxime benzoate and 6 g. of titanium dioxide (an inert filler). To the mixture was added a solution of 3 mg. of ferric acetylacetonate in 2.3 g. of pentaerythritol tetrakis (β-mercaptopropionate) available as Q43, with stirring. The mixture became hard and rubbery in approximately 30 seconds. After 24 hours the solid cured polythioether product was removed from the aluminum weighing dish. The polythioether product had a Shore A hardness of 60.

Example 3: Example 2 was repeated except that no cyclohexanone oxime benzoate was added to the mixture. The curing reaction required approximately 30 minutes to reach the state of cure which it had obtained in Example 2 in 30 seconds. After 24 hours the cured polythioether product had a Shore A hardness of 32.

Example 4: Example 2 was repeated using oxime esters other than cyclohexanone oxime benzoate. Oxime esters employed, each in the amount of 0.15 g. were glyoxime dibenzoate, dimethylglyoxime dibenzoate, 2-methylcyclohexanone oxime benzoate, cycloheptanone oxime benzoate and 3-phenyl-4,5-dihydro-6-oxo-1,2-oxazine. The results in all cases were substantially the same as in Example 2.

Example 5: The reactants of Example 2 were employed and were exposed to UV radiation from a 275 watt Sylvania sun lamp. After 2 minutes, a hard solidified cured polythioether product resulted. In a control run using the reactants and procedure herein except that the cyclohexanone oxime benzoate and ferric acetylacetonate were omitted, the reaction required over 2 hours of UV radiation under the same conditions to obtain a hard solidified cured polythioether. The solid cured polythioether polymer products resulting from the process have many and varied uses. Examples of some uses include, but are not limited to, adhesives; caulks, elastomeric sealants; liquid castable elastomers, thermoset resins; laminating adhesives, and coatings.

Alpha-Hydroxy Carboxylic Acid as Accelerator

A process described by W.R. Wszolek; U.S. Patent 3,578,614; May 11, 1971; assigned to W.R. Grace & Co. relates to accelerated curing of a liquid composition comprising a polyene containing at least 2 reactive unsaturated carbon to carbon bonds per molecule and a polythiol containing at least 2 thiol groups per molecule, the total combined functionality of (a) the reactive unsaturated carbon to carbon bonds per molecule in the polyene and (b) the thiol groups per molecule in the polythiol being greater than four, in the presence of a class of curing rate accelerators from the group consisting of oxalic acid and an α-hydroxy carboxylic acid of the following formula.

$$\begin{array}{c} \text{OH} \\ | \\ \text{R}-\text{C}-\text{R}' \\ | \\ \text{COOH} \end{array}$$

R and R' in the above formula are independently selected from the group consisting of $-CH_2COOH$, $-CHOHCOOH$, hydrogen, aryl, alkyl containing 1 to 18 carbon atoms. The following examples illustrate the process.

Example 1: Preparation of Polyene — 1 mol of a polyoxypropylene glycol, PPG 2025, was charged to a resin kettle equipped with a condenser, stirrer, thermometer and a gas inlet and outlet, 4 g. of dibutyl tin dilaurate as a catalyst was added to the kettle along with 348 g. (2.0 mol) of tolylene-2,4-diisocyanate and 116 g. (2 mol) of allyl alcohol. The reaction was carried out for 20 minutes at room temperature under nitrogen. Excess alcohol was stripped from the reaction kettle by vacuum over a 1 hour period. The thus formed $CH_2=CH-$ terminated liquid prepolymer had a molecular weight of approximately 2,400 and will hereinafter be referred to as Prepolymer A.

Example 2: 0.25 g. of benzilic acid was mixed with 10 g. of Prepolymer A from Example 1 in an aluminum dish for 5 minutes at 110°C. to dissolve the acid in the polyene. The polyene was then cooled to room temperature. 1 g. of pentaerythritol tetrakis (β-mercaptopropionate), Q43, was added to the liquid polyene and the thus formed composition was placed in an oven at 100°C. After 5 minutes the composition was removed from the oven as a solid cured mass. A control run using the same reactants and procedure as set out herein except that the 0.25 g. of benzilic acid was omitted from the adhesive composition failed to cure at 250°F. within 24 hours.

Example 3: 0.25 g. of benzilic acid were added to Prepolymer A from Example 1 and mixed for 5 minutes at 110°C. to dissolve the acid. The mixture was then cooled to room temperature and 1.0 g. of pentaerythritol tetrakis(β-mercaptopropionate) was mixed into the system. The composition was poured into an aluminum dish and exposed to UV radiation from a 275 watt Sylvania sun lamp. After 2 minutes, a hard solidified cured mass resulted. A control run using the reactants and procedure herein except that the 0.25 g. of benzilic acid was omitted, required over 2 hours of UV radiation to obtain a hard solidified cured mass.

Example 4: Example 3 was repeated except that 0.005 g. of benzilic acid was substituted for the 0.25 g. of benzilic acid. After 4 minutes a hard solidified mass resulted.

Example 5: The reactants and procedure of Example 2 were repeated except that curing of the composition was performed under an infrared lamp which heated the composition to 100°C. After five minutes a cured, solid mass resulted.

Curable Plasticizers Using Polythiols

In a process described by N.E. Davenport; U.S. Patent 3,652,733; March 28, 1972 the thermoplastic polymer compositions contain a plasticizer having addition-polymerizable double bonds and a polythiol. The compositions can be cured by exposure to free radicals, and it is believed that the cured product is a polythioether. Preferably the thermoplastic polymer is vinyl chloride and the plasticizer diallyl phthalate. The compositions are useful for forming protective or decorative coatings. The following examples illustrate the process.

Example 1: A plastisol made up of 100 parts of Breon 121 (a polyvinyl chloride emulsion resin of nominal specific gravity 1.40, specific viscosity 0.63 to 0.69 in 0.5% solution in cyclohexanone, and K value 70 to 74), 50 parts of diallyl phthalate, 45 parts of pentaerythritol mercaptopropionate, 5 parts of basic lead carbonate, 0.5 part of benzophenone and 0.4 part of Azosol Black MA was prepared as follows. The benzophenone was dissolved in the diallyl phthalate and the solution then added to the polyvinyl chloride. The basic lead carbonate and black dye were then added.

The resulting mixture, which was in the form of a paste, was then de-gassed, and the pentaerythritol mercaptopropionate was stirred in under vacuum. The mixture was spread onto a steel plate to form a film 0.51 mm. thick and fluxed at 200°C. for three minutes. The fluxed film was flexible, but on exposure to ultraviolet radiation for ten minutes became hard, tough and abrasion-resistant. It was insoluble in common organic solvents.

Example 2: Following the procedure of Example 1, a plastisol was prepared with 100 parts of Breon 121, 40 parts of diallyl phthalate, 40 parts of allyl-terminated polyester of molecular weight about 3,200, 44 parts of pentaerythritol mercaptopropionate, 5 parts of basic lead carbonate, and 1 part of benzophenone. The polyester used above was obtained by reacting hydroxy-terminated Polyester Glycol RC-S 106 with two mols of allyl isocyanate. A portion of this plastisol was cast on a steel surface into a film, fluxed and cured as in Example 1. The cured film was tough and moderately rigid.

Example 3: A composition was prepared with 100 parts of Geon 434 (a vinyl chloride-vinyl acetate copolymer suspension resin, having a specific viscosity of 0.31 to 0.36 in 0.5% solution in cyclohexanone and K value of 49 to 53), 165 parts of diallyl phthalate, 150 parts of pentaerythritol tetrakis thioglycollate, and 2.5 parts of benzophenone.

The Geon 434 was blended with the diallyl phthalate to form a smooth paste which was then heated until a molten solution was formed. The thiol and the benzophenone were then slowly stirred into the melt. A similar composition was also made containing 150 parts of Geon 434. Both these compositions were liquid at elevated temperatures and solidified on cooling to room temperature. They could be cured by ultraviolet radiation in either the solid or molten form. The cured compositions were tough, flexible and nontacky and did not melt when reheated after the curing. Other polymers can also be used in place of the Geon 434, in particular ethylene-vinyl acetate copolymers. The resulting compositions are of a similar nature to those described in Examples 1 and 2 except that a smaller amount of the polymer is used and the composition is heated so that the polymer dissolves in the plasticizer.

APPLICATIONS

Self-Sealing System for Spacecraft

In a process described by C.J. Benning; U.S. Patent 3,666,133; May 30, 1972; assigned to W.R. Grace & Co. a space vehicle containing a double-walled closure can be made self-sealing by placing a viscous photocurable material (can be slightly pressurized) in the hollow portion of the double-walled closure. When a meteorite pierces both walls, the internal pressure forces some of the photocurable material into the hole in the outer wall. A seal is rapidly formed as the photocurable material is photocured by ultraviolet light into a hardened substance.

U.S. Patent 3,291,333 describes a hollow-walled closure unit housing a two-component system where the components will be separately maintained until such time as the unit is penetrated, whereupon the components will intermix and react to form a seal of the opening formed by such penetration. One of such components is a fluid capable of polymerization by molecular cross-linking or molecular extension such as a liquid diisocyanate. The other component is a cross-linking agent, foaming agent or catalyst, likewise in liquid form, which, when exposed to the first component, will cause such first component to polymerize.

Such components are maintained in a double-walled enclosure and are held apart by a third wall or membrane interiorly thereof until at least this membrane is pierced by a meteoroid, at which time the two components will come in contact under predetermined conditions of rate and proportion so as to effect a coagulation or polymerization into a tough, nonflowing substance capable of sealing the puncture against loss of pressure from within the vehicle. The proportions to which and the rate at which the reactive components come together to form such a seal is controlled in a preferred embodiment by a relatively elastic and tear-resistant separating membrane so as to minimize the size of the opening that will be made and by the employment of a relatively high-viscosity fluid, usually the polymerization material, toward the outer or vacuum side of the vehicle and of a relatively low-viscosity material, usually the catalyst, inwardly or toward the pressurized side of the vehicle. Other systems consisting of balloons, compressed foams and liquid elastomers have been tested and show poor reliability. One system consists of polyvinyl acetate emulsion and expanded styrene beads contained in an elastomeric bladder. Also when a micrometeorite (1/32 to 1/8 inch in diameter) hits the structure at hypervelocities of up to 20,000 feet per second, the pellet can be trapped by urethane foam but the wall of the structure is left ruptured.

In this process, spacecraft (4) is partially illustrated in Figure 3.1a. Outer wall (8) and inner wall (12) are so positioned in relation to each other that hollow-wall unit (enclosure 16) is formed over the surface of spacecraft (4) (except where sensing devices, etc., protrude). Enclosure (16) can have any desired shape and, preferably, is constructed of a rigid metallic material. The photocurable composition is contained in enclosure (16) and is a medium- to high-viscosity liquid. Enclosure (16), as shown, is preferably divided into a series of subchambers (20) by means of dividers (24), which are preferably arranged in an interconnecting, criss-crossed arrangement forming a honeycomb structure.

In that manner, any opening in a chamber wall will expose only a relatively small quantity of the fluid within the entire chamber so that the dimensional stability of the closure unit and the spatial relationship of its components will be substantially maintained throughout the stresses and strains of its space flight. It is understood that interior portion (28) enclosed by the sealing wall construction will be pressurized in relationship to the surrounding environment. To provide a controlled atmosphere capable of comfortably supporting human life without spacesuits, the pressure will be on the order of 14.7 pounds per square inch absolute so as to approximate the conditions of the earth's atmosphere at sea level. When spacesuits are used, the cabin pressure will normally be on the order of 6 to 8 pounds per square inch absolute.

When the vehicle is passing through celestial space, however, it will be further understand that a substantially complete vacuum (e.g., less than 1×10^{-9} torr) will surround its exterior. Referring to Figure 3.1b and 3.1c, it can be seen that after high-velocity particle (32), e.g., a micrometeorite, passes through walls (8) and (12), the relatively greater atmospheric pressure within the vehicle and adjacent inner wall (12) will tend to escape through the wall opening toward the vacuum exterior of outer wall (8) and will start to force or carry photocurable composition (36) out of hole (40) in outer wall (8). Ultraviolet light encountered in orbit around the earth and the other planets of this solar system or celestial travel near this solar system rapidly photocures photocurable composition (36). The photocured material is represented as (44). Referring to Figure 3.1d, it is seen that photocured material (44) air-tightly seals hole (40). The photocuring is

FIGURE 3.1: RADIATION-INITIATED, SELF-SEALING SYSTEM FOR SPACECRAFT

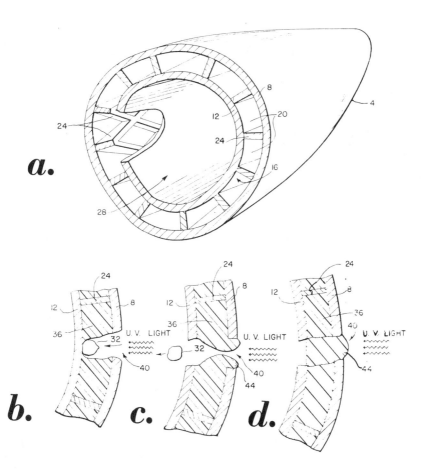

(a)   Space Vehicle with Self-Sealing Wall Construction
(b)(c)(d)   View of Hollow Wall Closure Unit Under Impact from Projectile

Source: C.J. Benning; U.S. Patent 3,666,133; May 30, 1972

extremely rapid provided that UV light is available. As time proceeds the seal becomes more substantial due to the fact that the photocure progressively occurs at a greater depth in the seal-forming material.

The crucial ingredients in the photocurable composition are: (1) about 2 to about 98 parts by weight of an ethylenically unsaturated polyene containing 2 or more reactive carbon to carbon bonds; (2) about 98 to about 2 parts by weight of a polythiol; and (3) about 0.0005 to about 50 parts by weight, based on 100 parts by weight of (1) and (2), of a photocuring rate accelerator. The reactive carbon to carbon bonds of the polyenes are preferably located terminally, near terminally, and/or pendant from the main chain. The polythiols, preferably, contain two or more thiol groups per molecule. The usable photocurable compositions are liquid at over the temperature range encountered by a spacecraft.

Example 1:   3,456.3 g. (1.75 mol) of poly(propylene ether) glycol, commercially available as PPG 2025, and 1.7 g. of di-n-butyl tin dilaurate were placed in a 5-liter, round-bottom, 3-neck flask. The mixture in the flask was degassed at 110°C. for 1 hour and was then cooled to 25°C. by means of an external water bath. 207 g. (3.50 mols) of allyl alcohol was added with stirring to the flask. 609 g. (3.50 mol) of an 80 to 20% isomer mixture of tolylene-2,4-diisocyanate and tolylene-2,6-diisocyanate, respectively, sold as Mondur TD 80, was charged to the flask. The mixture was stirred well. The flask was cooled by the water bath during this period.

Eight minutes after the Mondur TD 80 was added the temperature of the mixture was 59°C. After 20 minutes the NCO content was 12.39 mg. NCO per gram; after 45 minutes it was 9.87 mg. NCO per gram; and after 75 minutes it was 6.72 mg. NCO per gram. The water bath was removed 80 minutes after the Mondur RD 80 had been added, the temperature of the mixture being 41°C., and heat was applied until the mixture temperature reached 60°C. That temperature was maintained 105 minutes after the Mondur TD 80 was added, the NCO content was 3.58 mg. NCO per gram; after 135 minutes it was 1.13 mg. NCO per grams; and after 195 minutes it was 0.42 mg. NCO per gram.

At that point in time the resultant polymer composition was heated to 70°C. and vacuum-stripped for 1 hour. The resultant polymer composition was labeled Composition 1 and had a viscosity of 16,000 cp. as measured on a Brookfield Viscometer at 30°C. Unless otherwise stated, all the viscosity measurements were made on a Brookfield Viscometer at 30°C. The above procedure was repeated five times, and resultant compositions were labeled Compositions 2 through 6, respectively. The heating step lasted 180 minutes, 140 minutes, 140 minutes, 205 minutes, and 180 minutes, respectively. With Composition 2 the temperature was 60°C. after 8 minutes;

with Composition 3 the temperature was 57°C. after 6 minutes; with Composition 4 the temperature was 41°C. after 20 minutes, at which time the temperature was raised and held at 60°C.; with Composition 5 the temperature was 57.5°C. in 8 minutes, was 42°C. in 40 minutes, then taken up to 60°C. and lowered to 58°C. after 120 minutes; and with Composition 6 the temperature was 57°C. in 6 minutes, and was 41°C. after 60 minutes, at which time the temperature was immediately raised to 60°C. The viscosity of the resultant polymer compositions was 15,500 cp.; 16,000 cp.; 17,000 cp.; 16,800 cp.; and 16,200 cp., respectively.

Compositions 1, 2, 3, 4, 5 and 6 were placed in a 6 gallon container and stirred well. The resultant polymer composition had a viscosity of 16,600 cp. and the NCO content was 0.01 mg. NCO per gram. This composite polymer composition was labeled Polymer A. 100 parts of Polymer A, 10 parts of polythiol A, 15 parts of Hi-Sil 233 (a fine particle silica reinforcing filler and thickening agent), and 0.5 part of benzophenone were thoroughly mixed. This resulted in photocurable Compositon A. Polythiol A was pentaerythritol tetrakis(β-mercaptopropionate) which is available as 0-43 Ester.

Photocurable Composition A was placed in a hollow-walled enclosure similar to the one shown in Figure 3.1. The structure containing the hollow-walled enclosure was placed in a vacuum chamber. The internal pressure of the system was 14.7 pounds per square inch absolute, and the external pressure was 1×10^{-6} torr. The external wall of the system was irradiated with ultraviolet light from one Westinghouse RS 275 watt sunlamp at a distance of 3 inches. A particle (steel, having a diameter of 0.03 inch) was shot through the hollow wall. The initial velocity of the particle was about 500 feet per second. The particle punctured both walls. The photocurable composition entered the hole in the outer wall and completely formed a hardened, photocured plug in less than 30 seconds.

Example 2: Example 1 was repeated, except Polymer B was used. Polymer B was a polyester (MW = 2,000) made up of diethylene glycol, adipic acid, and 3% pentaerythritol (OH number is 60). Polymer B had a negligible acid number and was reacted with 2.2 equivalents of tolylene diisocyanate. The isocyanate-terminated resin was reacted with the stoichiometric amount of allyl alcohol relative to the isocyanate groups to produce a polymer with terminal double bonds. A hardened, photocured plug was obtained in about 15 to 20 seconds.

Examples 3 to 6: Example 1 was repeated four times, except that the pentaerythritol tetrakis(β-mercaptopropionate) was replaced with 10 parts of trimethylolpropane tris(β-mercaptopropionate); 15 parts of trimethylolpropane tris(thioglycolate); 15 parts of pentaerythritol tetrakis(thioglycolate);

and 5 parts of polypropylene ether glycol bis(β-mercaptopropionate), respectively. Hardened, photocured plugs were obtained.

Example 7:   Example 1 was repeated, except that half of the pentaerythritol tetrakis(β-mercaptopropionate) was replaced with 7.5 parts of ethylene glycol bis(β-mercaptopropionate). A hardened, photocured plug was obtained in about 30 to 40 seconds.

Example 8:   Example 1 was repeated except that 60 parts of Polymer C was used in place of Polymer A. Polymer C was prepared as follows: 485 g. (0.23 mol) of a commercially available liquid polymeric diisocyanate (Adiprene L-100) was charged to a dry resin kettle maintained under a nitrogen atmosphere and equipped with a condenser, stirrer, thermometer, and gas inlet and outlet. 37.8 g. (0.65 mol) of allyl alcohol was charged to the kettle and the reaction was continued for 17 hours with stirring at 100°C.

Thereafter the nitrogen atmosphere was removed and the kettle was evacuated 8 hours at 100°C. 50 cc dry benzene was added to the kettle and the reaction product was azeotroped with benzene to remove the excess unreacted alcohol. This allyl-terminated liquid polymer had a molecular weight of approximately 2,100 and was labeled Polymer C. A hardened, photocured plug was obtained in about 15 seconds.

Coated Roller

In a process described by C.B. Lundsager; U.S. Patent 3,656,999; April 18, 1972; assigned to W.R. Grace & Co. industrial rollers are prepared by photocuring a coating of photocurable composition on a metallic core. As the core cylinder is rotated, a thin layer of photocurable composition containing an epoxy polyene is fed intermittently or continuously onto the rotating cylinder, where it can optionally be smoothed by a doctor blade. That layer is applied as an adhesive layer. Then at least one complete layer of another photocurable composition is fed onto the coated rotating cylinder. The coating is photocured by an ultraviolet light source, which is preferably located on the cylinder side opposite the places where the photocurable compositions are applied, so that premature hardening does not occur in the feedstock.

Multiple, consecutive layers of the second photocurable composition can be built up on the coated core, each (after the first) being placed upon a partially hardened photocured sublayer. In this manner, the photocured material on the rigid core can be built up to any desired and practical thickness. The photocured surface of the roller can be ground and buffed to make a final product of accurately controlled dimensions.

An advantage of this process is that expensive molds or long heating cycles are not needed to coat a rigid core. The process is quick, convenient and economical, and produces a superior, fully cured product, which usually has an extremely smooth, glaze-like surface. Postfabrication curing or aging steps are not required, since the application and photocuring to completion (e.g., to constant final physical properties) of the photocurable composition is almost simultaneous. The most salient advantage is the tremendous adhesion of the coating to the core that is obtained.

Referring to Figure 3.2, roller core (4) rotates in a counterclockwise direction. The mounting and moving means for roller core (4) is any conventional device capable of rotating the core about its own axis (e.g., it is mounted on a lathe). Roller core (4) can be cleaned before coating, and, in case where metal rollers comprised of steel, can be sand-blasted prior to mounting to remove any rust and to expose a clean and slightly roughened coating surface. Reservoir (8), which may be heated if desired, contains photocurable composition (12) which contains a liquid epoxy polyene. After photocurable composition (12) is prepared, it must be stored in a dark area, i.e., in the absence of ultraviolet light.

Delivery tray (16) is in a slightly sloping position, with the lower end (delivery lip) about 2 to 20 mils from the surface of roller core (4) during operation and start up and with the upper end (receiving portion) positioned under delivery throat (20) of reservoir (8). A plurality of delivery throats, etc., can be used to insure that there is coverage over the entire roller length. Photocurable composition (12) is gravity fed down delivery tray (16). Delivery tray (16) is moveable in a horizontal manner in relationship to roller core (4). Also delivery tray (16) has side walls (not shown) to prevent lateral overflow. UV light source (24), e.g., one or several 275 watt RS sunlamps, is located so that its irradiating face is about 1.5 inches from the surface of roller core (4).

Several sunlamps can be used or a long tubular lamp could be used. Shield (28) encompasses light source (24), except that slit (32) allows the ultraviolet light to be beamed directly onto the surface of roller core (4). Reservoir (36), which may be heated if desired, contains photocurable composition (40). Delivery tray (44) is in a slightly sloping position, with the lower end (delivery lip) about 2 to 20 mils from the surface of roller core (4), after the adhesion layer has been coated, during operation and with the upper end (receiving portion) positioned under delivery throat (48) of reservoir (36). A plurality of delivery throats, etc., can be used to insure that there is coverage over the entire roller length.

Delivery tray (44) is shown positioned below delivery tray (16), but can be above it, etc. Photocurable composition (40) is gravity fed down delivery

FIGURE 3.2: COATED ROLLER APPARATUS

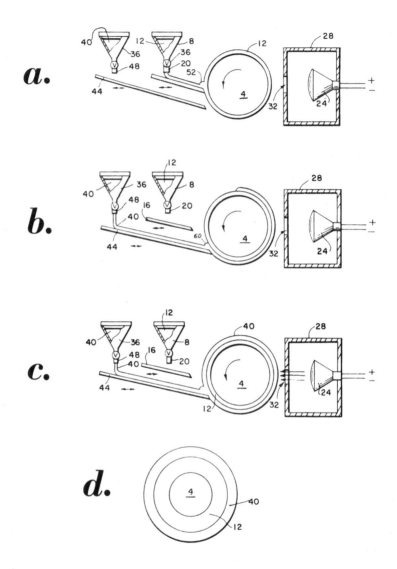

(a) Application of Adhesive Layer on Core
(b) Application of Roller Layer
(c) Activation by UV Source
(d) Cross-Sectional View of Completed Roller

Source: C.B. Lundsager; U.S. Patent 3,656,999; April 18, 1972

tray (44). Delivery tray (44) is movable in a horizontal manner in relationship to roller core (4). Also, delivery tray (44) has side walls (not shown)
to prevent lateral overflow. UV light source (24) is located so that it does
not expose photocurable composition (40) in delivery tray (44). In operation, photocurable composition (12) is coated onto rotating roller core (4)
to put an adhesive layer (12) thereon. The thickness of the adhesive layer
(12) can be any thickness, but is preferably between 0.0001 and 0.01 inches
in thickness. Any number of roller core passes or rotation can be used to
place the adhesive layer thereon. The edge of delivery tray (16) is slowly
moved away from roller core (4) as composition (12) is coated thereon.

During the coating operation, "rolling bank" (52) of viscous composition
(12) is preferably maintained in the end of delivery tray (16) against the
surface of core (4). After the adhesive layer (12) is applied, delivery tray
(16) is backed off, the valve having already been turned off. Delivery
tray (44) is advanced into delivery position and composition (40) is released
from reservoir (16). After one or so complete passes in coating composition
(40) on the adhesive layer (12), the photocuring of the entire coating on
core (4) is started by turning on UV light source (24). The contiguous
photocuring of the adhesive layer (12) and the photocurable layer (40) gives
a contiguous coating that will not separate.

During the coating operation, rolling bank (60) of viscous composition (40)
is preferably maintained in the end of delivery tray (44) against the coated
surface of core (4). The coated thickness can be increased by applying more
coatings (passes) of uncured photocurable composition (40) on photocured
composition (40) by moving the edge of delivery tray (16) slowly away
from coated roller core (4). Coating thickness as great as about 2.0 or
more inches can be obtained, and customary total thicknesses up to about
0.5 to 1.0 inches are easily obtained. After achieving a desired coating
thickness, delivery tray (44) is retracted away from the coated surface of
roller (4), after the valve has been turned off. The roller is rotated for
several seconds to several minutes thereafter to insure a complete photocure
of the coating of photocurable composition (40).

The coatings are essentially nontacky and are usually applied in less than
about 15 to 30 minutes of total operating time, depending on the ultimate
coating thickness, the core diameter, the intensity of the light source, etc.
The rollers have excellent characteristics and capabilities for the printing
and graphic arts industries and have coatings that are essentially contiguous
and that have good adhesion to core, particularly to metal cores and more
particularly to steel cores. The roller core can be constructed of nonmetallic substances, but this process is specifically and preferably directed to
metallic roller cores because the photocured sublayer containing the reacted epoxy polyene has so successfully cured the coating problems on

metallic roller cores. Cores can be constructed of steel, copper, aluminum. The roller core can be a hollow cylinder, a solid cylinder, a porous sintered cylinder, a porous polymeric structure, e.g., a filament wound spindle, etc. The following examples illustrate the process.

Example 1:   One mol diglycidyl ether of Bisphenol A (Epon 828) having a molecular weight in the range 370 to 384 and 2 mols of allyl hydrazine were dissolved in 500 ml. benzene in a beaker at room temperature (25°C.). The reaction was continued with stirring for 18 hours during which time the exotherm and reaction temperature was maintained below 80°C. The benzene solvent was removed by vacuum. The resultant liquid polyene was termed Polyene A.

Example 2:   Example 1 was repeated except that 2 mols of allyl amine was substituted for the 2 mols of allyl hydrazine. The resultant liquid polyene was termed Polyene B.

Example 3:   Example 1 was repeated except that 2 mols of diallyl amine was substituted for the 2 mols of allyl hydrazine in Example 1. In this instance no solvent was used in the synthesis reaction. The resultant liquid polyene was termed Polyene C.

Example 4:   Example 1 was repeated except that 2.mols of N, N-dimethyl-N-allyl amine hydro-p-toluenesulfonate was substituted for the 2 mols of allyl hydrazine in Example 1. The resultant liquid polyene was termed Polyene D.

Example 5:   3,456.3 g. (1.75 mol) of poly(propylene ether) glycol, (PPG 2055) and 1.7 g. of di-n-butyl tin dilaurate were placed in a 5-liter, round-bottom, 3-neck flask. The mixture in the flask was degassed at 110°C. for 1 hour and was then cooled to 25°C. by means of an external water bath. 207 g. (3.50 mols) of allyl alcohol, with stirring, were added to the flask. 609.0 g. (3.50 mols) of an 80 to 20% isomer mixture of tolylene-2,4-diisocyanate and tolylene-2,6-diisocyanate, respectively, (Mondur TD 80), was charged to the flask. The mixture was stirred well. The flask was cooled by the water bath during this period.

Eight minutes after the Mondur TD 80 was added, the temperature of the mixture was 59°C. After 20 minutes, the NCO content was 12.39 mg. NCO per gram; after 45 minutes, it was 9.87 mg. NCO per gram; and after 75 minutes, it was 6.72 mg. NCO per gram. The water bath was removed 80 minutes after the Mondur TD 80 had been added, the temperature of the mixture being 41°C., and heat was applied until the mixture temperature reached 60°C. That temperature was maintained. One hundred and five minutes after the Mondour TD 80 was added, the NCO content was 3.58 mg. NCO per gram; after 135 minutes, it was 1.13 mg. NCO per gram; and after 195 minutes, it was 0.42 mg. NCO per gram. At that point in time, the resultant polymer composition was heated to 70°C., and vacuum-stripped for one hour. The resultant polymer composition was labeled Composition 1, and had a viscosity of 16,000 cp. as measured on a Brookfield Viscometer at 30°C. Unless otherwise stated, all the viscosity measurements were made on a Brookfield Viscometer at 30°C. The above procedure was repeated 5 times, and resultant compositions were labeled Compositions 2 to 6, respectively. The heating step lasted 180 minutes, 140 minutes, 140 minutes, 205 minutes and 180 minutes, respectively.

With Composition 2, the temperature was 60°C. after 8 minutes; with Composition 3, the temperature was 57°C. after 6 minutes; with Composition 4, the temperature was 41°C. after 20 minutes, at which time the temperature was raised and held at 60°C.; with Composition 5, the temperature was 57.5°C. in 8 minutes, was 42°C. in 40 minutes, then taken up to 60°C. and lowered to 58°C. after 120 minutes; and with Composition 6, the temperature was 57°C. in 6 minutes, and was 41°C. after 60 minutes, at which time the temperature was immediately raised to 60°C. The viscosity of the resultant polymer compositions was 15,500 cp; 16,000 cp.; 17,000 cp.; 16,800 cp.; and 16,200 cp., respectively.

Compositions 1, 2, 3, 4, 5 and 6 were placed in a 6-gallon container and stirred well. The resultant polymer composition had a viscosity of 16,000 centipoise and the NCO content was 0.01 mg. NCO per gram. This composite polymer composition was labeled Polyene E, which has the following approximate structure:

$$CH_2{=}CH{-}CH_2{-}O\overset{O}{\overset{\|}{C}}{-}NH{-}\underset{CH_3}{\bigcirc}{-}NH{-}\overset{O}{\overset{\|}{C}}{-}O{-}(C_3H_6O)_{34}{-}\overset{O}{\overset{\|}{C}}{-}NH{-}\underset{CH_3}{\bigcirc}{-}NH{-}\overset{O}{\overset{\|}{C}}{-}OCH_2{-}CH{=}CH_2$$

Example 6: Three steel strips were sand blasted, wiped off and coated with photocurable Composition A, which contained 100 parts of Polyene C, 55 parts of Q-43 Ester and 0.5 part of benzophenone. Q-43 Ester is pentaerythritol tetrakis (β-mercaptopropionate). The strips were irradiated with ultraviolet light from one Westinghouse RS 275 watt sunlamp at a distance of 3 inches for 10 minutes. The photocured polymer was glass-like in appearance, and stuck well to the metallic surfaces.

Example 7: Example 6 was repeated, except that Polyene E was used in place of Polyene C in photocurable Composition A, and the amount of Q-43 was reduced to 10 parts per 100 parts of Polyene C. The photocurable composition was labeled photocurable Composition B. The composition was photocured for 5 minutes. The photocured polymer only lightly stuck to the metallic surfaces.

Example 8:   Photocurable Composition A was coated on a sand blasted steel strip (the layer was 2 mils thick). Photocurable Composition B was then coated lightly on the coating of photocurable Composition A (the layer was 20 mils thick). The doubly coated steel strip was then irradiated with UV light from one RS sunlamp at a distance of 3 inches for 10 minutes. The photocured coating had excellent adhesion to the steel strip, and the two photocured layers could not be separated.

Example 9:   Example 6 was repeated, except after the first layer was photocured, a second layer of photocurable Composition A was applied and similarly photocured. The photocured coating stuck to the metallic surface.

Example 10:   Example 9 was repeated, except that the first and second layer was prepared using photocurable Composition B. The photocured coating only lightly stuck to the metallic surface.

Example 11:   Example 9 was repeated, except that the second layer was prepared using photocurable Composition B. The photocured coating stuck to the metallic surface.

Example 12:   Example 8 was repeated, except that photocurable Composition A was diluted with 155.5 parts of toluene. The photocured coating had excellent adhesion to the steel strip, and the two photocured layers could not be separated.

Example 13:   An amount of photocurable Composition A was placed in a reservoir like (8) shown in Figure 3.2a. An amount of photocurable Composition A was placed in a reservoir like (36) shown in Figure 3.2a. The rest of the experimental set up was similar to that shown in Figure 3.2, and the accompanying write-up thereof above. The end of delivery tray (16) was placed about 10 mils away from steel roller core (4). Roller core (4) was rotated at 25 rpm. The valve on the throat of the reservoir (8) was opened. As the coating was applied to roller core (4) delivery tray (16) was slowly moved (manually) away from the core. A rolling bank similar to (52) was maintained.

When the coating thickness reached about 0.010 inch, delivery tray (16) was completely backed away from the coated roller after the flow of Composition A had been stopped by turning off the valve. The end of delivery tray (44) was placed about 20 mils away from rotating coated core (4).

The valve was turned on. As photocurable Composition B was applied to rotating coated core (4), delivery tray (44) was slowly moved away from the core. A rolling bank similar to (60) was maintained. After two rotations of core (4), ultraviolet light source (24) was turned on (outer coating thickness was less than 0.010 inch). The application continued, with the UV light source on, until the outer coating thickness was 0.525 inch.

Delivery tray (44) was completely backed away from the coated roller, after the flow of Composition B had been stopped by turning off the valve. The coated roller was further rotated for 3 minutes before the UV lamp was turned off. The final hardness of the coating was measured at Shore A 30 (ASTM). The photocured coating had excellent adhesion to the steel core, and the two photocured layers could not be separated.

Additional studies with the photocuring of coatings on rigid cores is described by C.B. Lundsager; U.S. Patent 3,637,419; January 25, 1972; assigned to W.R. Grace & Co.

Resin Impregnated Bandage and Surgical Cast

In a process described by D.W. Larsen and R.J. Ceresa; U.S. Patent 3,613,675; October 19, 1971 a lightweight strong cast for the repair of broken bones is made by impregnating a fibrous felt with a photocurable resin, wrapping the felt in bandage form around the injured member until a sufficient thickness is built up, then curing the resin by exposing the wrapped "cast" to actinic radiation for a time sufficient to convert the impregnated wrapping into a rigid substance. Photocurable compositions which are operable in the process are those obtained by mixing polyenes or poly-ynes containing two or more reactive unsaturated carbon to carbon bonds located terminally, near terminally, or pendant from the main chain with a polythiol containing two or more thiol groups per molecule. The following example illustrates the process.

Example: A strip of lofty, air-laid Dynel felt, 5 yards long, basis weight 3 oz. per square yard, was impregnated with a diene/tetrathiol blend prepared in the following manner: a solid polyesterdiol having a molecular weight of 3,200 was reacted with allylisocyanate to result in a solid diene. The diene then was melted and dissolved in an equal weight of a 50/50 mixture of toluene and ethoxyethylacetate. Pentaerythritol tetrakis (β-mercaptopropionate) was then added in the proportion of 1 g. per 13 g. of diene. 1% of benzophenone calculated on the weight of the mixture was then added. The Dynel fabric was impregnated by passing a web of the fabric through the above solution and then squeezing it sufficiently to permit 90% by weight of the resin to remain in the web. The web was then set aside to dry for a period of 12 hours, during which time the toluene and

ethoxyethylacetate evaporated and the resin crystallized. The web was
run through a slitter and cut into 3-inch widths. The slit webs were rolled
up and placed in tin cans to exclude light. Sheets of the material were
held in the dark and remained uncured for a period of several weeks. In
building up the cast, the usual techniques of protecting the patient's skin
may be followed: e.g., protective fabrics of medicated gauze or soft padding may first be placed on the skin, or a film of plastic may be applied
as a separating medium. The solid polyester composition should be melted
at 50°C. and may be cooled to room temperature before application. Thereafter, the bandage is wound on and the cast built up by the usual wrapping
techniques.

No times for exposures to actinic radiation can be given, for every cast
possesses substantial differences in thickness, number of plies, and size.
The exposure to actinic radiation should continue until substantially all of
the resin in the cast has been converted to a solid substance. As an example, the bandage may be cured under the light flux of an Ascorlux pulsed
xenon lamp placed 30 inches from the surface of the cast. Substantial
gellation will take place under these conditions through resin coats of 0.030
of an inch thick with 2 minutes of exposure. Thicker casts require longer
times of exposure. In all exposures to actinic radiation it is necessary to
protect the skin of the patient from burn and damage. This is easily done
by covering the patient with sheets or blankets impenetrable to actinic radiation. After curing, the polyester composition will recrystallize and impart greater rigidity to the cast.

The compositions to be cured, i.e., to be converted into the solid mass of
the cast may include materials to increase the rigidity of the cast or resin
extenders such as wood, flour, talc, etc. Heavier materials which are frequently used as fillers or loaders in resinous compounds, e.g., barytes, are
operative but are useful only if the weight of the cast is of no importance.

Accelerators of the reaction to actinic radiation are highly desirable, and
among the photoinitiators or sensitizers are the benzophenones, acetophenone, acenaphthene-quinone, acenaphthol-quinone, methylethyl ketone,
etc. They greatly increase the rate of hardening. Many of these compositions can be compounded so that they are pastes rather than flowable mixtures, and such paste compositions are highly useful when an already impregnated base must be top coated.

If the bandages are properly packaged in light-impermeable packs or can-
nisters, they may be stored for months without loss of efficacy, but once
they are opened and exposed to daylight, they should be used promptly.
The speed, ease of application, and the cleanly conditions surrounding the
use of these photoreactive bandages makes the application of a cast a very

much simpler and quicker process than formerly. Their rigidity after cure means that a patient, rather than having to put up with the great weight and clumsiness of a plaster of Paris cast, has a light, thin, individually created support which permits him to move arms or legs with much less effort. The surgeon, too, can begin at once to bind the limb in the set position, and does not have to work with a slippery, pasty, or generally messy material such as plaster of Paris.

Waterproofing Porous Support Layers

C.L. Kehr and W.R. Wszolek; U.S. Patent 3,676,195; July 11, 1972; assigned to W.R. Grace & Co. describe a process for preparing a water-resistant material having improved electrical and mechanical properties. The material is porous, fibrous, etc.; is usually in sheet form; and is termed a support layer or porous support layer. The support layer is preferably essentially transparent to actinic radiation and translucent to visible light and preferably has a thickness of less than 0.1 inch. Although if the support layer is comprised of an ultraviolet transparent fiber, the support layer can have a thickness of up to about 1/2-inch.

The water-resistant material is prepared from a photocurable element which includes a porous support layer and a photocurable composition absorbed into the support layer and onto one or both surfaces. The amount of photo-curable composition in the surface is normally extremely minimal, as it is normally wiped or scraped off. Preferably the photocurable composition is only absorbed throughout the entire thickness of the support layer. The term "porous support layer" includes a unitary layer or a laminated multiplicity of porous sheets, fibrous webs, fabrics and sheets (woven, compressed, nonwoven, bonded, and so forth). Useful porous support layers include fibrous glass; fibrous polyester; fibrous polyamide; paper; fibrous polystyrene; fibrous low density polyethylene; fibrous polypropylene; etc.

The photocurable composition is applied initially as a liquid, and, after absorption into the porous support layer, it may solidify to the extent that it may crystallize at ambient temperatures or that any solvent, etc., evaporates or is evaporated off. The process itself involves exposing the layer containing the photocurable composition to actinic radiation containing a substantial portion of ultraviolet, electron beam or gamma radiation, whereby the photocurable composition is hardened (preferably to an insoluble state). This process also includes the initial photocurable element and the resultant water-resistant element.

If the porous support (containing the photocurable composition) is flexible, the support can be stored and shipped in a compacted form, shaped to the desired form, and set to a rigid condition when exposed to ultraviolet light (or electron beam or gamma radiation). No substantial heat is generated and no postheating is required. This way, rigid structures can be easily made. Also, air inflated fabric structures used for temporary shelter could be coated (if not precoated, etc.), exposed to sunlight for about 10 minutes to several hours and then utilized as a self-supporting structure. The following examples illustrate the process.

Example 1:   3,456.3 g. (1.75 mol) of poly(propylene ether) glycol (PPG 2025) and 1.7 g. of di-n-butyl tin dilaurate were placed in a 5-liter, round-bottom, 3-neck flask. The mixture in the flask was degassed at 110°C. for one hour and was then cooled at 25°C. by means of an external water bath. 207 g. (3.50 mols) of allyl alcohol, with stirring, was added to the flask. 609.0 g. (3.50 mols) of an 80 to 20% isomer mixture of tolylene-2,4-diisocyanate and tolylene-2,6-diisocyanate, respectively, (Mondur TD 80), was charged to the flask. The mixture was stirred well. The flask was cooled by the water bath during this period. Six minutes after the Mondur TD 80 was added, the temperature of the mixture was 59°C. After 20 minutes, the NCO content was 12.39 mg. NCO per gram; after 45 minutes, it was 9.87 mg. NCO per gram; and after 75 minutes, it was 6.72 mg. NCO per gram.

The water bath was removed 80 minutes after the Mondur TD 80 had been added, the temperature was maintained. 105 minutes after the Mondur TD 80 was added, the NCO content was 3.58 mg. NCO per gram; after 135 minutes, it was 1.13 mg. NCO per gram; after 150 min., it was 1.13 mg. NCO/gram; after 195 min., 0.42 mg. NCO/gram. The resultant polymer composition was heated to 70°C., and vacuum-stripped for one hour. The resultant polymer composition was labeled Composition 1, and had a viscosity of 16,000 cp. as measured on a Brookfield Viscometer at 30°C. Unless otherwise stated, all the viscosity measurements were made on a Brookfield Viscometer at 30°C.

The above procedure was repeated five times, and resultant compositions were labeled Compositions 2 to 6, respectively. The heating step lasted 180 minutes, 140 minutes, 205 minutes, and 180 minutes, respectively. With Composition 2, the temperature was 60°C. after 8 minutes; with Composition 3, the temperature was 57°C. after 6 minutes; with Composition 4, the temperature was 41°C. after 20 min. then raised and held at 60°C.; with Composition 5, the temperature was 57.5°C. in 8 minutes, was 42°C. in 40 minutes, then taken up to 60°C. and lowered to 58°C. after 120 min.; and with Composition 6, the temperature was 57°C. in 6 minutes, and 41°C. after 60 minutes, at which time the temperature was immediately raised to 60°C. The viscosity of the resultant polymer compositions was 15,500 cp.; 16,000 cp.; 17,000 cp.; and 16,800 cp., respectively.

Compositions 1, 2, 3, 4, 5 and 6 were placed in a 6-gallon container and stirred well. The resultant polymer composition had a viscosity of 16,600 centipoise and the NCO content was 0.01 mg. NCO per gram. This composite polymer was labeled Polymer A. 100 parts by weight of Polymer A, 10 parts by weight of Polythiol A and 0.5 part by weight of benzophenone were mixed. This resulted in photocurable Composition A. Polythiol A was pentaerythritol tetrakis (β-mercaptopropionate), (Q-43 Ester). Photocurable Composition A was dissolved in an equal weight of methyl ethyl ketone (serving as a solvent) and used to impregnate a lightweight paper (Seafoam bond).

The excess photocurable composition solution was wiped off the surface of the paper (porous support layer). The solvent was removed by hanging the impregnated paper in a forced draft oven (at 35°C.) for 120 minutes. The impregnated paper was irradiated (exposed) with UV light from a Westinghouse sunlamp (Model No. RS 275 watt) at a distance of 12 inches for 2 minutes. The exposed photocurable composition had photocured to a rubbery material in the paper. The impregnated paper was translucent, mechanically strong and water-resistant.

Example 2: Example 1 was repeated, except that the porous support layer was a sheet of newsprint paper. The exposed photocurable composition had photocured to a rubbery material in the paper. The impregnated paper was translucent, mechanically strong and water-resistant.

Example 3: Example 1 was repeated, except that the porous support layer was a sheet of coarse porosity paper (Whatman Grade 541). The exposed photocurable composition had photocured to a rubbery material in the paper. The impregnated paper was translucent, mechanically strong and water-resistant.

Example 4: Example 1 was repeated, except that the porous support layer was a sheet of fine porosity paper (Whatman Grade 50). Exposure was made with 2 sunlamps positioned on both sides of the photocurable element. The exposed photocurable composition had photocured to a rubbery material in the paper. The impregnated paper was translucent, mechanically strong and water-resistant.

Example 5: Examples 1 was repeated, except that the porous support layer was a sheet of tightly woven cloth (cotton). The exposed photocurable composition had photocured to a rubbery material in the fabric. The impregnated fabric was translucent, mechanically strong and water-resistant.

PLASTICS AND COATINGS

PLASTICS

Glass Fiber-Polyester Laminates

P. Borrel and J. Lehureau; U.S. Patent 3,655,483; April 11, 1972; assigned to Progil, France describe a process of obtaining stratified or laminated materials of glass fibers and unsaturated polyesters, by photopolymerization. It has been found that if there is first applied, during a short enough time, an intense luminous radiation on the two faces of the stratified material to harden, in the presence of a photosensitizing agent and that thereafter irradiation intensity is decreased, there results an attraction or migration of the ethylenically unsaturated monomer of the polyester resin from the center up to the faces of the stratified material. So, for example, when starting, for impregnating the glass fiber or fabric from a composition containing 40% of styrene for 60% (by weight) of diol/diacid polycondensate, the styrene content on stratified material faces may pass from 40 to 42 and even 43% by the use of this process.

Such a composition variation provides composite materials having surface properties and the characteristics of which may be modified at will in using as a raw material a polyester resin having always the same initial content of polymerizable cross-linking monomer. This exposure of the stratified material at two successive times to different irradiation intensities, which is a characteristic feature of this process, may practically be accomplished by a convenient disposition of the used luminous sources. A convenient and especially advantageous way consists of passing the polyester/glass fibers mass between series of ultraviolet radiation lamps having given wave lengths, generally comprised between 1500 and 5000 angstroms, the first bank of lamps being placed between 5 and 25 cm. from the surface to be irradiated in order to obtain an intensity of between 1 to 5 watts/cm.^{2}, while the second bank is placed at a further distance, generally 30 to 100 centimeters, from such surface to obtain an intensity of between 0.9 and 0.01 watts/cm.^{2}. The radiation intensity is calculated by dividing the electrical force consumed at the lamp binding posts (power input) by the illuminated surface. The process may be applied according to the known continuous fabrication techniques for stratified materials. However, it is especially advantageous to use the following operative modes, the description of which is illustrated in Figure 4.1.

On a substrate-support, constituted for example of regenerated cellulose or a terephthalic polyester, there is placed a glass fiber mat (10) fed in such as through a hopper (2). The polyester resin with its initiating agent, and possibly its regulator, of photopolymerization is then fed to the mat (10) from the hopper (3). Then the impregnated mat is covered with a second sheet (4) of plastic material, for example of the same type as (1), and air bubbles are removed by passing of the whole assembly through pressing rolls (5). There is placed at location (6), at a distance near the material, for example about 5 to 20 cm., lamps emitting a light having a convenient wave length for the used photoinitiator. The distance and intensity of those lamps are regulated to luminous regard to the material passing speed, in order to provide a slight surface gelation of the resin while the inside of

FIGURE 4.1:   MANUFACTURE OF STRATIFIED MATERIALS OF GLASS
FIBERS AND POLYESTERS

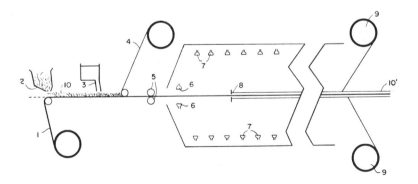

Source: P. Borrel and J. Lehureau; U.S. Patent 3,655,483; April 11, 1972

the laminated part remains fluid. Exposure time is generally short, from about 5 to 20 seconds. Other lamps (7), placed equally on both sides of the material to be polymerized, are at a greater distance, for example from 30 to 100 cm. and allow the progressive hardening of the sandwich-material whereas a diffusion of the monomer contained in the resin is made from the center up to surface proximity. During this second exposure to luminous radiations, which last generally from 5 to 20 minutes, the stratified material may be put in the desired form, for example in the form of corrugated plate with the help of a shaping machine (8) according to classical techniques. Then it is sufficient to recover the support-sheets (1) and (4) by winding them around spindles (9), and to proceed to sawing and cutting operations of final stratified substance (10').

According to a process variant it is possible first to coat the support-sheet (1) with polyester resin, then to proceed immediately to a pregelation by means of luminous radiations during a very short time. Then glass fibers mat resin and sheet (4) also coated with resin and pregelled as sheet (1) are set; the operations are pursued in the same way as previously, except the operations lamps (6) are not needed.

Example 1: An unsaturated polyester resin was prepared by dissolving 60 parts (weight) of a polycondensate of 1 mol of phthalic anhydride and 1 mol of maleic anhydride with 2.2 mols of propylene-glycol, in 40 parts of styrene. To the solution was added 0.02% of hydroquinone as a stabilizing agent and 0.2% of benzoin methyl ether as a photoinitiator. The resin was used to impregnate a glass fiber mat of 450 g./m.^{2} in order that a mixture by weight of 25% of glass and 75% of resin was obtained. The corresponding thickness was about 1.5 mm. The operation was carried out under the same conditions as set forth in the description of Figure 4.1 by adopting a passage speed of about 2 m./min. for the composite material.

Support sheets (1) and (4) were constituted of a terephthalic polyester (known as Mylar). The first mercury vapor lamps (6) providing radiation in the range of 1500 to 5000 angstroms were placed on each side of the laminated part at a distance from 10 cm. providing an intensity of 2.5 watts/cm.^{2} with an irradiation time of 10 seconds. Other series of lamps (7) were placed at 50 cm. to provide an intensity of 0.4 watts/cm.^{2} and the irradiation period was 10 minutes.

As reference operations a process was carried out exactly with the same quantities of materials and in the same conditions except that, in a first case (sample t_1) all the lamps were placed at 10 cm. with an irradiation time of 10 minutes, and in a second case (sample t_2) all the lamps were placed at 50 cm. of the composite material, in adopting the same irradiation time.

For the stratified material according to the process and the two samples (t₁ and t₂) there was measured on one hand styrene content at the surface, by measures of multiple reflection in infrared, then on the other hand flexion resistance and water absorption rate. Flexion resistance was measured on samples 1.5 mm. thick according to the indications of the American standard ASTM D 790-63 and in the conditions of this standard. Water absorption rate after sample immersion during 24 hours in water at 20°C. was valued in the conditions of the French standard NF 38 302. The obtained results are given in Table 1.

TABLE 1

Stratified material	Styrene content (percent) at the surface	Flexion resistance (kg./cm.²)	Water absorption [1]
i	42	1,510	0.2
t₁	40	1,220	0.5
t₂	40	1,290	0.4

[1] Percent of the weight of the stratified material.

Example 2:   Example 1 was repeated using the same conditions as in Example 1 by making also three tests, among which two as reference tests but adding in all the cases to the polyester solution in styrene 0.1% of ortho-hydroxybenzophenone as an absorber of ultraviolet radiations. This addition permitted a further increase in the effect of the styrene concentration in the superficial part of the stratified material by working according to the process (successive irradiations with lamps placed at unequal distance). The obtained results are given in Table 2.

TABLE 2

Stratified material	Styrene content (percent) at the surface	Flexion resistance (kg./cm.²)	Water absorption [1]
i'	43	1,550	0.1
t'₁	40	1,270	0.5
t'₂	40	1,310	0.4

[1] Percent of the weight of the stratified material.

Polyester Molding Compounds

K. Fuhr, H. Rudolph, H. Schnell and M. Patheiger; U.S. Patent 3,639,321; February 1, 1972; assigned to Farbenfabriken Bayer AG, Germany describe compositions which can be hardened by ultraviolet irradiation and have improved dark storability comprising an unsaturated polyester of an α, β-unsaturated dicarboxylic acid and a polyol, a copolymerizable monomeric compound and, as photosensitizer, a benzoin ether of a secondary alcohol. In the following Table 1 the values of stability in dark storage for benzoin are compared with those for benzoin ethers of primary and secondary alcohols. These sensitizers were dissolved in amounts of 2% by weight of the delivery form, stabilized in the usual manner with hydroquinone, of a typical unsaturated polyester obtained from maleic acid, phthalic acid and propylene glycol in admixture with styrene.

TABLE 1:   STORAGE STABILITY OF SENSITIZER*

> Benzoin: less than 1 day
> Benzoin-ethyl ether: less than 1 day
> Benzoin-isobutyl ether: less than 1 day
> Benzoin-isopropyl ether: 7 days
> Benzoin-sec.-butyl ether: 6 to 7 days
> Benzoin-cyclohexyl ether: 5 to 6 days

*Dark storage at 60°C. of typical polyester resin with
2% by weight of sensitizer

The materials according to the process are further characterized in that they have an extremely high reactivity which permits rapid hardening in the case of thin layers. In particular, they harden fast to form almost colorless moldings or coatings, even under ultraviolet and visible irradiation of low energy fluorescent lamps with an emission of 3000 to 5800 A. Examples of sensitizers to be used according to the process are the benzoin ethers of the following alcohols: propanol-2, butanol-2, pentanol-2, pentanol-3, 3-methylpentanol-2, 2,4-dimethyl-pentanol-3, and cyclohexanol.

Example 1:   An unsaturated polyester prepared by condensation of 152 parts by weight maleic acid anhydride, 141 parts by weight phthalic acid anhydride and 195 parts by weight propane-1,2-diol is mixed with 0.045 parts by weight hydroquinone and dissolved in styrene to give a 65% by weight solution. Portions of 100 parts by weight of the resultant resin delivery form are mixed with 20 parts by weight styrene, 1 part by weight of a 10% by weight solution of paraffin (MP 52° to 53°C.) in toluene and with various benzoin ethers of secondary alcohols.

The solutions so obtained are applied to glass plates by means of a film extruder (250μ) and illuminated with the radiation of a fluorescent tube (Osram L-lamp for tracing purposes, 40 watt, length 97 cm.) at a distance of 10 cm. The floating times of the paraffin are set out in Table 2. After a total illumination time of 10 to 15 minutes, the films have a pencil hardness of more than 6 H. They are almost colorless.

TABLE 2

Additive, referred to resin delivery form (% by wt.):	Floating time of paraffin after (minutes)
2 benzoin-isopropyl ether	1.9
2 benzoin-sec.-butyl ether	2.1
2 benzoin-2-pentyl ether	2.3
2 benzoin-3-pentyl ether	2.3
2 benzoin-2,4-dimethyl-3-pentyl ether	2.4

Example 2: 100 parts by weight of the resin delivery form described in Example 1 are mixed with 20 parts by weight styrene, 1 part by weight of a 10% by weight solution of paraffin (MP 52° to 53°C.) in toluene, 1 part by weight benzoin-isopropyl ether and, in addition, with various peroxides, acidic phosphoric acid esters and metal-containing compounds (see Table 3). These solutions are applied in layers of 250μ thickness to glass plates by means of a film extruder and illuminated at a distance of 10 cm. with the radiation of the fluorescent lamp described above.

TABLE 3

Additives, referred to resin delivery form (percent by weight)	Floating time of paraffin after (minutes)—	Pencil hardness >6H after (minutes)—
(1) 4 benzoyl peroxide paste (50% in plasticiser)	2.2	20
(2) 4 coumol hydroperoxide solutoion (70%)	2.2	20
(3) 4 methyl ethyl ketone peroxide (40% in plasticiser)	2.3	22
(4) 0.1 mixture of mono- and dibutyl-phosphoric acid ester	2.4	18
(5) 4 methyl ethyl ketone peroxide solution (40% in plasticiser); 0.1 mixture of mono- and dibutyl-phosphoric acid ester	2.3	24
(6) 1 cobalt napthenate solution (20% in toluene)	2.2	16
(7) 1 zirconium napthenate solution (6% zirconium in a toluene solution)	2.4	18
(8) 4 methyl ethyl ketone peroxide (40% in plasticiser); 1 cobalt napthenate solution 20% in toluene)	2.2	19
(9) 4 methyl ethyl ketone peroxide (40% in plasticiser); 1 cobalt napthenate solution (20% in toluene); 0.1 mixture of mono- and dibutyl phosphoric acid ester	2.3	19

After a total illumination time of 10 to 20 minutes, the pencil hardness amounts to 6 or more. If, after the floating of the paraffin, the polymerization is completed with the aid of infrared irradiation or hot air, then a pencil hardness of >6 is attained already after an illumination time of 10 to 13 minutes. The floating time of the paraffin in the presence of 1% by weight benzoin-isopropyl ether but without further additives amounts to 2.2 minutes; the hardness of >6 is attained after an illumination time of 20 min.

The films produced according to (1), (2), (4) and (7) are almost colorless, while those obtained according to (6), (8) and (9) are pale pink, and those prepared according to (3) and (5) are practically colorless.

Example 3: An unsaturated polyester obtained by condensation of 1,765 parts by weight maleic acid anhydride, 756 parts by weight glycol, 405 parts by weight 1,3-butane-diol and 1,540 parts by weight trimethylol-propane-diallyl ether in the presence of 0.83 part by weight hydroquinone is dissolved in styrene to give a 70% by weight solution. 100 parts by weight of the resultant delivery form and 1 part by weight of a cobalt naphthenate solution (20% by weight in toluene) are mixed with 2 parts by weight benzoin-isopropyl ether. The solution is photopolymerized in the form of a layer of 250μ thickness at a distance of 10 cm. by the fluorescent lamp described above. The film has gelled after illumination for 3 minutes. After an illumination time of 30 minutes, the film has a pencil hardness of >6 H. The same result is obtained when the benzoin-isopropyl ether is replaced with the same amount of benzoin-sec.-butyl ether.

Heat Shrinkable Polyolefin Films

R.C. Golike and G.J. Ostapchenko; U.S. Patent 3,663,662; May 16, 1972; assigned to E.I. du Pont de Nemours and Company describe a polyethylene film which exhibits improved heat-shrink and heat-sealing characteristics. Specifically, the process provides a cross-linked, oriented film of a blend of polymers comprising (a) about from 70% to 85% by weight, based on the total weight of the blend, of a low density polymer selected from the group consisting of polyethylene and copolymers of ethylene and olefinically unsaturated monomers, the polymer having a density of from about 0.91 to 0.93 gram/cc at 25°C., and (b) about from 30 to 15% by weight of a high density polymer selected from the group consisting of polyethylene and copolymers of ethylene and olefinically unsaturated monomers, the high density polymer having a density of about from 0.94 to 0.98 gram per cubic centimeter at 25°C.

The film has a shrinkage of at least 15% in each direction in the plane of the film at a temperature of 100°C., a shrink tension of at least 200 psi at 100°C. and zero strength temperature of at least 175°C. in each direction in the plane of the film. The process for the preparation of these films comprises forming a self-supporting film of the polymer blend indicated above, heating the film to a temperature of about from 90° to 115°C., stretching the film at least about 5 times in each of two mutually perpendicular directions in the plane of the sheet, cooling the film under tension, and irradiating the film for a time sufficient to raise the zero strength temperature ot the film to at least 175°C.

Examples 1 through 3: Irradiation with Ultraviolet Light — Three blended polymers are prepared by melt blending 75 parts branched polyethylene resin having a melt index of 4.1 and a density of 0.913 gram/cc at 25°C. and 25 parts linear polyethylene resin comprising an ethylene/1-octene copolymer having a melt index of 0.45 and a density of 0.956 gram/cc at 25°C. The linear polyethylene also contains 250 parts per million of an antioxidant, 2,6-di-t-butyl-4-methylphenol, 800 parts per million of mixed amide (approximately 533 parts of oleamide and 267 parts of stearamide) and 500 parts per million of silica, as well as ethylene/methacryloxybenzophenone/methyl methacrylate copolymer in amounts to provide levels of 0.1, 0.2 and 0.4% of benzophenone moiety in the resin in Examples 1, 2 and 3, respectively.

The blended resins are extruded through an apparatus described in U.S. Pat. 3,141,912, at a melt temperature of 200°C., and a die opening of 35 mils. The cast tubes, having a thickness of 20 mils, are stretched 5.8 times in the machine direction and 5 times in the transverse direction, at a temperature of 115°C. The extruded and stretched films are then exposed to ultraviolet light for a period of approximately 1 second. The light source is a 2,100 watt, 550 volt Hanovia 78 A lamp positioned 1 to 2 inches from the film.

The treated films are thereafter characterized with respect to zero strength temperature. The shrinkage of the film upon immersion in boiling water is measured, as well as the shrink tension developed in both the machine and transverse directions when the film is heated to 100°C. The results are shown in Tables 1, 2 and 3. A control film is similarly prepared and evaluated, except that the ethylene/methacryloxybenzophenone/methyl methacrylate copolymer sensitizer is omitted and the film is not irradiated. The test films show a large increase in zero strength temperature and a corresponding improvement in heat sealing performance on automatic packaging equipment. With the films of Examples 1, 2 and 3, effective sealing is attained over a temperature range of 120° to 200°C. in contrast to the performance with the control film, which has not been given the ultraviolet treatment, which is sealable only in a range of 115° to about 160°C.

TABLE 1: DYNAMIC ZERO STRENGTH TEMPERATURE

	Sensitizer level, percent	Zero strength temperature, °C.	
		MD	TD
Example Number:			
1	0.1	243	251
2	0.2	255	247
3	0.4	242	238
Control	None	145	142

TABLE 2:   SHRINKAGE AT 100°C.

	Percent shrinkage	
	MD	TD
Example Number:		
1	21	34
2	22	32
3	19	31
Control	20	32

TABLE 3:   SHRINK TENSION AT 100°C.

	Shrink force (p.s.i.)	
	MD	TD
Example Number:		
1	348	351
2	327	338
2	301	342
Control	346	386

Example 4:   A heat-shrinkable film is prepared as described in Examples 1 through 3, except that the photosensitizer is omitted from the resin blend. The film is irradiated to a dosage of two megarads by exposing it under the beam of a Van de Graaff accelerator. The properties of the irradiated film and an unradiated control film are summarized below. The film exhibits a marked increase in zero strength temperature and a broadening of the heat-sealing range as a result of the irradiation. This irradiated film shows excellent heat sealing performance on automatic packaging machinery.

Property	Control	Irradiated film
Zero strength temperature, ° C. MD/TD	140/040	260/260
Percent shrinkage (MD/TD) boiling water	23/34	23/34
Shrink tension (100° C.) (p.s.i. MD/TD)	293/426	266/457
Heat seal range, ° C., bar sealer	120-160	125-200

The above example is repeated, using heat-shrinkable films made from blends of low density and high density polyethylene resins in the ratio of 80/20 and 85/15, as described in U.S. Pat. 3,299,194. On exposure to high energy irradiation, these films exhibit a similar increase in zero strength temperature and improvement in heat sealing performance.

Polypropylene Dielectric Spacers

In a process described by W.R. Hendrix and S. Tocker; U.S. Patent 3,622,848; November 23, 1971; assigned to E.I. du Pont de Nemours and Company films of photocross-linked blends of linear polypropylene with 0.02 to 2 weight percent of a photosensitizing agent are used as dielectric spacers for capacitors. The sensitizing agent can be polymeric, e.g., a copolymer of ethylene and acryloxy benzophenone, or nonpolymeric, e.g., 2-methyl-anthraquinone. The films are advantageously oriented after extrusion and before cross-linking.

Example 1:   A copolymer of ethylene with acryloxy benzophenone is prepared by the high pressure synthesis of Example 1 of U.S. Pat. 3,214,492. The copolymer contains about 90% ethylene units. The copolymer is blended in a twin screw mixer with linear polypropylene having a melt index of 3 at 230°C. in the ratio of 10 to 90 parts, respectively. The mixture is extruded and tubularly oriented in accordance with Example 1 of U.S. Pat. 3,141,912. The film is about 0.5 mil thick. The film is cross-linked by exposure on a drum to a Hanovia ultraviolet source placed 4 inches from the film. The exposure time is about 13 seconds. The photocross-linked film has a gel content of about 40%, whereas an uncross-linked control film has a gel content of about 3%.

The loss factor is measured as a function of temperature. A liquid immersion container with a test cell (Balsbaugh Lab. Type 100) is used for single sheet measurement. The impregnating liquid was trichlorobiphenyl (Arochlor 1242). The cross-linked polymer maintains a desirably low loss factor and is still intact above 140°C. The loss factor for the control rises rapidly above 100°C. and the film fails by dissolution at 130°C. The behaviors of the control and the film of this process are essentially identical below 100°C. Film made in accordance with the above can be vacuum metallized and made into a capacitor as shown in Figure 1 of U.S. Pat. 3,271,642.

Example 2:   Capacitors are prepared (0.5 microfarad) by winding two dielectric spacer films prepared in example 1 with two 0.25 mil aluminum foils to form a structure such as in Figure 2 of U.S. Pat. 3,363,156. The film is 2 inches wide and the foil is 1.5 inches wide. Approximately 250 turns are made. The capacitors tested at room temperature typically withstand about 2,000 volts DC.

Example 3:   A capacitor of Example 2 is placed in a glass container and vacuum dried at 85°C. and 0.02 torr for 2 hours. The container is filled with trichlorobiphenyl and allowed to impregnate as described in U.S. Pat. 3,363,156 in the oven for 20 hours. The capacitor is sealed, removed from the oven and cooled. A control capacitor with uncross-linked dielectric and a cross-linked capacitor are given a thermal test. The capacitors are placed in an oven at 50°C. and heated at the rate of 1°C. per minute. The capacitor of this process is typically maintained for 1 hour at 130°C. without failure, whereas a control capacitor typically fails at 130°C. by shorting of the electrodes because of the softened film.

The following typical data are obtained by AC testing at 75 volts and 100 hertz:

	Dissipation Factor		
Sample	100°C.	125°C.	130°C.
Control	0.0040	0.00067	Shorted
Cross-linked	0.0043	0.00058	0.0092

Impregnation of Nylon with Cupric Chloride

A. Ishitani and K. Nukada; U.S. Patent 3,658,534; April 25, 1972; assigned to Toray Industries, Inc., Japan describe a photosensitive polymeric material comprising an oxygen, sulfur, phosphorus, nitrogen, halogen or coordination compound forming aromatic nucleus-containing polymer bonded by coordination bonding to an organic or inorganic salt of a metal from Groups IB, IIB, VIB, VIIB, and VIIIB of the Periodic Table. Typically, nylon or a polyester film is immersed in a solution of a metal salt, such as cupric halide, to effect molecular dispersion on the polymer and formation of a coordination bonded complex. Color or absorptivity change is then produced by irradiation. Some of the complexed polymer materials are heat sensitive and/or reversible in their photo and heat sensitivity. Color changes may be fixed in some materials after irradiation by other treatment steps. Selective treatment with chemical reagents or dyestuffs is also possible following irradiation.

Example 1: (a) A nylon-6 film approximately 100μ thick, was immersed in a 50% by weight aqueous solution of $CuCl_2$ at 80°C. for 30 minutes. An increase of weight, by coordination complexing with $CuCl_2$, of 15% was observed and the film was yellow. This polymer complex film showed three absorption bands at $930m\mu(\epsilon \sim 200)$, $270m\mu(\epsilon \sim 2,500)$ and $400m\mu$ ($\epsilon \sim 400$), indicating that a complex was formed between Cu^{++} and the amide group of the nylon-6. When this yellow film was irradiated with a light at a distance of 15 centimeters from a light source, comprising a 250 watt high pressure mercury lamp, for 30 minutes, the yellow color disappeared and the film became colorless.

At the same time, the aforesaid three absorption bands disappeared, showing that the light had cut the complex bond. From observation of the ESR spectrum of the Cu^{++} forming the complex bond and from the fact that the yellow color disappeared upon irradiation, it was understood that a light reduction reaction of Cu^{++} to Cu^+ took place. When the film thus rendered colorless by irradiation was left to stand in a dark place for several hours, the film reversibly returned to the original yellow color, demonstrating inverse photochromism.

(b) When the yellow film made in (a) was heated to more than 80°C., it changed to reddish brown. However, when the temperature was returned to room temperature, the film immediately returned to the original yellow, thus exhibiting thermochromism. It is possible to repeat this thermochromic reaction at temperatures up to 100°C.

(c) When the colorless film irradiated with the light in (a) was heated at 120°C. for 10 minutes, it became brown. Further, it was found that this brown color was fixed or permanent. Utilizing this property, the yellow film of (a) was contacted with a silver salt negative of a photograph. The film was then irradiated with light as in (a) and heated at 120°C. for 10 minutes to obtain a brown positive on a yellow ground. The film was then washed (either hot water or diluted sulfuric acid may be used) to remove the unreacted $CuCl_2$. The result was a brown positive on a colorless ground. In an alternative procedure, the film was first washed after irradiation and then heated. The same brown positive reproduction resulted. The image thus obtained had good resolution and intermediate color tone and could be used as a positive for slide projection.

(d) The yellow film of (a) was contacted with a silver salt photographic negative and irradiated. The film was then immersed in an aqueous solution of NaOH and heated. The resultant product was a negative good in contrast wherein only the nonirradiated part was colored a dark brown.

(e) When the film irradiated as in (d) was immersed in an aqueous solution of KI and heated, the nonirradiated part became orange in color. and a negative reproduction was thus produced.

(f) When the film irradiated as in (d) was washed with dilute sulfuric acid and treated with an aqueous solution of $K_2Fe(CN)_6$, the irradiated part became colored to form a permanent brown image.

(g) When the film irradiated as in (d) was immersed in an aqueous solution of Na_2S and heated, the nonirradiated part became dark green in color and the irradiated part became brown. An image with good contrast was thus made.

(h) When the film irradiated as in (d) was immersed in an aqueous solution of $Na_2S_2O_4$, the irradiated part became a permanent or fixed grey color.

(i) When the film irradiated as in (d) was immersed in an acetic acid solution of benzidine containing a small amount of KI, and thereafter immersed in dilute sulfuric acid, the irradiated part became brown in color.

(j) When the film irradiated as in (d) was immersed in an aqueous solution of Na_2CO_3, the nonirradiated part became dark brown in color and formed a negative image.

(k) When the film irradiated as in (d) was immersed in a dilute sulfuric acid solution of $Na_2S_2O_3$, the nonirradiated part became green in color and the irradiated part became blue.

(l) When a fabric woven from a nylon-6 yarn was immersed in a 50% by weight aqueous solution of $CuCl_2$ and treated at 80°C. for 2 to 3 minutes, 10 to 20% by weight of $CuCl_2$ was added to the fabric as a coordinately bonded complex and a yellowish green fabric was made. A mask was contacted with this fabric and the fabric was irradiated with the light under conditions as in (a) for 20 minutes, then heated at 130°C. for 10 minutes and washed with water. It was then possible to print brown letters and patterns on a white background on this fabric.

(m) When the film irradiated as in (d) was immersed first in a hot aqueous solution of NaOH(10 to 15 weight percent) and then in a dilute sulfuric acid, a layer of metallic copper was formed on the irradiated portion of the film. This copper layer was highly electrically conductive. (Wherein ϵ is molecular extinction coefficient, and ESR spectrum is Electron Spin Resonance Spectrum.)

Example 2: A 70μ thick nylon-12 film was immersed in a 50% by weight aqueous solution of ferric chloride at 85°C. for 30 minutes. The film weight increased by 6.10% and the film became yellow in color due to coordination bonding between the nylon film and the ferric salt. The film-metal salt complex had absorption maximums at $365m\mu$ and $315m\mu$. When a negative was placed in contact with the yellow film and the two were irradiated with an ultraviolet lamp, more specifically a high pressure mercury lamp, for 5 minutes, the film became colorless and transparent in the irradiated portion. At the same time, the aforesaid absorption bands disappeared. However, when this colorless transparent film was left to stand or heated to a temperature below 100°C., the color reversibly returned to the original yellow. When the colorless transparent film was treated at 130°C. for 5 minutes, the irradiated part assumed a yellowish brown color. A permanent and fixed image was thus obtained.

Example 3: In 30 ml. of dimethyl formamide, 1.2 g. of polyacrylonitrile and 35 mg. of cupric chloride were dissolved and from the resultant solution a 21μ thick film was cast and formed. As a result, 3.23% (based on polyacrylonitrile) by weight of the cupric salt coordinated with the polyacrylonitrile. This film was yellow in color, showing maximum absorptions at $260m\mu$, $400m\mu$ and $950m\mu$ in the near infrared region. When irradiated, these absorption maximums disappeared, as did the color of the film.

Example 4:   In 50 ml. of methanol, 0.93 g. of polymethacrylic acid and
26.3 mg. of cupric chloride were dissolved and from the resultant solution
a 20μ thick film was cast and formed. As a result, 2.8% (based on the
polymer) by weight of the cupric salt coordinated. The treated polymethacrylic acid film was green, showing absorption bands at 700mμ and, in
the ultraviolet region, also at 270mμ. When this film was irradiated, the
color changed to brown having an absorption position at 340 to 360mμ.
When the irradiation was discontinued, the color of the film returned to
green. When the green film was heated its color changed to brown. This
change also was reversible, however, indicating that the complex was
thermochromic as well as photochromic.

COATINGS

Nonair Inhibited Polyester Resin-Wood Coatings

A process described by A.C. Keyl and M.G. Brodie; U.S. Patent
3,669,716; June 13, 1972; assigned to The Sherwin-Williams Company
provides a method for producing cured polyester resin coatings of predetermined substantial thickness which are mar-resistant, scratch-resistant and
solvent-resistant. It has been found that improved results in the coatings
art can be obtained in preparing coatings having a thickness of 1 to 12 mils
(0.001 to 0.012 inch) by forming such coatings on a substrate from photopolymerizable nonair-inhibited polyester resins, preferably containing a
photosensitizer, and curing such resins by subjecting them to light waves
within the range of 1850 to 4000 angstroms. The process is useful in forming coatings on metal, wood or other substrate but is especially valuable
for producing cured resinous coatings on wood, e.g., plywood panels.

The resins which are effective are all characterized by the fact that they
are not inhibited by air or oxygen. They can also be described as air
drying polyesters. These resins, because of their chemical composition, are
capable of achieving good surface cure in the presence of air or oxygen.
Examples of the chemical types involved include allyl ether resins, benzyl
ether resins, tetrahydrophthalic anhydride resins, endomethylene tetrahydrophthalic anhydride resins, cyclopentadiene modified resins, acetal resins,
polyalkylene fumarate resins where there are at least three ethylene groups,
and tetrahydrofurfuryl resins. The nonair inhibited polyesters can also be
described as air drying unsaturated polyesters. Air drying unsaturated
polyesters which are employed in this process to produce polyester coatings
are especially modified to prevent air inhibition of cure.

Various groups can be introduced into the polyester most of which are subject to auto-oxidation and thus actively prevent the inhibiting action of

atmospheric oxygen which is normally dissolved at the surface of polyester coatings exposed to air. Examples of groups which may be introduced into the polyester formulation include aliphatic, cycloaliphatic and aromatic ethers, e.g., trimethylol propane (TMP) diallyl ether, tetrahydrofurfuryl alcohol, dioxane, dicyclopentyl ethers, TMP monobenzyl ether, and triethylene glycol; acetal type structures; cyclohexene type compounds, e.g., tetrahydrophthalic anhydride, and endomethylene tetrahydrophthalic anhydride; dicyclopentadiene derivatives such as 8-oxytricyclodecene-4-$(5,2,1,0^{2,6})$, and tricyclodecane-$(5,2,1,0^{2,6})$ dimethylol; and by direct modification with cyclopentadiene or dicyclopentadiene.

Example 1: Preparation of 30% Diallyl PE (Pentaerythritol) Diethylene Maleate Phthalate —

	Weight, g.
(a) Maleic anhydride	441
(b) Phthalic anhydride	222
(c) Diethylene glycol	490
(d) Diallyl PE	428
(e) Sulfonated styrene–divinyl benzene polymer (Dowex 50)	14.6
Total charge	1,595.6

Ingredients (a), (b) and (c) were charged to a 3 liter flask fitted with a water cooled condenser and separatory trap and heated gradually to 360°F. with agitation. 86 g. of toluene were added for refluxing and a light nitrogen blanket was introduced and maintained throughout the reaction. The temperature was held at 360°F. After the acid value had dropped to 141 to 142, ingredients (d) and (e) were added. The batch was reheated to 360°F. and held for a final acid value of 37 to 39. At a late stage in the reaction the azeotrope solvent (toluene) was removed by blowing with nitrogen. At an acid value of 37 to 39, the polyester resin was cooled and reduced with styrene to produce a final solution of 75% resin and 25% styrene. The final resin had an acid value of 38.6 and a viscosity of W–. Tertiary butyl hydroquinone was added as inhibitor at a concentration of 100 parts per million based on the weight of the total solution.

Example 2: Preparation of 30% Diallyl PE Ethylene Maleate Phthalate —

	Weight, g.
(a) Maleic anhydride	441
(b) Phthalic anhydride	222
(c) Ethylene glycol	285
(d) Diallyl PE	428
(e) Sulfonated styrene–divinyl benzene polymer (Dowex 50)	12.7
Total charge	1,388.7

This resin was prepared in the same manner as Example 1 except that the addition of (d) and (e) was made at an acid value of 173 to 176. The completed resin was reduced with styrene to produce a resin solution containing 25% styrene and 75% polyester resin. The final resin solution had a viscosity of Z_1+ and an acid value of 37.4. Inhibitor was added as in Example 1.

Example 3: A liquid unsaturated polyester coating composition was prepared by blending the ingredients listed below:

	Weight, g.
Polyester prepared according to Example 2	100
Styrene monomer	40
Silicone solution*	2
Nondrying capric acid alkyd	2
1-Chloromethyl naphthalene	1.4
2-Naphthalene sulfonyl chloride	1.4
Total	146.8

*1% silicone oil (Linde R-12) reduced in styrene monomer.

The above polyester blend was flow coated onto several wooden (maple) 4" x 6" panels primed with a primer or active ground coat as shown in the following table. After coating with polyester, the panels were aged at room temperature (or heat treated) prior to light irradiation with a type (b) lamp as shown in the table. The distance between the panels and the helical quartz flash tube during light irradiation was measured at 10 inches. A capacitance of 350 microfarads connected to a 3,000 volt power source was used to operate the flash tube at a power input of approximately 1,600 watt seconds per flash. Duration of each flash was of the order of 2 milliseconds and the off time between flashes was from 25 to 30 seconds. Ambient temperature under the flash tube was 84°F.

Panel No.	Primer	Pretreatment	Flashes	Results
A₁	Active ground coat.¹	Aged 4 min. at room temp. heat treatment for 3-4 min., at 201° F.	20 flashes or about 0.04 sec. light exposure.	Slightly tacky but fairly hard film
A₂	......do...........	Heat treatment 7 min. at 195° F. and aged 3 min. at room temp.	40 flashes or about 0.08 sec. light exposure.	Film harder than A₁. Very little tack.
A₃	Primer ³	Heat treatment 5-6 min. at 200 ° F	20 flashes or about 0.04 sec. light exposure.	Surface skin formed. Film not cured underneath.
A₄	Active ground coat.	Aged 19 min. at room temp	...do...	Surface cure. Film gel-like and soft.

¹ See the following table:

	Percent
60% solution of methyl ethyl ketone peroxide in dimethyl phthalate (Lupersol DDM)	6 5
Vinyl chloride-vinyl acetate copolymer (Vinylite VAGH).	13.2
Solvent: Xylene, methyl isobutyl ketone 2nd ethyl amyl ketone	80.3
Total	100. 0

³ See the following table:

	Percent
Vinylite VAGH	15. 0
Solvent: Xylene and methyl isobutyl ketone	85. 0
Total	100. 0

The various panels listed in the table which were subjected to heat treatment did not evidence any cure until exposure to light radiation. As shown in the table, light radiation alone was effective to partially cure the untreated panel, A_4, at low temperature (approximately 84°F.) but additional light flashes would have been necessary to obtain a complete cure. Panel A1 illustrates that a more complete cure may be obtained with equivalent light flashes by employing an active ground coat plus heat treatment. As shown by panel A2, a minimum of 40 flashes was required for effective cure. At an energy input per flash of 1,575 watt seconds the total energy used was 63,000 watt seconds. Heat treatment alone without the use of an active ground coat, is not effective as shown by the panel A3 data. In addition, data contained in the table illustrates the effect of increased light exposure upon the cure of polyester coatings.

Example 4:

	Weight, g.
Polyester prepared according to Example 1	67.00
Polyester prepared according to Example 2	33.00
Styrene monomer	35.00
Silicone solution*	2.00
Nondrying capric acid alkyd	2.00
1-Chloromethyl naphthalene	2.03
2-Naphthalene sulfonyl chloride	2.03
Total	143.06

*1% silicone oil (Linde R-12) reduced in styrene monomer.

The above ingredients were thoroughly mixed to obtain a liquid polyester coating. A wet film of the coating thus prepared was deposited on a wooden panel which had been primed with an active ground coat as in Example 3. Pretreatment of the coated panel consisted of baking in an oven for five minutes at 212°F. Thereafter the panel was placed 10 inches from the flash tube light source and subjected to light radiation as in Example 3. The flash tube was switched on immediately and the panel exposed to forty flashes of intense light irradiation. After exposure to the described radiation, the polyester coating was found to be hard with no after tack. Hardness of the film increased somewhat thereafter, indicating cure process continued after removal from the high intensity light source.

Fused Tetrafluoroethylene Polymer

A.N. Wright, V.J. Mimeault and E.V. Wilkus; U.S. Patent 3,673,054; June 27, 1972; assigned to General Electric Company describe a process for making laminated structures bonded by a fused tetrafluoroethylene polymer. The process for forming the tetrafluoroethylene polymer comprises providing tetrafluoroethylene vapor at a pressure of about 10 torr to 760 torr at a temperature of about 0° to 200°C. and subjecting the vapor to ultraviolet light having a wave length of about 1800 to 2400 Angstroms. The process is carried out satisfactorily with the temperature of the tetrafluoroethylene vapor at room temperature, i.e., 25°C., or at a temperature close to room temperature, and these temperatures are preferred. The pressure of the tetrafluoroethylene monomer may vary widely depending largely upon the particular rate of polymerization desired as well as upon the particular type of floc polymer desired to be formed. The process is operable with the tetrafluoroethylene monomer vapor pressure varying from about 10 torr to about 760 torr. The lower the monomer vapor pressure, the slower is the rate at which the polymer forms. For example, if the process is used to deposit a layer of the polymer on a substrate surface, the lower the tetrafluoroethylene vapor pressure, the slower is the rate of deposition. Likewise, the rate of polymerization or polymer deposition increases with increasing pressure. Pressures higher than atmospheric pressure are not useful due to the danger of explosion.

In addition, pressures higher than atmospheric would tend to produce high molecular weight polymers which are less thermoplastic, and at pressures significantly higher than atmospheric, the polymer formed would no longer be fusible. With increasing tetrafluoroethylene monomer vapor pressure, a tetrafluoroethylene floc polymer of higher thermal stability is formed. Specifically, the polymer formed at higher tetrafluoroethylene monomer vapor pressures requires higher fusion temperatures than the polymer formed at lower monomer vapor pressures. In addition, during fusion, the polymer formed at lower pressures undergoes more weight loss, i.e., has a lower percent retention, than the polymer formed at higher pressures indicating that larger amounts of lower molecular weight polymer are formed at the lower tetrafluoroethylene vapor pressures.

In carrying out the polymer process, the tetrafluoroethylene monomer vapor is exposed to ultraviolet light having a wave length of 1800 to 2400 A. to form the polymer. Since no sensitizers are used, wave lengths outside the 1800 to 2400 A. range are not effective. Light having a wave length of 1840 to 2200 A. produces a particularly satisfactory rate of polymerization, especially at room temperatures and atmospheric pressure. The rate of polymerization is also proportional to the intensity of the light as well as the pressure of the tetrafluoroethylene monomer, i.e., the more intense the light, the faster is the rate of polymerization. These factors can be readily controlled so as to obtain a satisfactory rate of polymerization.

The process is illustrated by the following examples. In all of these examples the procedure was as follows unless otherwise noted: Ultraviolet light was provided by a 700 watt Hanovia lamp, Model No. 674A, which emitted light of wave length ranging from about 1849 A. to about 13,673 A. Specifically, it emitted ~ 17 watts of light of wave length of about 1849 to 2400 A. and ~ 131 watts of wave length of 2400 A. to 3360 A. in the ultraviolet. The lamp was provided with a reflector and was capable of heating the reactor system to about 200°C. In all of the runs the temperature of the tetrafluoroethylene monomer vapor was below room temperature when introduced into the reactor, but shortly after introduction into the reactor, usually less than within about 1 minute, the vapor equilibrated to room temperature, and during the run its temperature was raised by the ultraviolet light to above room temperature. Dry N_2 gas, when used, was at room temperature when introduced into the reactor. The reactor used was essentially a rectangular vacuum reaction chamber, approximately 13 cm. wide, 29 cm. long, and 7 cm. high. In the chamber top was an 8 x 20 cm. quartz window which was situated directly above, and 3 cm. from, an 8 x 20 cm. copper cooling block inside the chamber.

The block was cooled by the internal flow of fluid, cooled and driven outside the chamber. Those laminae placed in the reactor were placed on the cooling block. Ultraviolet input was provided by the Hanovia lamp which was aligned over, and 5 cm. away from, the window. The quartz window used was transparent to light of wave length greater than about 1800A. Tetrafluoroethylene monomer pressure inside the chamber was obtained and maintained through a valved connection to a temperature-regulated source of tetrafluoroethylene monomer. The tetrafluoroethylene monomer gas used in all of the examples was free of inhibitor.

Specifically, if the monomer source were at room temperature in the shipping cylinder, the gas was passed through a dry ice trap to remove inhibitor. Otherwise, a liquid source of tetrafluoroethylene, prepurified from inhibitor by distillation, was used as a direct source of monomer vapor at low temperatures. Thermocouples inside the chamber enabled the recording of the temperature of the block as well as the specimens. Heating to fuse the polymer was carried out in a forced air oven. A standard bridge technique was used to determine capacitance. All glass slides and evaporated aluminum coated glass slides used as laminae were standard microscope slides, i.e., 1 inch wide, 3 inches long and about one thirty-seconds inch thick. All aluminum coupons used as laminae were 1 inch wide, 3 inches long and about one-eighth inch thick.

Example 1:   In this example, the laminae were glass slides. The slides were placed in the reactor, and with the tetrafluoroethylene monomer pressure at 300 torr and nitrogen gas pressure at about 400 torr, the tetrafluoroethylene floc polymer was deposited on the surfaces of the slides for 60 minutes. During the deposition the slides were cooled and maintained at a temperature of 5°C. The floc polymer deposited on the slides in a substantially uniform manner and formed a continuous coating thereon.

The floc coated slides were then placed together with the polymer intermediate the slides to form a lap joint with one square inch of overlap. Under a load of about 7 oz./in.^{2}, the lap assembly was heated at a temperature of 330°C. for 100 minutes and then allowed to cool to room temperature under the load. The resulting laminated structure could not be pulled apart manually and Instron measurements indicated a shear strength greater than 50 lbs./in.^{2}.

Example 2: In this example, the laminae were aluminum frying pan coupons which were precleaned with a trichloroethylene dip. The coupons were placed in the reactor, and with the tetrafluoroethylene monomer pressure at 300 torr and nitrogen gas pressure at about 400 torr, the tetrafluoroethylene floc polymer was deposited on the laminae surfaces for 50 minutes. During the deposition the coupons were cooled and maintained at a temperature of 25°C. The floc polymer deposited on the surfaces of the coupons in a substantially uniform manner and formed a continuous coating. The floc coated coupons were placed in a furnace and heated separately for 45 min. at a temperature of 350°C. to give continuous clear films, each having a thickness of about 54,000 A. The fused polymer coated coupons were placed back in the reactor with the fused coatings exposed to the reactor atmosphere and were maintained at 25°C. during floc deposition.

With the tetrafluoroethylene monomer pressure at about 500 torr and no other gas present, tetrafluoroethylene floc polymer was deposited on the fused coatings for 50 minutes. The floc polymer deposited on the fused coatings in a substantially uniform manner forming a continuous deposit. The floc coated coupons were then placed together with the polymer intermediate the coupons to form a lap joint which was two inches square. Under a load of 7 oz./in.^{2} the lap assembly was heated at a temperature of 350°C. for 45 minutes and then allowed to cool to room temperature under the load. The resulting laminated structure was held together by an insulating dielectric layer as determined by the standard bridge technique. The laminated structure was tested on an Instron tester and showed a shear strength of 2 lbs./in.^{2}.

Example 3: In this example a laminated structure was formed comprised of a silicon wafer bonded to a glass slide. The glass slide was placed in the reactor where it was maintained at a temperature below 50°C. during floc deposition. With the tetrafluoroethylene monomer gas pressure at 400 torr and with no other gas present in the reactor, floc polymer was deposited on the glass for 55 minutes in a substantially uniform manner and formed a continuous deposit.

A silicon wafer about the size of a 50 cent piece, but thinner, was placed on the deposited floc polymer and the assembly was heated at a temperature of 350°C. for 45 minutes under no compressive load. When the laminated structure had cooled to room temperature, it showed sufficient adhesion to support the weight of the glass slide.

Chlorosulfonyl-tert-Butylisocyanate for Urethanes

In a process described by D. Arlt; U.S. Patent 3,580,942; May 25, 1971; assigned to Farbenfabriken Bayer AG, Germany 2-isocyanato-2,2-dimethyl-ethane-sulfonic acid chloride is prepared by reacting isobutylene and cyanogen chloride with chlorosulfonic acid or sulfur trioxide. The process is illustrated by the following reaction schemes. The primary reaction product shown in brackets in the reaction schemes, which is first obtained by reacting butylene and cyanogen chloride with chlorosulfonic acid or sulfur trioxide, is generally not isolated.

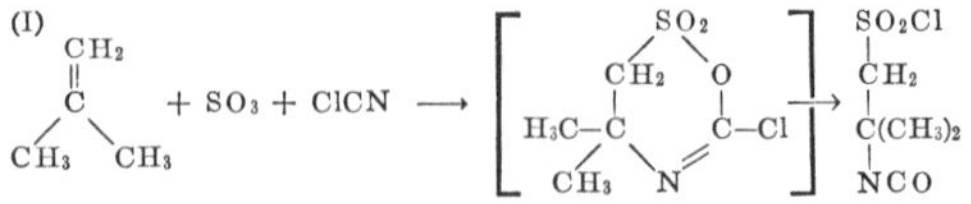

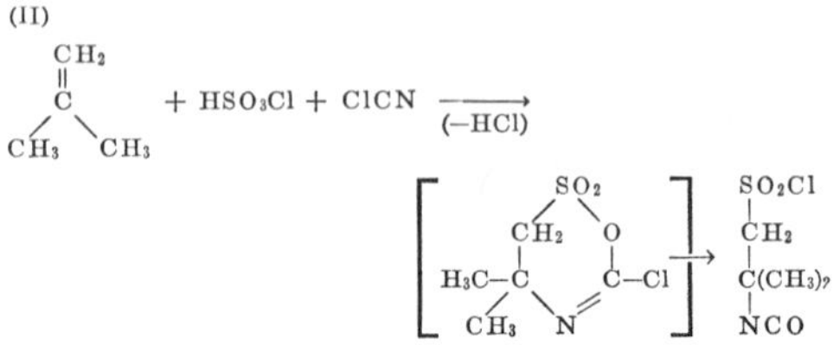

(I)

$$CH_2=C(CH_3)_2 + SO_3 + ClCN \longrightarrow \left[\begin{array}{c} SO_2 \\ H_3C-C(CH_3)(CH_2) \diagdown O \diagup C-Cl \\ N \end{array} \right] \rightarrow \begin{array}{c} SO_2Cl \\ | \\ CH_2 \\ | \\ C(CH_3)_2 \\ | \\ NCO \end{array}$$

(II)

$$CH_2=C(CH_3)_2 + HSO_3Cl + ClCN \xrightarrow[(-HCl)]{}$$

$$\left[\begin{array}{c} SO_2 \\ H_3C-C(CH_3)(CH_2) \diagdown O \diagup C-Cl \\ N \end{array} \right] \rightarrow \begin{array}{c} SO_2Cl \\ | \\ CH_2 \\ | \\ C(CH_3)_? \\ | \\ NCO \end{array}$$

The product of the process is especially useful as a cross-linker for photo lacquers. For this reason the product is transformed into the corresponding sulfazide and brought into the reaction with lacquer compositions which contain hydrogen atoms reactive against isocyanates and unsaturated bondings and cross-linking of these compositions is carried out by the action of light or radiation.

Example: About 580 parts of chlorosulfonic acid are dissolved in about 750 ml. of cyanogen chloride at about 0°C. with stirring and cooling. About 300 parts of isobutylene are then introduced over the course of about 2 hours, the reaction mixture being kept at a temperature of from about 0°C. to about 5°C. during the reaction by cooling. The reaction mixture is then heated to from about 50° to about 60°C., and excess cyanogen chloride distills off with simultaneous vigorous evolution of hydrogen chloride. The product remaining behind is further heated in a vacuum of about 1 mm. Hg, a distillate passing over.

The temperature is raised, as soon as distillation slows down, to a maximum of about 170°C. About 765 parts of a distillate which consists mainly of 2-isocyanato-2,2-dimethyl-ethane-sulfonic acid chloride and 2-methyl-propene-2-sulfonic acid chloride are obtained. About 450 parts, which corresponds to about 46% of the theoretical, of 2-isocyanato-2,2-dimethyl-ethane-sulfonic acid chloride of boiling point 83° to 85°C./0.2 mm. Hg are obtained by fractional vacuum distillation. Analysis — $C_5H_8ClNO_3S$ (197.7). Calculated (percent): C, 30.4; H, 4.1; Cl, 17.9; N, 7.1; O, 24.3; S, 16.2. Found (percent): C, 30.4; H, 4.2; Cl, 18.0; N, 7.1; O, 24.4; S, 16.0.

Acid-Hardening Varnishes

A process described by H.-J. Rosenkranz, H. Rudolph and H.-J. Kreuder; U.S. Patent 3,692,560; September 19, 1972; assigned to Farbenfabriken Bayer AG, Germany relates to acid-hardening resin compositions containing a benzophenone compound each of the two benzene nuclei of which is substituted by a halogenated methyl group. If the compositions are irradiated by ultraviolet light, the benzophenone compound splits off hydrogen halide which, in turn, catalyzes the hardening of the resin composition.

The following compounds can be used in the process, for example: p-benzoyl-benzyl chloride, p-benzoyl-benzal chloride, p-benzoyl-benzotrichloride, p-benzoyl-benzyl bromide, p-benzoyl-benzotribromide, 4,4'-bis-dichloromethyl-benzophenone, and 4,4'-bis-bromomethyl-benzophenone. The compounds are added to the resins in amounts of about 0.1 to about 10, preferably of about 1 to about 6, percent by weight, referring to the resin to be hardened which may be dissolved in the usual solvents. Mixtures of this type have virtually unlimited storage stability in the dark, but when they are illuminated with ultraviolet light which can be generated with the usual light sources, the desired gelling sets in and hardening takes place either immediately or after the usual stoving time, depending on the composition of the resin.

Example 1: An acid-hardening varnish consists of the following components: 60 parts by weight of a 60% solution of an alkyd resin in butanol (the alkyd resin was prepared by condensing 90.2 parts by weight castor oil, 128.4 parts by weight soya bean oil, 95.1 parts by weight trimethylolpropane, 76.3 parts by weight pentaerythritol, 14.3 parts by weight benzoic acid and 196.8 parts by weight phthalic acid anhydride up to acid number 8 and hydroxyl number 170); 40 parts by weight of a commercial 60% solution of an urea-formaldehyde condensate in butanol; 7 parts by weight ethyl glycol; 7 parts by weight butanol; 7 parts by weight ethanol; and 1 part by weight of a 1% solution of silicone oil in xylene.

This varnish is mixed with additives according to the following table. The mixtures are applied to glass plates by means of a film extruder (100μ) and then further treated according to the table. The high pressure burner used for illumination is an apparatus which acts on the films from a distance of 20 cm. When the mixtures (1) to (5) are hardened according to the table, they still show the same course of reaction after dark storage at room temperature for 5 months.

Initiator	Additive as parts by weight, referred to the varnish	Time required for complete hardening (min.) when illumination with—		
		Daylight	High press. burner	2 min. high press. burner, then daylight
(1) 4,4'-bis-bromomethyl-benzophenone....	2.2	>180	6	40
(2) p-benzoyl-benzyl bromide..............	1.6	>180	8	40
(3) p-benzyl-benzal bromide..............	2.1	>180	6.5	40
(4) p-benzoyl-benzal chloride.............	1.6	>180	5.5	30
(5) p-benzoyl benzotrichloride...........	2.3	>180	6.5	30

Example 2:   88 parts by weight of a polyester which contains terephthalic acid radicals as acid radicals and bis-ethoxylated bisphenol A and glycerol as alcoholic components and which has been condensed to a viscosity of 50 seconds (measured in a DIN beaker or a 40% solution of the polyester in ethyl glycol acetate), 12 parts by weight hexa-bis-(methoxymethyl)-melamine and 3 parts by weight p-benzoyl benzotrichloride are dissolved in 500 parts by weight methylene chloride.

A metal plate is evenly coated with this solution with a thickness of 100μ and allowed to dry in air. A firm tack-free film is obtained. This film is illuminated through a negative for 30 seconds with the light of a high pressure mercury lamp placed at a distance of 20 cm. After heating the coated plate at 120°C. for 20 minutes, the nonexposed parts of the film can easily be detached with the aid of methylene chloride, whereas a firm film of good adhesion has formed on the illuminated areas.

POLYOLEFIN SURFACE TREATMENTS FOR ADHESION

Organic Isocyanate and Sensitizer

A process described by R.A. Bragole; U.S. Patent 3,607,536; Sept. 21, 1971; assigned to USM Corporation relates to adhesive processes and particularly processes including the treatment of low-energy polymeric resin bodies to enable strong bonding and to resin bodies so treated. Polyalkylene plastic materials, particularly polyethylene and polypropylene possess many desirable characteristics including inertness to most chemicals and solvents at ordinary temperatures, resistance to electricity, toughness and flexibility. By reason of these and other properties it has been desired to employ such materials in numerous relationships where the bonding of the material to itself or to other surfaces is required. These materials present a waxy, sometimes paraffinlike surface character, i.e., have a low critical surface

tension of wetting, which interferes with adhesion by the commonly employed adhesive or coating agents. Also it appears that many polyalkylene
plastics have at their surfaces a weak boundary layer developed in the
course of molding or other shaping. In many relationships, for example, in
the use of flexible polyethylene sheet material or the lamination of polyethylene to flexible sheet materials, hot melt adhesives which operate to
fuse and integrate with the polyethylene surface may be used to bond the
surface. However, there are many relationships where because of the rigidity of the materials to be combined or because of special contours or
other factors, such hot melt adhesive systems are not usable.

In this process an organic isocyanate and a photosensitizer are provided at
the surface of a body of low surface tension of wetting polymer resin material and the surface is subjected to controlled ultraviolet radiation treatment to produce a chemical linkage between the resin material and the
organic isocyanate to form a urethane stratum integral with the resin body
and capable of being bonded by adhesives. Adhesive bonds to the surface
are formed by use of adhesives known to be capable of bonding to polyurethane or rubber surfaces and assembling the adhesive-coated surface
against the surface of a second body to form an adhesive joint.

The first step in the process is the application to the surface of the material
to be bonded of a deposit of an organic isocyanate and a photosensitizer.
Useful isocyanate compounds may have an -NCO functionality of one or
more than one. Among isocyanates which have been used are triphenyl-
methane triisocyanate, polyarylene polyisocyanate (-NCO functionality
of 2.8), tolylene diisocyanate, methylene-bis-4-phenyl isocyanate and
phenyl isocyanate. Other compounds having at least one active -NCO
group, such as those from reaction of compounds having an active hydrogen,
e.g,, an alcohol, glycol or amine with an isocyanate, may be used. These
materials may be applied in organic solvent solution.

No special conditions of temperature or time are necessary for the contact
and it has been found that the desired results are obtained by merely spread-
ing a solution over the surface to be ultraviolet radiated and wiping off the
excess, the whole treatment being carried out at room temperature. Any
convenient solvent may be used for the isocyanate. In general it is preferred
to use volatile solvents which evaporate rapidly. Useful solvents include
methylene chloride, methyl ethyl ketone, and tetrahydrofurane; but other
solvents may be used which do not interfere with the activity of the isocyanate. Solutions containing as little as 0.25% by weight of the isocyanate may be used, but it is preferred to use concentrations of from about
1 to about 10%. Greater percentages of isocyanate are not necessary, and
aside from the unnecessary cost may be less effective than the lower percentages.

The -NCO groups of the applied compound will not react per se with the untreated surfaces of such materials as polyethylene and there is provided at the surface to be bonded at least a minor amount of a photosensitizer effective under the action of ultraviolet radiation to generate groups chemically binding the surface with the applied isocyanate-containing compound. Such ultraviolet radiation sensitizers may be halogenated hydrocarbons such as methylene chloride, trichloroethylene, and chloroform, ketone materials such as benzophenone, acetophenone, benzoin, 2-acetonaphthone or other known photosensitizers such as acenaphthene and fluorine.

In general, these materials are excited by ultraviolet radiation and, in excited state, interact with the resin substrate to initiate reactions producing groups which will react with isocyanates; sensitizers in which carbon atoms are linked to other atoms by multiple bonds as in benzophenone and trichloroethylene are believed to be engrafted into the molecular chain of the resin substrate and may themselves provide groups which will react with isocyanate. As little as 2% by weight of photosensitizer in an applied solution may be effective; but higher percentages may be used without adverse results.

When the organic isocyanate and the photosensitizer have been supplied to the surface of the material to be bonded, the surface is exposed to ultraviolet radiation. The ultraviolet radiation bombards the surface of the plastic and the photosensitizer with photons which excite the molecules to cause chemical and electronic charges. It has been found ultraviolet radiation at a wavelength of from 2000 to 3500 A. for as little as 300 watt seconds per square foot and preferably from about 350 to about 7,000 watt seconds per square foot generates groups which react with the isocyanate to form a strong urethane surface stratum. This stratum is integral with the body of material to be bonded and is easily wet and strongly adhered by adhesives effective to bond polyurethanes or rubbers.

Example 1:   A series of strips of low density polyethylene one inch wide and 0.060 inches in thickness were dipped in a 1% solution of polyarylene polyisocyanate having an -NCO equivalent of 2.8 (PAPI) in trichloroethylene and were subjected to radiation by disposing surfaces of the strip at a distance of 3 inches from a 1,500 watt ultraviolet lamp (major wavelength 2537 A.) for time periods calculated to give the dosage listed in Table 1. After radiation, the treated surfaces were given an adhesive coating of a 25% solids solution in methylene chloride of a copolyester obtained by condensation of 27 parts by weight of terephthalic acid, 6.6 parts by weight of hexahydrophthalic anhydride and 17.65 parts by weight of sebacic acid with 15.7 parts by weight of cyclohexane dimethanol and 26.1 parts by weight of 1,4-butane diol, the copolyester having a melting point of 120° to 130°C.

The adhesive coating was allowed to dry for about an hour at room temperature. Thereafter, the adhesive coatings were heated to 135° to 140°F. for
about 30 seconds and pairs of strips were assembled in a lap joint to give
1-square inch bonded area and subjected to a 240 lbs. per square inch pressure for 10 seconds. The joined strips were then mounted in the jaws of a
tester which were separated at a rate of 2 inches per minute to determine
the shear strength of the bonds. In Table 1 under "Failure" the term "Interfacial" means that the failure was a separation of the adhesive from the
polyethylene at the interface. This is an indication that the adhesive failed
to establish adequate wetting and strong adhesive engagement with the polyethylene. As shown in the table, the radiation produced a four-fold or
better increase in the strength of the bond and in fact the failure was only
the failure of the material itself.

TABLE 1

Irradiation	Shear Strength of Bond	Failure
None - control	16.0	Interfacial
858 watt-sec./sq. ft.	62.0	Severe stretching of plastic with-out any opening of the joint
2,574 watt-sec./sq. ft.	78.0	Severe stretching of plastic with-out any opening of the joint

Example 2: Strips of high density polyethylene 1 inch wide and 0.060 inch
thickness were dipped in a series of solutions of polyarylene polyisocyanate,
i.e., PAPI in different liquids prior to radiation and then irradiated, ce-
mented and tested as in Example 1. Results are listed in the following table.

TABLE 2

Dipping Medium	Irradiation, watt-sec./sq. ft.	Shear Strength of Bond (psi)	Failure
1% PAPI in methylethyl ketone	858	120.0	Interfacial
1% PAPI in trichloro-ethylene	858	188.0	Stock failure - without any opening of joint
1% PAPI/5% benzo-phenone/methylethyl ketone	858	180.0	Stock failure - without any opening of joint
1% PAPI in trichloro-ethylene	2,574	195.0	Stock failure - without any opening of joint
1% PAPI/5% benzo-phenone/methylethyl ketone	2,574	193.0	Stock failure - without any opening of joint

It can be seen that a markedly superior bond is obtained where trichloro-
methylene or benzophenone was present in the dipping solution over the
type of bond obtained where these materials were not present. This illustrates the importance of the presence of photosensitizer at the time of radiation.

Reactive Aldehyde Resin

In a process described by R.A. Bragole; U.S. Patent 3,627,609; Dec. 14,
1971; assigned to USM Corporation surfaces of substrates which are difficult to bond strongly, e.g., polyethylene, are subjected to ultraviolet
radiation in the presence of a photosensitizer and bonded with an adhesive
comprising an elastomer and a reactive aldehyde resin to form a structurally
strong joint. The preferred aldehyde-type resins are formed by condensation of alkyl or aryl substituted phenols with aldehydes under alkaline conditions and with an excess of aldehyde over the stoichiometric amount required for reaction with the phenol. In general, there may be combined
from slightly more than 1 mol to as high as 2 mols of aldehyde to 1 mol of
the substituted phenol. Formaldehyde is the customary aldehyde used in
forming this resin but other aldehydes may be used.

The phenol may be a lower alkyl substituted phenol such as para-tertiary
butyl phenol or para-tertiary amyl phenol, or an aryl substituted phenol, for
example, para-phenyl phenol. The phenol formaldehyde resin condensate
is preferably then reacted with a metal oxide such as magnesium oxide or
lead oxide. This reaction may be carried out in organic solvent solution
by addition to the solution of an excess of the metal oxide over the amount
which will combine with the resin, the excess being preferably physically
removed as by filtering or decantation thereafter.

The reaction product of the metal oxide and the resin retains its solubility
in volatile organic solvents but has become substantially infusible, i.e., it
does not melt even when heated to temperatures sufficient to initiate decomposition. The elastomer component of the adhesive may be a natural
or synthetic material. For example, natural rubber, polychloroprene,
butadiene acrylonitrile copolymer rubbers and butadiene styrene copolymer
rubbers may be used. The elastomer material and the reacted resin material are combined in amount such that there is at least about 5% of resin
by weight based on the weight of the elastomer and the amount of resin may
be 100% or more by weight based on the weight of the elastomer.

Example 1: Strips of low density polyethylene 1 inch wide, 3 1/2 inches
long and 0.060 inch in thickness were dipped in trichloroethylene and were
subjected to ultraviolet radiation by disposing surfaces of the strips at a
distance of 3 inches from a 1,500 watt ultraviolet lamp (major wavelength

2537 A.) for 15 seconds while trichloroethylene was still present on the polyethylene. After radiation, the treated surfaces were given a coating of the following adhesive.

Component	Parts by Weight
Neoprene	100
Stearic Acid	1
Magnesium Oxide	8
Pentaerythritol Ester of Rosin (Pentalyn K)	55
Acetic Acid (Glacial)	1
Alkali Catalyzed Condensation Resin from para-tertiary Butyl Phenol and Formalde- hyde (Bakelite Resin CKR 5360)	45
Toluene	25
Textile Spirits	300
Methyl Ethyl Ketone	200

The adhesive coating was dried for 1 1/2 hours and the surfaces were then subjected to radiant heat activation at 170°F. for 30 seconds. Strips were assembled with the adhesive coated surface of one against the adhesive coated surface of the other to provide a 1-inch overlap and the assembly was pressed. One day after completion of the bonds, the bonds were tested in shear. At a value of 191 lbs. per sq. inch, stock failure occurred and the joint remained intact. Bonds between strips of polyethylene which had not been radiated but which had been coated with the same adhesive and assembled as in the foregoing test showed failure of the bond by separation of the adhesive from the polyethylene at the interface at a value of only 64 lbs. per sq. inch.

Example 2:   A series of strips of polypropylene one inch wide, 3 1/2 inches long, and 0.060 inch in thickness were dipped in trichloroethylene, radiated, coated with adhesive, activated and assembled as in Example 1. At a value of 252 lbs. per sq. inch, the stock failed and the joint remained intact.

In a control test in which the polypropylene strips had not been dipped in trichloroethylene and radiated, the bond failed at 35 lbs. per sq. inch with interfacial failure, i.e., separation of the adhesive from the poly-propylene at the interface.

Example 3:   Using strips of polyethylene and of polypropylene and treat-ment prior to application of adhesive as listed in the following table, the bond strengths noted were obtained.

TABLE: RESULTS OF BOND STRENGTH TESTS

Radiation at 3" Distance	Pretreatment	Bond Strength, lbs./in.^{2}	Type of Failure
Polyethylene:			
5 seconds	None	72	Interfacial
15 seconds	None	81	Interfacial
5 seconds	Trichlorethylene dip	131	Cohesive
Polypropylene:			
5 seconds	None	58	Interfacial
15 seconds	None	54	Interfacial
5 seconds	Trichloroethylene dip	201	Stock

Mercapto-Substituted Polysiloxanes

In a process described by W.G. Gowdy and J.W. Keil; U.S. Patent 3,632,715; January 4, 1972; assigned to Dow Corning Corporation the surface characteristics of vinylic polymers are altered by the application of an organosilicon compound containing at least one HSR'Si— moiety, where R' is a divalent hydrocarbon radical free of aliphatic unsaturation, followed by application of energy. Alternatively, the organosilicon compound can be added to the molten vinylic polymer and this composition is further processed, for example, by extruding into fibers. Exemplary is an acrylonitrile-butadiene-styrene surface which has been treated with the mercapto-substituted polysiloxane which exhibits improved lubricity.

This process provides a means for making polymer surfaces which are oleophobic, hydrophobic, hydrophilic, receptive to inks and dyes, having high lubricity, or of low lubricity, as desired. For example, considering the vinylic polymer polyethylene: solvent-resistant containers can be made from polyethylene which has been provided with an oleophobic surface.

Polyethylene having an aminofunctional, hydrophilic surface is receptive to permanent printing and dyeing. Additionally, polyethylene sheets can have their surfaces altered so as to have a high coefficient of friction against other treated polyethylene sheets, making the sheets more easily handleable in bulk.

Example 1: A rubber formulation was made up of the following ingredients with amounts shown being in parts by weight: styrene-butadiene rubber stock, 23.8; polybutadiene rubber stock, 16.2; high surface area silica, 20.2; zinc oxide, 1.7; coumarone-indene resin, 3.4; sulfur, 0.8; stearic acid, 0.34; "Altax" rubber additive, 0.4; "Methyl Tuads" additive, 0.07.

A polysiloxane of the formula

$$
\begin{array}{ccc}
\text{CH}_3 & \text{CH}_3 & \text{CH}_3 \\
| & | & | \\
\text{H S CH}_2\text{SiO}(\text{SiO})\!\sim_{24}\!\text{SiCH}_2\text{S H} \\
| & | & | \\
\text{CH}_3 & \text{CH}_3 & \text{CH}_3
\end{array}
$$

having a viscosity of about 50 to 60 cs. at 25°C., was used in the following experiment. The lubricity of the rubber samples was tested by the "Crock test" where a 1/2 inch ball bearing is rubbed back and forth in a 6 inch path at about 60 strokes per minutes, the weight of the ball bearing on the rubber being 1.5 pounds. The time of failure is the time until the rubber begins to crack and abrade.

(a) The above rubber formulation was vulcanized for 45 minutes at 160°C. and 5,000 psi. It was then wiped with isopropyl alcohol. After 5 minutes of the Crock test, the rubber showed severe abrasion.

(b) The above rubber formulation was vulcanized as in (a), and then the polysiloxane was applied. The treated rubber withstood 4 hours on the Crock tester without abrasion. However, similarly treated rubber was cleaned with isopropyl alcohol and tested on the Crock tester. After 5 minutes, there was severe abrasion.

(c) The above rubber formulation was vulcanized as in (a), but the poly-siloxane was used as a mold release agent. The vulcanized rubber was cleaned with isopropyl alcohol and tested on the Crock tester. After 4 hours there was moderate abrasion on the rubber.

(d) The above rubber formulation was vulcanized for 1 hour at 175°C. and 5,000 psi, using the above siloxane as a mold release agent. The vulcanized rubber was cleaned with isopropyl alcohol and treated on the Crock tester. After 4 hours there was no abrasion on the rubber.

(e) The rubbers of (a) and (c) were tested for their physical characteristics with the following results:

Rubber	Durometer	Tensile strength at break point	Elongation percent
(a) _____	80	2,650	530
(c) _____	85	2,880	700

Example 2: The following vulcanized rubber samples were thinly coated with mercapto-siloxanes and treated with ultraviolet light from a standard UV lamp. The irradiated products were washed in isopropyl alcohol and tested by the Crock test of Example 1. The results follow.

Test Results of Mercapto-Siloxane Coated Rubber Samples

Rubber	Siloxane used	Irradiation	Crock test time to failure
(a) Natural	None	None	20 sec.
(b) Natural	CH_3 CH_3 CH_3 $(HS\,CH_2CHCH_2SiO_{1/2})_2(SiO)_n$ CH_3 CH_3	440 cs. at 25° C	5 min. of ultraviolet, 6 in. from U.V. source.	15 min.
(c) Natural	The siloxane of (b)	15 min. of ultraviolet, 6 in. from U.V. source.	4 hrs.
(d) Natural	CH_3 CH_3 $(HS\,CH_2SiO_{1/2})_2(SiO)_n$ CH_3 CH_3	51 cs. at 25° C	15 min. of ultraviolet, 6 in. from U.V. source.	4 hrs.
(e) Styrene-butadiene rubber	None	None	45 sec.
(f) Styrene-butadiene rubber	The siloxane of (b)	15 min. of ultraviolet, 6 in. from U.V. source.	5 min.
(g) Styrene-butadiene rubber	The siloxane of (b)	30 min. of ultraviolet, 6 in. from U.V. source.	30 min.
(h) Ethylene propylene-cyclohexadiene terpolymer.	None	None	30 sec.
(i) Ethylenepropylene cyclohexadiene terpolymer.	The siloxane of (b)	15 min. of ultraviolet, 6 in. from U.V. source.	10 min.
(j) Ethylenepropylene cyclohexadiene terpolymer.	The siloxane of (b)	30 min. of ultraviolet, 6 in. from U.V. source.	>16 hrs.
(k) Ethylenepropylene cyclohexadiene terpolymer.	The siloxane of (d)	30 min. of ultraviolet, 6 in. from U.V. source.	1 hr.
(l) Styrene-butadiene rubber with polybutadiene.	None	None	1 min.
(m) Styrene-butadiene rubber with polybutadiene.	The siloxane of (b)	15 min. of ultraviolet, 3 in. from U.V. source.	5 min.
(n) Styrene-butadiene rubber with polybutadiene.	The siloxane of (b)	30 min. of ultraviolet, 3 in. from U.V. source.	1 min.[1]
(o) Styrene-butadiene rubber with polybutadiene.	The siloxane of (b)	30 min. of ultraviolet, 6 in. from U.V. source.	30 min.
(p) Styrene-butadiene rubber with polybutadiene.	The siloxane of (b)	1 hr. of ultraviolet, 6 in. from U.V. source.	2 min.[1]
(q) Butyl	None	None	20 sec.
(r) Butyl	The siloxane of (d)	30 min. of ultraviolet, 6 in. from U.V. source.	1 min.[2]

[1] This indicates that there has been an overdose of ultraviolet.
[2] The pull needed to remove Scotch tape from the rubber of (q) was 750 gms. The pull needed to remove Scotch tape from the rubber of (r) was about half of this 400 gms. This clearly indicates the presence of a surface lubricant.

Synthetic Hydrocarbon Elastomers with Improved Tack

In a process described by R.E. Tarney and J.J. Verbanc; U.S. Patent 3,657,203; April 18, 1972; assigned to E.I. du Pont de Nemours and Company synthetic hydrocarbon elastomers are tackified by uniformly dispersing in N,N'-disubstituted-p-arylene diamine, a 1,2-dihydro-2,2,4-trialkyl quinoline, 1,4-di-2,4-cyclopentadien-1-yl-butene or a diester of 5-norbornene-2-methanol and a dicarboxylic acid, and exposing the resulting mixture to a greater than ambient concentration of ozone or to ultraviolet light in the presence of oxygen. Optionally, the elastomer can contain selected resins uniformly dispersed which enhance the generation and retention of building tack.

Part II.

Electron Beam Curing

COATINGS

REACTIVE MONOMERS

Divinyl Compounds from Monoepoxides

In a process described by S.S. Labana; U.S. Patent 3,586,528; June 22, 1971; assigned to Ford Motor Company a substrate is coated with a film-forming composition consisting essentially of a divinyl compound and an alpha-beta olefinically unsaturated paint binder resin having a molecular weight in excess of about 1,000, preferably in the range of about 2,000 to about 20,000 and the coating is converted into a tenaciously adhering, solvent-resistant, wear and weather-resistant coating by exposing the coated substrate to ionizing radiation, preferably in the form of an electron beam.

This divinyl compound is formed by first reacting a monoepoxide with acrylic acid and/or methacrylic acid and subsequently reacting the resultant monovinyl ester condensation product with a vinyl unsaturated acyl halide. The monoepoxides employed as starting materials for preparing the divinyl compounds are C_4 to C_{12} monoepoxides. In the preferred case the monoepoxide is a C_7 to C_{10} monocyclic monoepoxide in accordance with the formula

$$R-\overset{H}{\underset{H}{C}}\overset{O}{\overbrace{}}\overset{}{\underset{H}{C}}-H$$

where R is an aryl, alkylaryl, arylalkyl, aryloxy, cycloaliphatic or heterocyclic radical, e.g., phenylglycidyl ether, vinyl cyclohexene epoxide, vinyl cyclopentene expoxide, styrene epoxide, etc. The monoepoxides have a molecular weight in the range of 112 to 151. The vinyl unsaturated acyl halides used are preferably acryloyl chloride and/or methacryloyl chloride.

Example 1: A divinyl compound is prepared in a manner below set forth from the materials hereinafter named: (1) To a reaction vessel equipped with a condenser stirrer, nitrogen inlet and thermometer are charged the following materials.

Materials	Parts by Weight
Vinylcyclohexene epoxide	126
Methacrylic acid	85
Toluene (solvent)	500
Dimethylbenzylamine (catalyst)	2

(2) The vinylcyclohexene epoxide, the methacrylic acid and the dimethy-benzylamine are intimately mixed and incrementally added to the toluene which is at 90°C. in a nitrogen atmosphere. (3) The reaction mixture is maintained at 90°C. until reaction of the epoxide groups is essentially complete as measured by product acid number of less than about 10. (4) The solvent is removed under vacuum.

(5) The reaction product of (4) in the amount of 210 parts by weight is dissolved in 500 parts by weight of toluene and 95 parts by weight of methacryloyl chloride are added dropwise with the reaction mixture maintained at 65°C. until HCl evolution ceases. (6) The solvent is removed under vacuum and the divinyl compound is recovered. An alpha-beta olefinically unsaturated vinyl resin, Resin A, is prepared in the following manner:

Starting Materials	Parts by Weight
Xylene	600
Methyl methacrylate	196
Ethyl acrylate	333
Glycidyl methacrylate	71
Azobisisobutyronitrile	6
Hydroquinone	0.12
Methacrylic acid	42
Triethylamine	0.96

The solvent, xylene, is charged to a flask fitted with a stirring rod, an addition funnel, a thermometer, a nitrogen inlet tube and a condenser. The amount of xylene is equal to the total weight of vinyl monomers to be added. The xylene is heated to reflux, nitrogen is bubbled through the solution during heat up and throughout the reaction. The combined monomers, excepting the methacrylic acid and initiator (azobisisobutyronitrile) is added to the refluxing solution evenly over 2 hours. The initiator weight is 10 parts by weight per 1,000 parts by weight of vinyl monomers. The reaction solution is refluxed until the conversion of monomers to polymers is greater than about 97%. In the second step, hydroquinone is added as an inhibitor and then the methacrylic acid is added to react with the epoxy groups on the polymer.

Triethylamine is used as a catalyst. This esterification reaction is carried out at reflux temperatures until about 80% esterification is accomplished (determined by residual acid number). The xylene is then removed by vacuum distillation and the polymer recovered.

Substrates of wood, glass, metal and polymeric solid, i.e., polypropylene and acrylonitrile-butadiene-styrene copolymer, are coated with a paint binder consisting essentially of this divinyl compound and Resin A using the following procedure: (1) 20 parts by weight of the tetravinyl compound is mixed with 80 parts by weight of Resin A and diluted to spraying viscosity with acetone. This solution is sprayed upon the substrates to an average depth of about 1 mil (0.001 inch) and the solvent flashed off.

The coated substrate is passed through a nitrogen atmosphere and at a distance of about 10 inches below the electron emission window of a cathode ray type, electron accelerator through which an electron beam is projected upon the coated surface until the wet coating is polymerized to a tack-free state. The electrons of this beam have an average energy of about 275,000 electron volts with a current of about 25 milliamperes.

(2) A second group of substrates are coated in the manner above using the same conditions and materials except for the single difference that the paint binder solution used consists of 60 parts by weight of the divinyl compound, 40 parts by weight of Resin A, and the acetone and the coating is applied to a depth of about 3 mils.

(3) A third group of substrates are coated in the manner above using the same conditions and materials except for the single difference that the paint binder solution used consists of 10 parts by weight of the divinyl compound, 90 parts by weight of Resin A, and acetone.

(4) A fourth group of substrates are coated in the manner above using the same conditions and materials except for the single difference that the paint binder solution used consists of 80 parts by weight of the divinyl compound, 20 parts by weight of Resin A, and acetone.

Example 2: The procedure of Example 1 is repeated with the following difference: (1) Resin A is replaced with a polyester resin, Resin B described below, (2) the irradiation atmosphere is helium, and (3) the electron beam used has an average energy of about 350,000 electron volts. The preparation of Resin B is as follows.

Starting Materials	Parts by Weight
Maleic anhydride	14.7
Tetrahydrophthalic anhydride	72.3
Neopentyl glycol	75.0
Dibutyltin oxide, catalyst	7.06

To a reaction vessel, the reactants are charged and heated to about 340°F. and held at this temperature for 1 hour. The temperature of the charge is then raised to 440°F. and maintained at such temperature until the acid number of the resin is below about 20. The excess glycol and water are removed by vacuum and when the acid number is below about 10, there are added 0.03 part by weight hydroquinone.

Example 3:   The procedure of Example 2 is repeated with the only difference that Resin B is replaced with an equal amount by weight of Resin C, a polyester prepared by the procedure used to prepare Resin B except that an equimolar amount of phthalic anhydride is substituted for the tetrahydrophthalic anhydride.

Example 4:   The procedure of Example 2 is repeated with the only difference that Resin B is replaced with an equal amount by weight of Resin D, a polyester prepared by the procedure used to prepare Resin B except that an equimolar amount of ethyleneglycol is substituted for the neopentyl glycol.

Example 5:   The procedure of Example 2 is repeated with the only difference that Resin B is replaced with an equal amount by weight of Resin E, a polyester prepared by the procedure used to prepare Resin B except that an equimolar amount of trimellitic anhydride is substituted for the tetrahydrophthalic anhydride.

Example 6:   The procedure of Example 2 is repeated with the only difference that Resin B is replaced with an equal amount by weight of Resin F, a polyester prepared by the procedure used to prepare Resin R except that an equimolar amount of pentaerythritol is substituted for the neopentyl glycol.

Example 7:   The procedure of Example 2 is repeated with the only difference that Resin B is replaced with an equal amount by weight of Resin G, a polyester prepared by the procedure used to prepare Resin B except that an equimolar amount of 1,6-hexamethylene glycol is substituted for neopentyl glycol.

Example 8:   The procedure of Example 2 is repeated with the only difference that Resin B is replaced with an equal amount by weight of Resin H, a polyester prepared by the procedure used to prepare Resin B except that an equimolar amount of fumaric acid is substituted for the maleic anhydride.

Example 9:   The procedure of Example 2 is repeated with the only difference that Resin B is replaced with an equal amount by weight of Resin I, a polyester prepared by the procedure used to prepare Resin B except that an equimolar amount of 2-butene-1,4-diol is substituted for neopentyl glycol.

Example 10:   The procedures of Examples 1 and 2 are repeated in those embodiments where a nonpolymerizable solvent is employed with the only difference that the acetone is replaced by toluene.

Example 11:   The procedure of Example 1 is repeated with the only difference that Resin A is replaced with a different vinyl monomer comprising resin, Resin J. The preparation of Resin J is as follows.

Starting Materials	Parts by Weight
Methyl methacrylate	400
Ethyl acrylate	400
Hydroxyethyl methacrylate	195
Toluene	1,000
Benzoyl peroxide	30

The procedure for Step 1 is as follows. The benzoyl peroxide is dissolved in a solution of methyl methacrylate, ethyl acrylate and hydroxyethyl methacrylate and one-half of the toluene. This solution is added incrementally to the remainder of the toluene at reflux over a 7 hour period with a final part temperature of about 138° to 140°C. Reflux is maintained for another 3 hours and the solution cooled. Step 2 is as follows.

500 parts by weight of the solution from Step 1 was added to 33.8 parts by weight of acryloyl chloride and 30 parts by weight of toluene. The solution of Step 1 is heated to 60°C. in a solution of the acryloyl chloride and toluene is added dropwise over a 4 hour period while the temperature is allowed to rise to about 90°C. After heating for another 2.5 hours the polymer is recovered by vacuum distillation. Additional studies with monoepoxides to prepare divinyl compounds are described by S. Labana; U.S. Patent 3,595,687; July 27, 1971; assigned to Ford Motor Company.

Divinyl Compounds from Diepoxides

In a process described by E.J. Aronoff, S.S. Labana and E.O. McLaughlin;

U.S. Patent 3,586,529; June 22, 1971; assigned to Ford Motor Company a substrate is coated with a film-forming composition comprising a divinyl compound and the coating is converted to a tenaciously adhering, solvent-resistant, wear and weather-resistant film by exposure to ionizing radiation, preferably in the form of an electron beam. This divinyl compound is formed by first reacting a diepoxide with acrylic and/or methacrylic acid and then reacting the ester condensation product with a saturated acyl halide.

FIGURE 5.1: REACTION SEQUENCE FOR PREPARING DIVINYL COMPOUND

a.

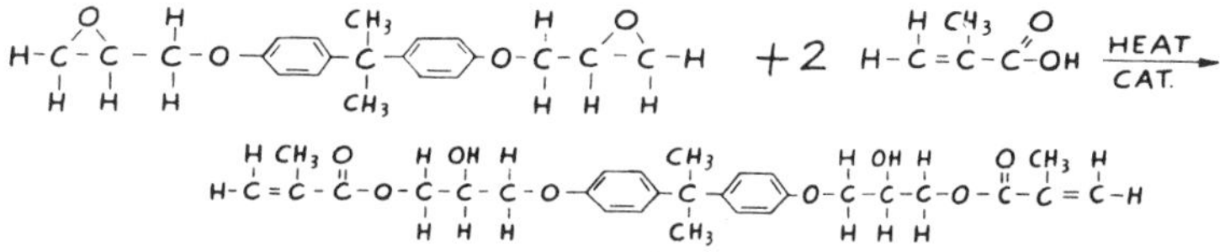

(Reaction scheme showing the diepoxide reacting with 2 molecules of methacrylic acid, HEAT/CAT.)

Reaction Step 1

b.

(Reaction scheme showing the product of step 1 reacting with 2 Cl–C(O)–R, HEAT/CAT., yielding product + 2 HCl)

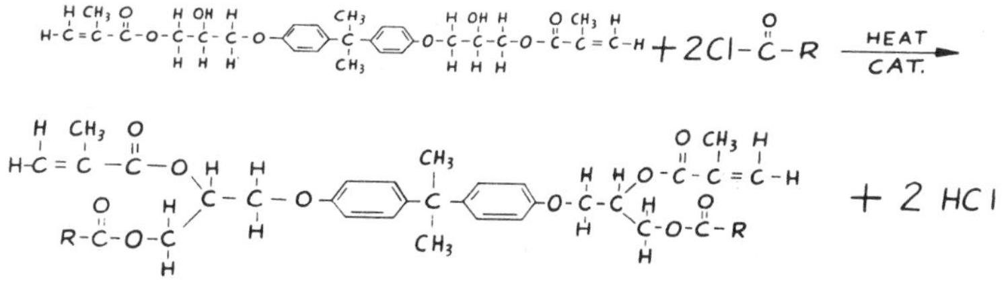

Reaction Step 2

Source: E.J. Aronoff, S.S. Labana and E.O. McLaughlin; U.S. Patent 3,586,529; June 22, 1971

The first reaction step in preparing the divinyl compounds is illustrated by the representative reaction shown in Figure 5.1a. The second reaction step is illustrated by the representative reaction shown in Figure 5.1b. The following examples illustrate the process.

Example 1: A divinyl compound is prepared in the following manner from the material indicated. To the reaction vessel equipped with condenser, stirrer, nitrogen inlet and thermometer are charged the following materials: 192 parts by weight of diepoxide (as shown by the formula on the next page); 86 parts by weight of methacrylic acid; 500 parts by weight of toluene (solvent) and 1 part by weight of dimethylbenzylamine (catalyst).

The diepoxide, the methacrylic acid and the dimethyl benzylamine are intimately mixed and incrementally added to the toluene which is at 90°C. in a nitrogen atmosphere. The reaction mixture is maintained at 90°C. until reaction of the epoxide groups is essentially complete as measured by product acid number of less than about 10. The solvent is removed under vacuum and a solid reaction product (softening point 45°C.) is recovered.

The solid reaction product as above in the amount of 280 parts by weight is dissolved in 500 parts by weight toluene, and 95 parts by weight of butyric acid chloride are added dropwise with the reaction mixture maintained at 65°C. until HCl evolution ceases. The solvent is removed under vacuum and a divinyl compound is recovered.

Substrates of wood, glass, metal and polymeric solid, i.e., polypropylene and acrylonitrile-butadiene-styrene copolymer are coated with this divinyl compound using the following procedure: The above prepared divinyl compound is diluted to spraying viscosity with xylene and the paint film is sprayed on the substrates to an average depth of about 1 mil (0.001 inch) and the solvent flashed off.

The coated substrate is passed through a nitrogen atmosphere and at a distance of about 10 inches below the electron emission window of a cathode ray type, electron accelerator through which an electron beam is projected upon the coated surface until the wet coating is polymerized to a tack-free state. The electrons of this beam have an average energy of about 275,000 electron volts with a current of about 25 milliamperes.

Example 2: The procedure of Example 1 is repeated except that the diepoxide employed is 3,4-epoxy-6-methyl cyclohexyl-methyl-3,4-epoxy-methyl cyclohexanecarboxylate and the paint film is applied to an average depth of about 3 mils.

Example 3: The procedure of Example 1 is repeated except that the diepoxide employed is 1-epoxyethyl-3,4-epoxy cyclohexane.

Example 4: The procedure of Example 1 is repeated except that the diepoxide employed is dipentene dioxide.

Example 5:   The procedure of Example 1 is repeated except that the di-epoxide employed is dicyclopentadiene dioxide.

Example 6:   The procedure of Example 1 is repeated with the only difference that the electrons of the electron beam have an average energy of about 350,000 electron volts.

Example 7:   The procedure of Example 1 is repeated with the only difference that the atmosphere of irradiation is helium.

Example 8:   A divinyl compound is prepared as in Example 1 and a different divinyl compound is prepared using the same procedure with the single exception that cinnamic acid chloride is substituted for the second step reactant butyric acid chloride. Substrates are then coated as in Example 1 using a paint binder composition consisting of 51 parts by weight of the divinyl compound prepared with butyric acid chloride, 49 parts by weight of the divinyl compound prepared with cinnamic acid chloride, and toluene in an amount sufficient to provide the composition with a good spraying viscosity.

Substrates are coated with this composition and toluene flashed off prior to irradiation. Irradiation conditions are the same as in Example 1. Additional substrates are coated in like manner except that the paint binder composition consists of 99 parts by weight of the divinyl compound prepared with butyric acid chloride, one part by weight of the divinyl compound prepared with cinnamic acid chloride and toluene.

Additional substrates are coated in like manner except that the paint binder composition consists of 75 parts by weight of the divinyl compound prepared with butyric acid chloride, 25 parts by weight of the divinyl compound, prepared with cinnamic acid chloride and toluene.

Additional polymerization studies with the divinyl compounds and other vinyl-containing monomers or unsaturated resins are described by E.J. Aronoff, S.S. Labana and E.O. McLaughlin; U.S. Patent 3,586,531 and U.S. Patent 3,586,530; June 22, 1971; assigned to Ford Motor Company.

Tetravinyl-Unsaturated Resin Compositions

In a process described by E.J. Aronoff and S.S. Labana; U.S. Patent 3,586,527; June 22, 1971; assigned to Ford Motor Company a substrate is coated with a film-forming composition consisting essentially of a tetravinyl compound having a molecular weight below about 350, preferably in the range of about 220 to about 1,100, and an alpha-beta olefinically unsaturated paint binder resin having a molecular weight in excess of about

1,000, preferably in the range of about 2,000 to 20,000, is converted into a tenaciously adhering, solvent-resistant, wear and weather-resistant coating by exposing the coated substrate to ionizing radiation, preferably in the form of an electron beam. This tetravinyl compound is formed by first reacting a diepoxide with acrylic acid and/or methacrylic acid and subsequently reacting the resultant ester condensation product with a vinyl unsaturated acyl halide.

FIGURE 5.2:   REACTION SEQUENCE FOR PREPARING TETRAVINYL COMPOUND

a.

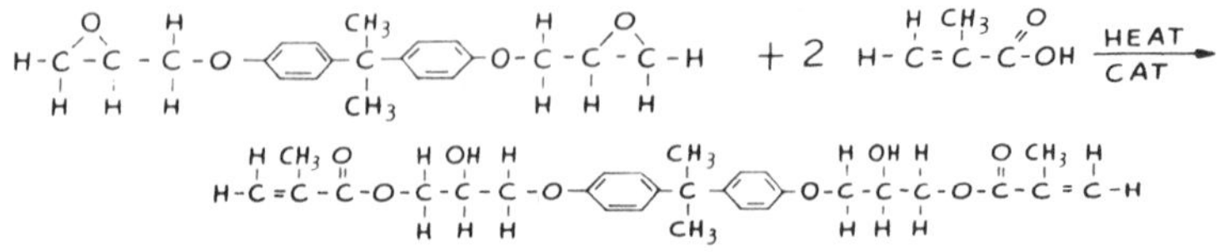

Reaction Step 1

b.

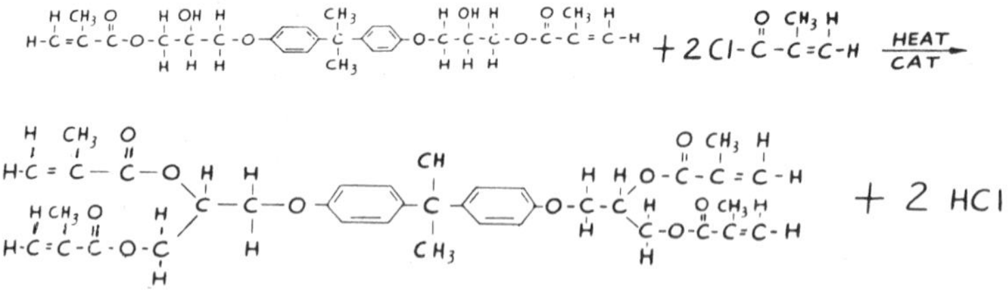

Reaction Step 2

Source: E.J. Aronoff and S.S. Labana; U.S. Patent 3,586,527; June 22, 1971

The first reaction step in preparing the tetravinyl compounds is illustrated by the representative reaction shown in Figure 5.2a. The second reaction step is illustrated by the representative reaction shown in Figure 5.2b. The following examples illustrate the process.

Example 1:   A tetravinyl compound is prepared in the following manner from the materials named. To a reaction vessel equipped with condenser, stirrer, nitrogen inlet and thermometer are charged the following materials:

	Parts by Weight
Diepoxide (as shown in the formula below)	192
Methacrylic acid	86
Toluene (solvent)	500
Dimethyl benzylamine (catalyst)]	1

The diepoxide, the methacrylic acid and the diemthyl benzylamine are
intimately mixed and incrementally added to the toluene which is at 90°C.
in a nitrogen atmosphere. The reaction mixture is maintained at 90°C.
until reaction of the epoxide groups is essentially complete as measured by
a product acid number of less than about 10. The solvent is removed under
vacuum and a solid reaction product (softening point 45°C.) is recovered.

The solid reaction product of above in the amount of 280 parts by weight is
dissolved in 500 parts by weight toluene; 110 parts by weight of methacryloyl chloride are added dropwise with the reaction mixture maintained at
65°C. until HCl evolution ceases. The solvent is removed under vacuum
and a tetravinyl compound is recovered in the form of a viscous liquid.
An alpha-beta olefinically unsaturated vinyl resin, Resin A, is prepared
in the following manner:

Starting Materials	Parts by Weight
Xylene	600
Methyl methacrylate	196
Ethyl acrylate	333
Glycidyl methacrylate	71
Azobisisobutyronitrile	6
Hydroquinone	0.12
Methacrylic acid	42
Triethylamine	0.96

The solvent, xylene, is charged to a flask fitted with a stirring rod, an
additional funnel, a thermometer, a nitrogen inlet tube and a condenser.
The amount of xylene is equal to the total weight of vinyl monomers to be
added. The xylene is heated to reflux, nitrogen is bubbled through the

solution during heat up and throughout the reaction. The combined monomers, excepting the methacrylic acid, and initiator (azobisisobutyronitrile) is added to the refluxing solution evenly over a 2 hour period. The initiator weight is 10 parts by weight per 1,000 parts by weight of vinyl monomers. The reaction solution is refluxed until the conversion of monomer to polymer is greater than about 97%.

In the second step, hydroquinone is added as an inhibitor and then the methacrylic acid is added to react with the epoxy groups on the polymer. Triethylamine is used as a catalyst. This esterification reaction is carried out at reflux temperatures until about 80% esterification is accomplished (determined by residual acid number). The xylene is then removed by vacuum distillation and the polymer dissolved in methyl methacrylate so that the weight ratio of polymer to solvent is two.

Substrates of wood, metal and polymeric solid, i.e., polypropylene and acrylonitrile-butadiene-styrene copolymer, are coated with a paint binder consisting essentially of this tetravinyl compound and Resin A using the following procedure: 20 parts by weight of the tetravinyl compound is mixed with 80 parts by weight of Resin A and diluted to spraying viscosity with acetone. This solution is sprayed upon the substrates to an average depth of about 1 mil (0.001 inch) and the solvent flashed off.

The coated substrate is passed through a nitrogen atmosphere and at a distance of about 10 inches below the electron emission window of a cathode ray type, electron accelerator through which an electron beam is projected upon the coated surface until the wet coating is polymerized to a tack-free state. The electrons of this beam have an average energy of about 275,000 volts with a current of about 25 milliamperes.

A second group of substrates are coated in the manner above using the same conditions and materials except for the single difference that the paint binder solution used consists of 60 parts by weight of the tetravinyl compound, 40 parts by weight of Resin A, and acetone.

A third group of substrates are coated in the manner above using the same conditions and materials except for the single difference that the paint binder solution used consists of 10 parts by weight of the tetravinyl compound, 90 parts by weight of Resin A, and acetone.

A fourth group of substrates are coated in the manner above using the same conditions and materials except for the single difference that the paint binder solution used consists of 80 parts by weight of the tetravinyl compound, 20 parts by weight of Resin A, and acetone.

Example 2: The procedure of Example 1 is repeated with the following differences: (1) Resin A is replaced with a polyester resin, Resin B described below, (2) the irradiation atmosphere is helium, and (3) the electron beam used has an average energy of about 350,000 electron volts. The preparation of Resin B is as follows.

Starting Materials	Parts by Weight
Maleic anhydride	14.7
Tetrahydrophthalic anhydride	72.3
Neopentyl glycol	75.0
Dibutyltin oxide, catalyst	7.06

To a reaction vessel, the reactants are charged and then heated to about 340°F. and held at this temperature for 1 hour. The temperature of the charge is then raised to 440°F. and maintained at such temperature until the acid number of the resulting resin is below about 20. The excess glycol and water are removed by vacuum and when the acid number is below about 10, there are added 0.03 part by weight hydroquinone.

Example 3: The procedure of Example 2 is repeated with the only difference that Resin B is replaced with an equal amount by weight of Resin C, a polyester prepared by the procedure used to prepare Resin B except that an equimolar amount of phthalic anhydride is substituted for the tetrahydrophthalic anhydride.

Example 4: The procedure of Example 2 is repeated with the only difference that Resin B is replaced with an equal amount by weight of Resin D, a polyester prepared by the procedure used to prepare Resin B except that an equimolar amount of ethylene glycol is substituted for the neopentyl glycol.

Example 5: The procedure of Example 2 is repeated with the only difference that Resin B is replaced with an equal amount by weight of Resin E, a polyester prepared by the procedure used to prepare Resin B except that an equimolar amount of trimellitic anhydride is substituted for the tetrahydrophthalic anhydride.

Example 6: The procedure of Example 2 is repeated with the only difference that Resin B is replaced with an equal amount by weight of Resin F, a polyester prepared by the procedure used to prepare Resin B except that an equimolar amount of pentaerythritol is substituted for the tetrahydrophthalic anhydride.

Example 7: The procedure of Example 2 is repeated with the only difference that Resin B is replaced with an equal amount by weight of Resin G, a polyester prepared like Resin B except that an equimolar amount of 1,6-hexamethylene glycol is substituted for the neopentyl glycol.

Other studies with tetravinyl paint compositions are described by E. Aronoff and S.S. Labana; U.S. Patent 3,679,447; July 25, 1972 and U.S. Patent 3,586,526; June 22, 1971; assigned to Ford Motor Company.

Acryloxypivalyl Acryloxypivalate

In a process described by R. Dowbenko and R.M. Christenson; U.S. Patent 3,645,984; February 29, 1972; assigned to PPG Industries, Inc. an acrylic monomer is prepared by reacting a diol such as

$$\underset{\underset{CH_3}{\mid}}{\overset{\overset{CH_3}{\mid}}{HOCH_2\,C}}CH_2OCOC\underset{\underset{CH_3}{\mid}}{\overset{\overset{CH_3}{\mid}}{C}}CH_2OH$$

with an acrylic or methacrylic acid or their anhydrides or acid chlorides. The reaction product is a monomer which may be polymerized by subjecting it to ionizing irradiation, actinic light, or to free-radical catalysts, and the resulting polymer is a hard, mar-resistant material. The compounds produced in accordance with this process are acrylic monomers having the formula

$$CH_2=C-COOC-(CH_2)_k\left(\begin{array}{c}R_1\\\mid\\C\\\mid\\R_2\end{array}\right)_l-(CH_2)_m-O-CO(CH_2)_n-\left(\begin{array}{c}R_3\\\mid\\C\\\mid\\R_4\end{array}\right)_q-(CH_2)_p-COOC-C=CH_2$$

where R_1, R_2, R_3, and R_4 are selected from the group consisting essentially of H, alkyl, aryl, and cycloalkyl, substituted alkyl, substituted aryl and substituted cycloalkyl groups. The radicals R_5, R_6, R_7 and R_8 are selected from the group consisting of H, alkyl, aryl, and cycloalkyl.

The radicals R_9 and R_{10} are selected from the group consisting of H, alkyl groups containing from 1 to 2 carbon atoms, halo-substituted alkyl groups containing from 1 to 2 carbon atoms, and halogen. Examples are methyl, ethyl, bromo-ethyl, and chlorine. k, l, m, n, q and p are whole numbers having values from 0 to 5.

The most preferable compound having this structure is acryloxypivalyl acryloxypivalate, in which R_1, R_2, R_3, R_4 are CH_3 radicals and R_5, R_6, R_7, R_8, R_9, and R_{10} are H and m, l, and q are 1 and k, n and p are 0. The following examples illustrate the process.

Example 1: An acrylic monomer was prepared as follows: A reactor was
charged with 7,140 g. of hydroxypivalyl hydroxypivalate having the
formula

$$HOCH_2\overset{\overset{\displaystyle CH_3}{|}}{C}CH_2OCO-\overset{\overset{\displaystyle CH_3}{|}}{C}-CH_2OH$$
$$\underset{CH_3}{}\qquad\underset{CH_3}{}$$

1,190 g. of cyclohexane, 1,324 g. of acrylic acid, 62.2 g. of sulfuric
acid, and 124.4 g. of hydroquinone. The reactants were heated to reflux
at 200°F. and 3,972 g. of acrylic acid were added dropwise over a period
of 30 minutes at 208°F. The reaction was run for an additional 4 hours
during which time 2,730 g. of cyclohexane were added and 1,235 g. of
water were distilled off. The final product, acryloxypivalyl acryloxypivalate, was obtained in 90% yields after purification by washing.

Example 2: A radiation-sensitive monomer was prepared as follows: A reaction vessel was charged with 204 g. of hydroxypivalyl hydroxypivalate,
47.4 g. of methacrylic acid, 3.9 g. of hydroquinone, 1.9 g. of sulfuric
acid, and 50 g. of cyclohexane. The reactants were heated to reflux at
100°C., and 142.1 g. of methacrylic acid were added dropwise over a
period of 35 minutes. The reaction was run for an additional 4 1/2 hours
during which time 100 cc of cyclohexane were added and 28.6 g. of water
were distilled out. The final product, after purification, was methacryl-
oxypivalyl methacryloxypivalate.

Example 3: A radiation-sensitive monomer was prepared as follows: A reaction vessel was charged with 204 g. of hydroxypivalyl hydroxypivalate,
19.8 g. of acrylic acid, 23.7 g. of methacrylic acid, 3.8 g. of hydroquinone, 1.9 g. of sulfuric acid, and 50 g. of cyclohexane. The reactants
were heated to 99°C. and 59.5 g. of acrylic acid and 71.1 g. of methacrylic acid were added dropwise over a period of 30 minutes. The reaction
was continued for 5 hours during which time 85 cc of cyclohexane were
added and 34.5 g. of water were collected.The resulting acrylate–methacrylate was isolated and purified in a manner similar to the monomers
described in Examples 1 and 2.

Example 4: Acryloxypivalyl acryloxypivalate formed by the method of
Example 1 was cured by the following method: A steel plate was covered
with a composition comprising 100 parts of the acryloxypivalyl acryloxypivalate of Example 1 (after the hydroquinone had been removed) and 1 part
of cumene hydroperoxide. The composition was heated in a nitrogen atmosphere at 170°F. for 30 minutes. The resulting cured product was a hard,
mar-resistant and stain-resistant film.

Example 5:   The acrylic monomers of Examples 1, 2 and 3 were cured by
subjecting them to ionizing irradiation in the following manner: The monomers were applied to aluminum panels and subjected to electron beam
impingement at an accelerating potential of 400 kilovolts and a tube current of 16 milliamps. The films received a total dosage of 4 megarads. The
cured films were found to have excellent mar-resistance, were extremely
hard and were resistant to staining by ink, thimerosal, and mustard.

Example 6:   A mixture of 75 parts of the acryloxypivalyl acryloxypivalate
prepared as in Example 1 and 25 parts of 2-ethyl-hexyl acrylate were copolymerized by subjecting the mixture to electron beam impingement at an
accelerating potential of 400 kilovolts and a tube current of 16 milliamps.
The total dosage received by the mixture was 4 megarads. The resulting
copolymer was hard, mar-resistant and was only very slightly stained by
ink, thimerosal, and mustard.

In related work R. Dowbenko and R.M. Christenson; U.S. Patent 3,647,737;
March 7, 1972; assigned to PPG Industries, Inc. have found that when the
diols and acrylic acid are coreacted with a monofunctional aliphatic, alicyclic, or aromatic acid, then the resulting monomers upon being polymerized, form polymers having the outstanding properties of the acrylic polymers described above in U.S. Patent 3,645,984 and are more flexible.

Acryloxy Esters of Anhydrides

A process described by G.M. Parker and R.C. Heuser; U.S. Patent
3,690,927; September 12, 1972; assigned to PPG Industries, Inc. involves
acryloxy (or methacryloxy)-alkyl (or alkyloxyalkyl) organic dicarboxylic
esters prepared by reacting an organic anhydride with a hydroxyalkyl acrylate (or methacrylate) and then reacting this half-ester with a terminally-
saturated glycidyl ether or ester. These compounds are useful in coating
compositions and are particualrly useful in coating compositions curable
by radiation.

The compounds of this process include 3-acetyloxy-2-hydroxypropyl 2-acryl-
oxyethyl phthalate; 3-butyryloxy-2-hydroxypropyl 3-acryloxypropyl tetra-
hydrophthalate; 3-ethoxy-2-hydroxypropyl 2-methacryloxyethyl hexahydro-
phthalate; 3-butoxy-2-hydroxypropyl 3-methacryloxypropyl hexachloro-
phthalate; 3-hexanoyloxy-2-hydroxypropyl 2-(2-acryloxyethoxy)-ethyl
chlorendate; 3-hexyloxy-2-hydroxypropyl 3-(3-acryloxypropoxy-propoxy)
propyl maleate; 3-benzoyloxy-2-hydroxypropyl 2-acryloxyethyl chloro-
maleate and 3-decanoyloxy(caprinoyloxy)-2-hydroxypropyl 3-acryloxy-
propyl citraconate. In the following examples, parts given are by weight.

Example 1:   In a reactor were charged 50.9 parts of 2-acryloxy-ethyl hydrogen phthalate, 0.1 part hydroquinone and 0.1 part m-methyl morpholine. The mixture was heated to 110°C. and there was added over a period of 15 minutes 72.3 parts of glycidyl stearate, maintaining a temperature between 90° and 110°C. After the addition was complete, the reaction mixture was held between 115° to 120°C. for 3 hours. The final reaction mixture contained 2-acryloxyethyl 3-stearoyloxy 2-hydroxypropyl phthalate.

Example 2: Into a reactor were charged 75 parts of 2-acryloxyethyl hydrogen phthalate, 0.15 part of hydroquinone and 0.15 part of triethylamine. This mixture was heated to 110°C. and there was added incrementally over a period of 20 minutes, 75.1 parts of a mixture of C_8 and C_{10}, normal alkyl monoglycidyl ethers. (Epoxide No. 7.) After addition was complete, the reaction mixture was held at 115°C. for 9 hours. The product contained a mixture of 3-octyloxy and decyloxy 2-hydroxypropyl 2-acryloxyethyl phthalate.

Example 3: Into a reactor was charged 185 parts of 2-acyloxyethyl hydrogen phthalate, 0.4 part of hydroquinone and 0.4 part of triethylamine. This mixture was heated to 115°C. and there was then added incrementally 207.5 parts of a C_{15} monoglycidyl ether. (Epoxide No. 8.) The addition was made over a 1 hour period. The reaction mixture was then maintained at 115°C. for 17 hours. The final reaction mixture contained 3-pentadecyloxy-2-hydroxypropyl 2-acryloxyethyl phthalate.

Example 4: Into a reactor were charged 245 parts Cardura E (a C_8 to C_{10} coke acid glycidyl ester); 264 parts of 2-acryloxyethyl hydrogen phthalate, 0.5 part of hydroquinone and 0.5 part of n-methyl morpholine. The reaction mixture was heated at 115° to 120°C. for 3 hours. The final mixture had an acid value of 1.8 and contained no measurable epoxy groups. The reaction mixture contained the corresponding phthalic diester.

Example 5: A composition made up of 47.5 parts 3-acryloxy-2-hydroxypropyl 2-acryloxyethyl phthalate; 5.0 parts butyl acrylate and 47.5 parts of the compound of Example 4 (all parts by weight) was cured under an electron beam in a nitrogen atmosphere at 400 kv. with a total dose of 2.3 megarads. The coating was cured and was solvent resistant and had good mar-resistance. The same composition was drawn down on a panel and heated for 5 minutes at 300°F., then electron cured under the same conditions. This panel had a direct and a reverse impact resistance of 50 foot pounds.

Example 6: The following compositions were electron cured at 400 kv. under nitrogen with a total dose of 2.6 megarads. All of the following samples were diluted with 10% butyl acrylate.

Parts by Weight		
Acryloxy Ester*	Expoxide No. 8	Properties
95	5	Hard, mar-resistant film
90	10	Hard, mar-resistant film
80	20	Hard, mar-resistant film
70	30	Hard, mar-resistant film
60	40	Hard, mar-resistant film
50	50	Soft, readily marred film
0	100	Tacky, adherent film

*2-acryloxyethyl 3-acryloxy-2-hydroxypropyl phthalate

Bis(Acryloxyalkyl) Carbonates

In a process described by G.M. Parker; U.S. Patent 3,619,260; November 9, 1971; assigned to PPG Industries, Inc. phosgene is reacted with acrylates or methacrylates containing one hydroxyl group in the presence of an acid acceptor to form bis(acryloxyalkyl) carbonates. The product is highly radiation sensitive so that it may be polymerized by ionizing radiation and forms a coating which is hard and stain resistant. The following examples illustrate the process.

Example 1: A flask fitted with a dry ice condenser was cooled to 0°C. and to the flask were added 232 g. of 2-hydroxyethyl acrylate, 150 g. of trimethylamine, and 250 ml. of benzene. Phosgene gas was bubbled through the mixture at the rate of approximately 2 g. per minute. At the end of 46 minutes, the temperature of the reactants was 8°C. and 99 g. of phosgene had been added.

The product was filtered and washed with 3% hydrochloric acid solution and rewashed with water, and then washed with dilute, aqueous sodium hydroxide and finally washed with water. The pH was 7.0. The product was stripped and the resulting bis(acryloxyethyl) carbonate monomer (a water white liquid of low viscosity) had a hydroxy value of 0. The yield was 39%.

Example 2: A flask was charged with 226 g. of methacrylic acid and 1.2 g. of hydroquinone and heated to 110°C. To this mixture were added 1.8 g. of N-methylomorpholine and 355 g. of glycidyl methacrylate and the reactants were heated to 110°C. The reactants were heated until an acid number of 20 was reached. The resulting glycerol 1,3-dimethacrylate had a hydroxyl value of 254. A flask was charged with 222 g. of the above prepared diacrylate, 100 g. of pyridine and 400 ml. of benzene and brought to a temperature of -8°C. Phosgene gas was bubbled through the mixture at a rate of approximately 1.5 g. per minute. At the end of 32 minutes, the temperature of the reaction was -5°C. and 55 g. of phosgene had been added. The product was washed and stripped. The resulting tetramethacrylate had an acid number of 1.40, a hydroxyl value of 13.91, and a Gardner-Holdt viscosity of K-L.

Example 3: A 3-mil thick coating of the product of Example 1 was applied to an aluminum panel by the drawn-down method. The coated panel was then subjected to electron beam impingement in nitrogen atmosphere at an accelerating potential of 400 kv. and a tube current of 16 milliamps. The total dose given the coating was 2.5 megarads.

The coating was a hard film which was tested for stain and solvent resistance by staining for 4 hours with ink, mustard and thimerosal. The coating was found to be completely resistant to these materials. The same procedure was used to coat substrates of wood, Lexan, vinyl plastic, and Plexiglas, at dosages of 0.4 megarad to 5.5 megarads. All of the resulting coatings were hard and had excellent adhesion and were stain and solvent resistant.

Example 4: To 5 g. of the composition of Example 2 was added 0.05 g. of diacetyl. A 1-mil thick coating was drawn down on an aluminum panel and cured under ultraviolet light. The curing of the film was carried out in a nitrogen atmosphere and exposed to the ultraviolet light for 3 minutes using a 450 watt medium pressure mercury lamp. The coating cured to a hard, tack-free film.

Example 5: A 3-mil thick coating of the composition of Example 2 was applied to an aluminum panel by the drawn-down method. The coated panel was then subjected to electron beam impingement in a nitrogen atmosphere at an accelerating potential of 400 kvs. and a tube current of 14 milliamps. The total dose given the coating was 2.3 megarads. The coating cured to a hard, stain-resistant and mar-resistant film.

Diesters of N-Acryliminodiacetic Acids

In a process described by D.E. Jefferson and N.S. Marans; U.S. Patent 3,598,792; August 10, 1971; assigned to W.R. Grace & Co., polymers and copolymers of diesters of N-acrylyliminodiacetic acids are prepared by irradiating such diesters or mixtures with methyl acrylate, methyl methacrylate, a vinylpyridine, N-vinylpyrrolidone, styrene, acrylonitrile, or vinyl acetate.

These polymers can be hydrolyzed (or saponified and acidified) to convert them from the ester form to the carboxylic form. A preferred form of the process is directed to preparing the polymer described above by irradiating a monomeric diester having the formula as shown on the following page.

$$CH_2=C-\overset{\overset{\displaystyle O}{\|}}{C}-N\overset{\displaystyle CH_2COOR'}{\underset{\displaystyle CH_2COOR'}{\diagup}}$$

R is H or lower alkyl and R' is lower alkyl (i.e., about 1 to 7 carbon atoms). The diester is irradiated while maintaining the temperature within the range of about 0° to 100°C. (preferably about 20° to 80°C.), in an inert atmosphere with about 0.1 to 10 megarads (preferably about 0.5 to 5 megarads) of high energy ionizing radiation (e.g., a high energy electron beam), maintaining the irradiated diester at about 30° to 100°C. (preferably about 45° to 75°C.) for about 10 to 300 minutes (preferably 60 to 150 minutes), and recovering the resulting polymer.

In an alternative and substantially fully equivalent procedure the radiation (preferably using a total dose of about 0.01 to 10 megarads, more preferably about 0.1 to 5.0 megarads) can be conducted over a period of about 10 to 600 minutes while maintaining the monomer at an elevated temperature, e.g., about 30° to 120°C. (preferably about 50° to 100°C.). When this alternative procedure is used polymerization of the diester occurs during the irradiation and the subsequent step of maintaining the irradiated diester at about 30° to 100°C. can be eliminated. The process is shown as follows.

Example 1: A solution of 0.25 mol (40.29 g.) of dimethyliminodiacetate in 200 ml. of diethyl ether (ether) was added dropwise to a solution of acrylyl chloride (0.125 mol, 11.31 g.) in 100 ml. of ether. The temperature of the solution was adjusted to about 0° to 25°C. and the resulting product mixture was cooled as the dimethyliminodiacetate solution was added dropwise to maintain the temperature of the product mixture within the range of about 10° to 30°C.

A white crystalline precipitate formed immediately upon the addition of the acrylyl chloride. The chloride was added over a period of about 30 minutes. The solid by-product (dimethyliminodiacetate hydrochloride) was filtered from the liquid phase of the product mixture, and the filtrate was concentrated under vacuum (i.e., volatile constituents were evaporated therefrom) to yield a solid material which was recovered.

The solid material was crystallized from ether using conventional techniques. The crystallized solid (MP, 60° to 61°C.) was recovered and analyzed. The solid (which was labeled Monomer No. 1) was identified by its infrared spectrum, by NMR (nuclear magnetic resonance), and by functional group analysis as the dimethyl ester of N-acrylyliminodiacetate. Conversion (1 pass yield) based on the weight of the recrystallized material was 56% theory.

Example 2: The general procedure of Example 1 was repeated. However, in this instance the acrylyl chloride was replaced with methacrylyl chloride using 0.125 mol (13.1 g.) in 100 ml. of diethyl ether. Also, in this instance after evaporating the diethyl ether, the residue (crude product) was purified by distilling under vacuum and collecting the fraction boiling between about 117° and 120°C. at 0.75 mm. of mercury absolute pressure.

The distilled product (obtained in a conversion of 85% of theory) was Monomer No. 2 and identified as the dimethyl ester of N-methacrylyliminodiacetic acid by its infrared spectrum and by NMR (nuclear magnetic resonance). Similar results have been obtained where using ethyl, propyl and other esters of iminodiacetic acid and where using substituted acrylyl halides having the formula

$$CH_2=C-\overset{\overset{\displaystyle O}{\|}}{C}-X$$
$$\underset{\displaystyle R}{|}$$

where X is Cl or Br and R is an ethyl or butyl or other lower alkyl group.

Example 3: A sample of Monomer No. 1 described in Example 1 was placed in a tube and evacuated to a pressure of about 0.3 mm. of mercury absolute. The tube was sealed while maintaining the vacuum therein. This sample was irradiated with high energy electrons from a 2 mev Van de Graaff electron accelerator at 25°C. at a rate of 1 megarad per pass for 4 passes.

No visual evidence of polymerization could be detected. The irradiated tube was then placed in a water bath at about 50° to 55°C. for about 2 hours. The tube was then opened and the material in it was removed. Upon examination and analysis it was found that substantially all of the monomer had polymerized to form a polymer consisting of about 200 repeating units having the formula

$$\left[\begin{array}{c} \overset{\displaystyle H}{|} \\ -CH_2-C- \\ \underset{\displaystyle |}{} \\ C=O \\ \underset{\displaystyle |}{} \nearrow CH_2COOCH_3 \\ N \\ \searrow CH_2COOCH_3 \end{array}\right]$$

Substantially identical results were obtained with samples irradiated at 2 and 3 megarads per pass for 1 pass and for 4 passes, and aged at about 50° to 55°C. subsequent to irradiation. Similar results, except that the methyl groups of the diester moiety of the polymer were replaced with other lower alkyl groups, were obtained when Monomer No. 1 was replaced with monomers having the formula

$$CH_2=CH-\overset{\overset{\displaystyle O}{\|}}{C}-N\overset{\nearrow CH_2COOR'}{\searrow CH_2COOR'}$$

where R' was ethyl, n-propyl, isopropyl, n-butyl, isobutyl and other lower alkyl groups.

Example 4:   The general procedure of Example 3 was repeated. However, in this instance Monomer No. 1 was replaced with Monomer No. 2 (described in Example 2) and the monomer, in a sealed evacuated tube, was irradiated with high energy electrons at 2 megarads per pass for 5 passes. Subsequent to irradiation the tube was placed in a water bath at about 70°C. for about 100 minutes. Then the tube was removed from the bath, opened, and its contents recovered. Unreacted monomer was separated from the polymer product by vacuum distillation, and the product was recovered. Examination and analysis of the polymer product established that it was a polymer consisting of about 2,000 repeating units with the formula

$$
\left[-CH_2-\underset{\underset{\underset{\underset{CH_2COOCH_3}{\displaystyle N-CH_2COOCH_3}}{\displaystyle |}}{\overset{\overset{\overset{\displaystyle CH_3}{\displaystyle |}}{\displaystyle C}}{\underset{\displaystyle C=O}{|}}} \right]
$$

This polymer was labeled "Polymer A." Similar results, except that the methyl groups of the diester moiety and the methacrylyl moiety of the polymer were replaced with other lower alkyl groups were obtained when Monomer No. 2 was replaced by monomers having the formula

$$
CH_2=C-\overset{\overset{\displaystyle O}{\|}}{C}-N\overset{\diagup CH_2COOR'}{\underset{\diagdown CH_2COOR'}{}}
$$
$$
\quad\ \underset{R}{|}
$$

where R and R' were other lower alkyl groups such as n-propyl, isopropyl, n-hexyl, isohexyl, and the like including monomers in which R and R' were identical, and those in which R and R' were not identical.

Example 5:   A 5 g. sample of Polymer A in particulate form, the particle size being about 0.5 to 2 mm., was maintained in an aqueous system having a temperature of about 40°C. and a pH of about 8 for about 30 minutes. Diluted (ca. 3 molar) sulfuric acid was then added to the system to bring the pH thereof to about 3. This treatment converted substantially all of the ester groups of the polymer to carboxyl groups and formed a polymer consisting of about 2,000 repeating units, the units having the formula

$$
\left[-CH_2-\underset{\underset{\underset{\underset{CH_2COOH}{\displaystyle N}}{\displaystyle |}}{\overset{\overset{\overset{\displaystyle CH_3}{\displaystyle |}}{\displaystyle C}}{\underset{\displaystyle C=O}{|}}}\overset{\diagup CH_2COOH}{\underset{\diagdown CH_2COOH}{}} \right]
$$

Example 6:   The general procedure of Example 4 was repeated. However,
in this instance the procedure was modified by using a mixture of monomers
in place of Monomer No. 2 which had been used in Example 4. The mixture of monomers was a mixture of Monomer No. 2 and styrene in 1 to 1
mol ratio. The results were substantially the same as those obtained in
Example 4 except that the product consisted of a copolymer of about 3,000
repeating units per polymer molecule, the units having the formulas

where X is —H and Z is

the unit ratio being 1 to 1. This polymer (copolymer) was labeled "Copolymer B." Similar results (except for the structure of "X" and "Z") were
obtained when the styrene was replaced by acrylic acid, methyl acrylate,
methyl methacrylate, 2-vinylpyridine, 3-vinylpyridine, 4-vinylpyridine,
N-vinylpyrrolidone, acrylonitrile, and vinyl acetate.

Example 7:   The general procedure of Example 5 was repeated. However,
in this instance the procedure was modified by replacing Polymer A with
Copolymer B. The final product was a copolymer consisting of about 3,000
repeating units, the units having the formulas

where X is —H and Z is

the unit ratio being 1 to 1.

Acrylate-Capped Polycaprolactone Compositions

In a process described by O.W. Smith, J.E. Weigel and D.J. Trecker; U.S. Patent 3,700,643; October 24, 1972; assigned to Union Carbide Corporation unsaturated acrylate-capped polycaprolactone polyol deriva- tives are produced having terminal acrylyl groups and at least one poly- caprolactone polyol chain residue in the molecule. In one of its simplest forms the final product can be the reaction product of a polycaprolactone diol, an organic isocyanate and hydroxyethyl acrylate. These derivatives can be used to produce coating compositions that are readily cured to solid protective films. The following examples illustrate the process.

Example 1: A polycaprolactone polyol (98 g.), produced as described in U.S. Patent 3,169,945 by the reaction of diethylene glycol and epsilon- caprolactone and having an average molecular weight of about 530, was placed in a 500 ml. flask that was equipped with a stirrer, thermocouple and two dropping funnels.

After the addition of 1 drop of dibutyltin dilaurate the mixture was heated to 80°C. in an oil bath and 128 g. of bis(2-isocyanatoethyl)-5-norbornene- 2,3-dicarboxylate and 52 g. of 2-hydroxypropyl acrylate were co-fed in a dropwise manner while stirring vigorously. The addition of the hydroxy- propyl acrylate was completed in 2 hours; thereafter the mixture was stirred for an additional one-half hour at 80°C. The acrylate-capped polycapro- lactone urethane produced had the basic structure:

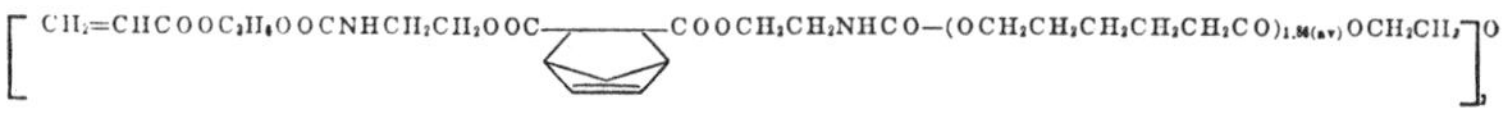

$$\left[\ \text{CH}_2=\text{CHCOOC}_3\text{H}_4\text{OOCNHCH}_2\text{CH}_2\text{OOC}\ \underset{\bigcirc}{}\ \text{COOCH}_2\text{CH}_2\text{NHCO}-(\text{OCH}_2\text{CH}_2\text{CH}_2\text{CH}_2\text{CH}_2\text{CO})_{1.5(av)}\text{OCH}_2\text{CH}_2\ \right]\text{O}$$

This product was a clear, straw yellow liquid. A solution containing 73 weight percent of the acrylate-capped polycaprolactone and 27 weight percent of 2-butoxyethyl acrylate had a Brookfield viscosity of 1,780 cp. at 23°C. In a similar manner the acrylate-capped polycaprolactone of the following structure is produced by the use of 2-hydroxyethyl methacrylate and a polycaprolactone polyol having an average molecular weight of about 786 that was produced by the reaction of hexanol and epsilon-caprolactone

$$\overset{\text{CH}_3}{\underset{|}{\text{CH}_2=\text{CCOOC}_2\text{H}_4\text{OOCHNCH}_2\text{CH}_2\text{OOC}}}\ \underset{\bigcirc}{}\ \text{COOCH}_2\text{CH}_2\text{NHCO}-(\text{OCH}_2\text{CH}_2\text{CH}_2\text{CH}_2\text{CH}_2\text{CO})_6\text{OC}_6\text{H}_{11}$$

Example 2:   A solution of 98 g. of the polycaprolactone polyol used in
Example 1, 52 g. of 2-hydroxypropyl acrylate and 1 drop of dibutyltin
dilaurate was heated to 70°C. The solution was stirred while 69.6 g. of
an 80/20 mixture of 2,4- and 2,6-tolylene diisocyanates was added in a
dropwise manner over a period of 2 hours. The mixture was stirred for an
additional one-half hour and then 0.002 g. of 4-methoxyphenol was added
as a stabilizer. The acrylate-capped polycaprolactone urethane produced
had the basic structure:

$$
\left[CH_2=CHCOOC_3H_6OOCNH-\underset{}{\overset{CH_3}{\bigcirc}}-NHCO(OCH_2CH_2CH_2CH_2CH_2CO)_{1.8(av.)}OCH_2CH_2 \right]_2 O
$$

This product was a clear liquid. A solution containing 73 weight percent
of the acrylate-capped polycaprolactone and 27 weight percent of 2-butoxy-
ethyl acrylate had a Brookfield viscosity of 2,270 cp. at 23°C.

Example 3:   In a method similar to that described in Example 1, a solution
of 98 g. of the same polycaprolactone polyol, 73.2 g. of 2-butoxyethyl
acrylate as solvent, 1 drop of dibutyltin dilaurate and 0.01 g. of 4-meth-
oxyphenol was heated to 70°C. and 69.6 g. of an 80/20 mixture of 2,4-
and 2,6-tolylene diisocyanate and 52 g. of 2-hydroxypropyl acrylate were
simultaneously co-fed as separate streams over a period of 2 hours. The
mixture was stirred an additional one-half hour at 65°C. The light yellow
solution of the acrylate-capped polycaprolactone urethane had a Brookfield
viscosity of 20,000 cp. at 23°C.

Example 4:   A solution containing 98 g. of the same polycaprolactone
polyol used in Example 1, 52 g. of 2-hydroxypropyl acrylate, 81.2 g. of
2-butoxyethyl acrylate and 1 drop of dibutyltin dilaurate was heated to
70°C. and then 69.6 g. of an 80/20 mixture of 2,4- and 2,6-tolylene
diisocyanates were added in a dropwise manner over a period of about 2
hours.

The reaction mixture was stirred another half-hour at about 70°C. and then
0.002 g. of 4-methoxyphenol was added. The solution of the acrylate-
capped polycaprolactone urethane had a Brookfield viscosity of 13,600 cps.
at 23°C. The acrylate-capped polycaprolactone polymer produced in
Examples 3 and 4 had the same basic structure as that of Example 2.

Acrylic Esters of Halogenated Acetylenic Diols

G.F. D'Alelio; U.S. Patent 3,664,992; May 23, 1972 describes esters
which comprise polymerizable acrylic esters of the halogenated acetylenic

alcohols, represented by the following structures,

$$\text{HO}\!-\!\!\left(\text{CHRCHRO}\right)_{n}\!\!\left(\text{CR}_2\right)_{y}\!\!-\!\text{CX}\!=\!\text{CX}\!-\!\left(\text{CR}_2\right)_{y}\!\!\left(\text{OCHRCHR}\right)_{n}\!\!-\!\text{OH}$$

$$\text{HO}\!-\!\!\left(\text{CHRCHRO}\right)_{n}\!\!\left(\text{CR}_2\right)_{y}\!\!-\!\text{CX}_2\text{CX}_2\!-\!\left(\text{CR}_2\right)_{y}\!\!\left(\text{OCHRCHR}\right)_{n}\!\!-\!\text{OH}$$

where R represents hydrogen preferably or a monovalent hydrocarbon containing 1 to 10 carbon atoms, X is a halogen selected from the class consisting of chlorine and bromine, y represents an integer having a value of 1 or 2, and one n represents zero and the other n represents an integer having a value of 0 to 3, preferably zero. These acrylic monomers are useful for the preparation of self-extinguishing homopolymers and copolymers; the diesters are also particularly useful as cross-linking agents for other monomers and polymers. The following examples illustrate the process.

Example 1: (A) To 86 parts of 2-butyne diol, $\text{HOH}_2\text{CC}\!\equiv\!\text{CCH}_2\text{OH}$, in 350 parts of CCl_4 is added slowly with stirring at 0° to 25°C. 161 parts of bromine in 625 parts of CCl_4 and the reaction allowed to continue after the bromine addition, for 3 hours at 25° to 40°C. The reaction mixture is then washed first with aqueous 10% Na_2CO_3, then with distilled water following which, the CCl_4 layer is separated, dried over anhydrous sodium sulfate, decolorized with activated carbon and filtered.

The CCl_4 is removed by distillation at reduced pressure, leaving an almost quantitative yield of $\text{HOCH}_2\text{CBr}\!=\!\text{CBrCH}_2\text{OH}$, melting point on recrystallization, 116.5° to 117°C. The elemental analysis yields values of 19.47% C and 65.06% Br, which are in good agreement with the theoretical values.

(B) When procedure (A) above is repeated using 322 parts of bromine instead of 161 parts the tetrabromo derivative, $\text{HOCH}_2\text{CBr}_2\!-\!\text{CBr}_2\text{CH}_2\text{OH}$, (78.7% bromine), is obtained which is in close agreement with the theoretical value.

(C) Into a stirred solution of 86 parts of $\text{HOH}_2\text{CC}\!\equiv\!\text{CCH}_2\text{OH}$ in 500 parts of CCl_4, maintained at 10° to 15°C., is slowly passed a stream of chlorine until 71 parts of Cl_2 are reacted. Then the solution is purified by the procedure of Example 1(A) above, yielding $\text{HOCH}_2\text{CCl}\!=\!\text{CClCH}_2\text{OH}$, MP 78°C., which on analysis for chlorine yields a value of 45.26%, which is close to the theoretical value for the compound.

$$\underset{\overset{|}{\text{Br}}\quad\overset{|}{\text{Br}}}{\text{HOCH}_2\text{CCl}\!-\!\text{CClCH}_2\text{OH}}$$

(D) The above procedures are typical of those used to prepare the following dihalo- and tetrahalo-diols from which the acrylic esters are prepared (Table 1). Bromination of this compound by the procedure given in Example 1(A) yields the product.

TABLE 1

Dihalo-diols	Tetrahalo-diols
1. $HOCH_2CBr=CBrCH_2OH$	a. $HOCH_2CBr_2-CBr_2CH_2OH$
2. $HOCH_2CCl=CClCH_2OH$	b. $HOCH_2CCl_2-CCl_2CH_2OH$
3. $HOCH_2CH_2CBr=CBrCH_2CH_2OH$	c. $HOCH_2C(Br)(Cl)-C(Br)(Cl)CH_2OH$
	$\quad\quad\quad CH_3 \quad\quad\quad CH_3$
4. $HOCH(CH_3)CH_2CBr=CBrCH_2CH(CH_3)OH$	d. $HOC-CBr_2CBr_2C-OH$
	$\quad\quad\quad CH_3 \quad\quad\quad CH_3$
5. $HOCH(CH_3)CBr=CBrCH(CH_3)-OH$	e. $HOCH_2CH_2CBr_2-CBr_2CH_2CH_2OH$
	$\quad\quad\quad CH_3 \quad\quad\quad CH_3$
6. $HOCH(C_6H_5)CBr=CBrCH(C_6H_5)OH$	f. $HOC-CCl_2CCl_2C-OH$
	$\quad\quad\quad CH_3 \quad\quad\quad CH_3$
	$\quad\quad\quad CH_3 \quad\quad\quad CH_3$
7. $HOC(CH_3)_2CBr=CBrC(CH_3)_2OH$	g. $HOC-HCBr_2CBr_2C-HOH$
	$\quad\quad\quad CH_3 \quad\quad\quad CH_3$
8. $HOC(CH_3)_2CCl=CClC(CH_3)_2OH$	h. $HOC-HCCl_2CCl_2C-HOH$
9. $HOOCH_2Cbr=CBrCH_2OOH$	
10. $\left[HOOCH(CH_3)CH_2OCBr= \right]_2$	
11. $HOC(CH_3)(C_2H_5)CBr=CBrC(CH_3)(C_2H_5)OH$	

Example 2: The following is a typical procedure for preparing diesters of the halogenated diols. To a mixture of 800 parts of dry benzene, 1 part of tertiary butyl catechol, 1 mol of halogenated diol and 2 mols of triethylamine, cooled to 5°C., there is added slowly with stirring 2 mols of the acid chloride (209 parts of $CH_2=C(CH_3)COCl$ or 181 parts of $CH_2=CHCOCl$) over a period of 2 hours. The mixture is then filtered to remove triethylamine hydrochloride and hexane is added to the filtrate until the solution becomes turbid. The solution is again filtered, decolorized with activated carbon and further purified by passing the solution through a column of chromatographic alumina or silica. The solvent is then removed from the solution at reduced pressure of 1 to 50 mm. Hg leaving the diester as a clear, viscous oil.

The diesters so prepared can be used directly, but if further purification is desired or required, such purification is readily accomplished in a falling film evaporator. The anhydrides of acrylic and methacrylic acid, $[CH_2=C(R)CO]_2O$, can be used instead of the acid chlorides. By the use of the above procedure, the acrylic esters and the methacrylic esters of dihalo-diols, (1) to (11) inclusive, and of the tetrahalo-diols, (a) to (h) inclusive, are readily prepared, whose elemental analyses for halogen are in good agreement with the calculated values, as illustrated by the analysis of some typical diesters. The analyses are given in Table 2 in which A refers to the acrylic ester and MA indicates the methacrylic ester. Similarly, the chloroacrylic esters and the bromoacrylic esters are prepared by using the halogenated acid chlorides, $CH_2=C(Cl)COCl$ and $CH_2=C(Br)COCl$, respectively.

TABLE 2

Number	Structure	Percent halogen
A.1	$[CH_2=CHCOOCH_2CBr]_2$	45.26
A.2	$[CH_2=CHCOOCH_2CCl]_2$	26.76
A.3	$[CH_2=CH-COOCH_2CH_2CBr]_2$	41.85
A.7	$[CH_2=CHCOOC(CH_3)_2CBr]_2$	38.48
A.8	$[CH_2=CHCOOC(CH_3)_2CCl]_2$	22.02
MA.1	$[CH_2=C(CH_3)COOCH_2CBr]_2$	41.90
MA.2	$[CH_2=C(CH_3)COOCH_2CCl]_2$	24.30
MA.3	$[CH_2=C(CH_3)COOCH_2CH_2CBr]_2$	38.53
MA.7	$[CH_2=C(CH_3)COOC(CH_3)_2CBr]_2$	36.80
MA.8	$[CH_2=C(CH_3)COOC(CH_3)_2CCl]_2$	20.21
MA.10	$[CH_2=C(CH_3)COOCH(CH_3)CH_2OCH(CH_3)CH_2OCBr]_2$	30.50
A.a	$[CH_2=CHCOOCH_2CBr_2]_2$	62.01
A.b	$[CH_2=CHCOOCH_2CCl_2]_2$	42.14
A.d	$[CH_2=CHCOOC(CH_3)_2CBr_2]_2$	55.75
A.f	$[CH_2=CHCOOC(CH_3)_2CCl_2]_2$	36.06
MA.a	$[CH_2=C(CH_3)COOCH_2CBr_2]_2$	58.58
MA.b	$[CH_2=C(CH_3)COOCH_2CCl_2]_2$	38.60
MA.d	$[CH_2=C(CH_3)COOC(CH_3)_2CBr_2]_2$	53.00
MA.f	$[CH_2=C(CH_3)COOC(CH_3)_2CCl_2]_2$	33.70

Example 3: To a mixture of 500 parts of dry benzene, 1 part of tertiary butyl catechol, 59 parts of trimethylamine, and 250 parts of the formula $HOCH_2CBr=CBrCH_2OH$, cooled to 5°C., there is added slowly with stirring over a period of 4 hours, 90 parts of $CH_2=CHCOCl$. The mixture is then filtered to remove precipitated trimethylamine hydrochloride, and hexane is added to the filtrate until the solution becomes turbid, is then refiltered, decolorized with activated carbon and passed through a chromatographic column of silica.

The benzene is then removed at 1 to 5 mm. Hg pressure leaving the product as a clear viscous oil. Its infrared spectrum shows strong bands for a free hydroxyl in the 3 micron region along with the typical bands expected for the acrylic ester linkages in the 12.34μ region. Vapor phase chromatography indicates that the product consists primarily of monoester and about 11.2% of diester which corresponds to A.1 of Example 2.

Separation of the mono- and diesters is accomplished readily in a falling film evaporator at 0.05 mm. Hg pressure. The elemental analysis of 39.43% bromine for the fraction showing the free-hydroxyl group, is in good agreement with the calculated value for the compound, with the formula $CH_2=CHCOOCH_2CBr=CBrCH_2OH$. By the above procedure the acrylic-type esters of the dihalo-diols, (1) to (11) inclusive, and the tetrahalo-diols, (a) to (h) inclusive, are readily prepared.

These are converted to the ethylene glycol, diethylene glycol and triethylene glycol derivatives by reacting the hydroxy group with 1, 2 and 3 mols of ethylene oxide in the presence of a small amount of sodium hydroxide. These monohydroxy derivatives are then converted by reaction of the respective terminal hydroxy groups with acrylyl chloride to diacrylates in which the second n of the formula is 1, 2 or 3. Various typical derivatives are illustrated as follows:

$$CH_2=CHCOOCH_2CBr=CBrCH_2(OCH_2CH_2)_2OH$$

$$CH_2=CHCOOCH_2CBr_2-CBr_2CH_2OCH_2CH_2OH$$

$$CH_2=CHCOOCH_2CBr=CBrCH_2OCH_2CH_2OOCCH=CH_2$$

Example 4: Under a nitrogen atmosphere, 50 parts of a product with the formula $(CH_2=CHCOOCH_2CBr)_2$ containing 0.5 parts of azoisobutyronitrile is added to 20 parts of poly(cis-butadiene) and the mixture stirred at 25°C. until it is homogeneous; then the mixture is heated at 70°C. for 10 hours and at 120°C. for 16 hours and there is obtained a cross-linked self-extinguishing polymerizate which is tough and exhibits impact properties.

Example 5: The following mixtures are first prepared and then irradiated with ultraviolet light from a 100-watt mercury lamp until the mixtures becomes solid and hard.

		Parts
(A)	Methyl methacrylate	75.0
	$\left[CH_2=CHCOOCH_2CBr_3 \right]_2$	25.0
	Benzophenone	0.2
(B)	Methyl methacrylate	65.0
	$CH_2=C(CH_3)COOCH_2CCl=CClCH_2OOCC(CH_3)=CH_2$	25.0
	$PO(OCH_2CH=CH_2)_3$	10.0
	Benzophenone	0.2

In both cases, self-extinguishing polymers are obtained.

Example 6: The following mixtures are first prepared:

		Parts
(A)	$\left[CH_2=CHCOOCH_2CBr_3 \right]_2$	20
	$\underset{CH_2=C-COOCH_3}{\overset{CH_3}{|}}$	80
(B)	$\left[CH_2=CHCOOC(CH_3)_2CBr_3 \right]_2$	20
	$\underset{CH_2=C-COOC_2H_5}{\overset{CH_3}{|}}$	80
(C)	$\left[CH_2=C(CH_3)COOC(CH_3)_2CBr_3 \right]_2$	10
	$\underset{CH_2=C-COOCH_3}{\overset{CH_3}{|}}$	80
	$CH_2=CHCOOCH_2PO(OC_2H_5)_2$	10
(D)	Alkyd Resin B	60
	$\left[CH_2=CHCOOCH_2CCl_2 \right]_2$	20
	$CH_2=CHCOOCH_2CH_2OH$	5
	$\underset{CH_2=C-COOCH_3}{\overset{CH_3}{|}}$	15

Samples of mixtures (A), (B), (C) and (D) are placed in glass vials which are swept out with nitrogen and sealed, and then each is exposed to the beam of a 1 mev Van de Graaff accelerator and insoluble, infusible, nonburning, self-extinguishing polymers are obtained at dosages varying from 4 to 8 megarads. Similar results are obtained when other sources of ionizing radiation are used, such as from natural or synthetic radioactive material, e.g., from Cobalt 60 or from the Varian type travelling wave linear accelerators or the types of accelerators described in U.S. Patent 2,763,609 and British Patent 762,953.

When 20 mil wood ply is impregnated and saturated with mixtures (A), (B), (C) and (D), then covered with 0.1 mil polyethylene sheet, irradiated to

6 megarads, and the polyethylene barrier sheet removed, the resulting wood is found to be dense, water-resistant, self-extinguishing and nonburning.

Additional studies with acrylic esters of halogenated acetylenic diols are described by G.F. D'Alelio; U.S. Patent 3,664,993; May 23, 1972.

BINDER RESINS, FORMULATIONS

Allyl Glycidyl Ether Modified Vinyl Compositions

In a process described by J.F. Fellers, J.E. Hinsch and E.O. McLaughlin; U.S. Patent 3,641,210; February 8, 1972; assigned to Ford Motor Company a film-forming, radiation-polymerizable, paint binder solution of vinyl monomers and an olefinically unsaturated vinyl monomer-comprising polymeric binder is applied as a liquid coating to an external surface of an article of manufacture and cured with ionizing radiation.

The binder polymer is characterized by having olefinic unsaturation between the terminal carbons of the side chains, i.e., alpha-beta unsaturation, with the unsaturation being separated from the principal carbon-to-carbon chain by two ether linkages. The binder polymer is formed by reacting an allylic glycidyl ether with a polymer formed by reacting an allylic alcohol with at least two different vinyl monomers at least one of which is an ester of acrylic or methacrylic acid. The following examples illustrate the process.

Example 1: A vinyl monomer-comprising polymer is prepared from the following reactants: 39 parts ethyl acrylate; 24 parts methyl methacrylate; 36 parts allyl alcohol; 1 part benzoyl peroxide and solvent xylene (all parts by weight). To a reaction vessel provided with a condenser, thermometer, agitator, and dropping funnel there are charged an amount of xylene equal in weight to the reactants to be added in the first reaction step. The xylene is heated to about 100° to 120°C.

The four reacting materials are thoroughly mixed and added slowly with a dropping funnel to the heated xylene over a period of 4 hours. The reaction is held at this temperature for 1 to 2 hours after addition is complete and then allowed to cool to room temperature.

A binder polymer is formed in a second reaction step from the following materials: 69 parts copolymer from Step 1; 3.18 parts allyl glycidyl ether and 0.2 parts potassium hydroxide (all parts by weight). A solution of the allyl glycidyl ether and potassium hydroxide is added to the copolymer at room temperature. The mixture is then heated to a temperature of 100° to 120°C. This temperature is maintained for about 7 hours and allowed to cool. The binder polymer reaction mixture is heated to about 60°C. and the xylene and the excess reactant are removed by vacuum distillation. At a temperature of about 60°C. styrene and hydroquinone are added to the polymer to form a film-forming solution having the following composition: 67 parts polymer from Step 2; 32.93 parts styrene and 0.07 parts hydroquinone (all parts by weight).

The film-forming solution is applied to a metal substrate to an average depth of about 0.3 mil and cured by electron beam irradiation. The conditions of irradiation are as follows: a voltage of 270 kv., a current of 25 milliamperes, a total dosage of 15 Mrads, under an atmosphere of nitrogen.

Example 2: The procedure of Example 1 is repeated with the single difference that an equivalent amount of methyl methacrylate is substituted for the styrene monomers in the film-forming solution.

Example 3:   The procedure of Example 1 is repeated with the single difference that an equivalent amount of a mixture of styrene, ethyl acrylate and 2-ethyl hexyl acrylate is substituted for the styrene monomers in the film-forming solution.

Example 4:   The procedure of Example 1 is repeated except that an equimolar mixture of methyl methacrylate and vinyl toluene is substituted for the styrene monomers in the film-forming solution, the substrate is wood, and the film depth is about 1 mil.

Example 5:   The procedure of Example 1 is repeated with the following changes: (a) The reactants of the first step reaction are 39.0 parts butyl acrylate; 24.0 parts styrene; 36.0 parts 1-penten-5-ol and 1.0 parts benzoyl peroxide (all parts by weight).

(b) The reactants of the second step reaction are 70 parts copolymer from (a); 29.8 parts allyl glycidyl ether and 0.2 parts KOH (all parts by weight).

(c) Composition of film-forming solution is 67 parts polymer from (b); 32.93 parts methyl methacrylate and 0.07 parts hydroquinone (all parts by weight).

Additional work with these allyl glycidyl ether-modified vinyl compositions is described by J.F. Fellers, J.E. Hinsch and E.O. McLaughlin; U.S. Patent 3,642,939; February 15, 1972; assigned to Ford Motor Company.

Siloxane Modified Polyesters

In a process described by W.J. Buriant and I.H. Tsou; U.S. Patent

3,632,399; January 4, 1972; assigned to Ford Motor Company a film–forming, radiation–polymerizable, paint binder solution of vinyl monomers and
an alpha–beta olefinically unsaturated, polysiloxane-modified polyester is
applied as a liquid coating to an external surface of an article of manufacture and cured thereon with ionizing radiation. The binder resin contains
about 0.5 to 5, preferably about 0.5 to 3, alpha–beta olefinic unsaturation
units per 1,000 units molecular weight.

In one example, the binder resin is formed by reacting a siloxane bearing
at least two functional groups selected from hydroxy groups and hydrocarbonoxy groups with a polyhydric alcohol and subsequently reacting the
siloxane–containing intermediate with a dicarboxylic acid which provides
the polymer with alpha–beta olefinic unsaturation and a second carboxylic
acid that provides no additional alpha–beta olefinic unsaturation to the
resulting polymer.

The films formed from the paints of this process are advantageously cured at
relatively low temperatures, e.g., between room temperature (20° to 25°C.)
and the temperature at which significant vaporization of its most volatile
component is initiated, ordinarily between 20° and 70°C. The radiation
energy is applied at dose rates of about 0.1 to 100 Mrad per second upon
a preferably moving workpiece with the coating receiving a total dose in
the range of about 0.1 to 100, ordinarily between about 1 and 25, and
most commonly between 5 and 15 Mrad. The films can be converted by the
electron beam into tenaciously bound wear and weather resistant coatings
which meet the following specifications.

Substrate Applicability	Type of Exposure	Requirements of Test
Wood or metal	Room temperature water soak	Withstand 240 hours in water at 20° to 25°C. without loss of gloss or film integrity
Wood	Cyclic boiling and baking	Withstand 25 cycles of 4 hours in boiling water followed by 15 hours drying at 62° to 63°C. without loss of gloss or film integrity
Metal	Elongation	Withstand 25% elongation without rupture (1 to 2 mil coating, 1/8 inch mandrel)
Wood or metal	Ultraviolet	Withstand 2,000 hours in Standard Atlas Ultraviolet Carbon Arc Weatherometer test without chalking or loss of gloss or film integrity

Example 1: A siloxane–modified polyester, paint binder resin is prepared
in the following manner: To a reaction vessel are charged 1,330 lbs. of
neopentyl glycol and 1,080 lbs. of a commercially available methoxylated

partial hydrolysate of monophenyl and phenylmethyl silanes consisting essentially of dimethyltriphenyltrimethoxytrisiloxane (Sylkyd 50) and have the following typical properties: an average molecular weight of 470, a combining weight of 155, a specific gravity at 77°F. of 1.105 and a viscosity at 77°F. of 13 cs.

The charge is heated to about 345°F. (174°C.) until about 215 lbs. methanol are removed overhead. The charge is cooled to about 250°F. (121°C.) after which there is added 196 lbs. maleic anhydride, 964 lbs. tetrahydrophthalic anhydride, 2.2 lbs. dibutyl tin oxide and 150 lbs. xylene. The temperature of the charge is raised slowly to about 420°F. (215°C.) and this temperature is maintained until the resulting resin has an acid number of 10. A vacuum is pulled to remove the xylene and 61 lbs. hydroquinone are charged and the charge is cooled to 200°F. and dumped into a mixing tank with 780 lbs. styrene.

A white mill base is then prepared by mixing 3,050 lbs. of TiO_2, 1,805 lbs. of resin, prepared as in the preceding paragraph, 146 lbs. of styrene, 507 lbs. of methyl methacrylate, and 20 lbs. of Bakers M.P.A., a waxlike, high molecular weight castor oil derivative to facilitate the grinding through viscosity adjustment and assist in retention of pigment dispersion in the grind, and passing the foregoing mixture through a conventional sand grinder.

This mill base is further diluted with styrene and methyl methacrylate in amounts such as to provide a paint comprising about 40% resin, 30% styrene and 30% methyl methacrylate. A film of the resulting paint is sprayed upon wood and metal panels and irradiated by an electron beam with a potential of 295 kv., a current of 25 ma., a distance (emitter to workpiece) of 10 inches, a line speed of 1.6 cm./sec. with 2 passes for a total dosage of 10 Mrad.

Example 2: The procedure of Example 1 is repeated except that an equivalent amount of ethylene glycol is substituted for neopentyl glycol.

Example 3: The procedure of Example 1 is repeated except that an equivalent amount of hexylene glycol is substituted for the neopentyl glycol.

Example 4: The procedure of Example 1 is repeated except that the methyl methacrylate component of the vinyl monomers is replaced with styrene and the tetrahydrophthalic acid component is replaced with an equivalent amount of trimellitic anhydride in the preparation of the resin.

Example 5: The procedure of Example 1 is repeated except that the styrene component of the vinyl monomers is replaced with an equimolar amount of methyl methacrylate.

Example 6:   The procedure of Example 1 is repeated except that one-half of the methyl methacrylate component of the vinyl monomers is replaced with an equimolar amount of butyl acrylate.

Example 7:   The procedure of Example 1 is repeated except that one-fourth of the styrene component of the vinyl monomers is replaced with an equimolar amount of 2-ethyl bexyl acrylate and the relative proportions of maleic anhydride and tetrahydrophthalic anhydride are adjusted to provide in the resin an amount of maleic anhydride which provides the resin with 5 alpha-beta olefinic unsaturation units per 1,000 molecular weight.

Example 8:   The procedure of Example 1 is repeated except that one-third of the methyl methacrylate component of the vinyl monomers is replaced with an equimolar amount of ethyl acrylate.

Example 9:   The procedure of Example 1 is repeated except that one-fifth of the styrene component of the vinyl monomers is replaced with an equimolar amount of alpha-methyl styrene.

Example 10:   The procedure of Example 1 is repeated except that the maleic anhydride component of the vinyl monomers is replaced with an equimolar amount of fumaric acid.

Example 11:   The procedure of Example 1 is repeated using the following materials to prepare the paint concentrate:

	Parts by Weight
White mill base from Example 1	5,680
Siloxane-modified polyester resin prepared as in Example 1	1,676
Styrene	1,050
Methyl methacrylate	1,527
Green mill base	267

The green mill base above referred to is prepared by first admixing 120 parts by weight of 50% xylene solution of a conventional acrylic paint binder resin (Gardner-Holdt Vis. at 77°F., WX), 60 parts by weight phthalocyanine green, 61 parts by weight xylene.

This mix is ground 48 hours in a steel ball mill to 8 Heg. after which 240 parts by weight of the acrylic polymer above mentioned and 60 parts by weight xylene are added and the mix is ground for an additional 2 hours. To the latter grind is added 264 parts by weight of the same acrylic polymer, 130 parts by weight of butylated melamine formaldehyde resin (60%

solids, 20% xylene, 20% butanol) and 65 parts by weight xylene. The paint concentrate thus prepared is diluted with styrene and methyl methacrylate to provide paints containing the following:

Resin, percent	Styrene, percent	Methyl Methacrylate, percent
60	20	20
50	20	30
50	30	20
40	30	30

These paints are sprayed on wood and on metal panels and cured by irradiation as in the preceding examples.

Example 12:   A siloxane-modified polyester, paint binder resin is prepared in the following manner: To a reaction vessel are charged 70 lbs. of neopentyl glycol, 10 lbs. of xylene, and 35 lbs. of a commercially available (Dow Corning Z-6018) hydroxy-functional, cyclic, polysiloxane having the following properties:

Hydroxy content, Dean Stark	
percent condensable	5.5
percent free	0.5
Average molecular weight	1,600
Combining weight	400
Refractive index	1.531 to 1.539
Softening point, (Durran's Hg Method, 60% solids in xylene), °F.	200
Specific gravity at 77°F.	1.075
Viscosity at 77°F.	
cp.	33
Gardner-Holdt	A-1

The charge is heated to about 345°F. (174°C.) for 2 1/2 hours, after which there is added 13.7 lbs. maleic anhydride, 54.2 lbs. of tetrahydrophthalic anhydride and 100 g. of dibutyl tin oxide. The temperature of the charge is raised slowly to about 430°F. (221°C.) and this temperature is maintained until the resulting resin has an acid number of about 10. Some of the xylene and water of reaction are removed during the cook and the excess is then removed by vacuum. To the charge is added 12.5 g. hydroquinone and the charge is cooled to 180°F. (82.5°C.) and diluted with 40 lbs. of styrene.

A white mill base is prepared by mixing 3,050 lbs. of TiO_2, 1,805 lbs. of resin, prepared as in the preceding paragraph, 146 lbs. of styrene, 507 lbs.

of methyl methacrylate, and 20 lbs. of Bakers M.P.A., a waxlike, high molecular weight castor oil derivative to facilitate the grinding through viscosity adjustment and assist in retention of pigment dispersion in the grind, and passing the foregoing mixture through a conventional sand grinder.

This mill base is further diluted with styrene and methyl methacrylate in amounts such as to provide a paint comprising about 40% resin, 30% styrene and 30% methyl methacrylate. A film of the resulting paint is sprayed upon wood and metal panels and irradiated by an electron beam under the following conditions:

Potential	295	kv.
Current	1	milliampere
Distance, emitter to workpiece	10	inches
Line speed	1.6	cm./sec.
Passes	2	
Total dosage	10	Mrad

Siloxane-Acrylic Polymers

J.D. Nordstrom and C.B. Zelek; U.S. Patent 3,650,812; March 21, 1972; assigned to Ford Motor Company describe an acrylic-siloxane paint binder resin that is cross-linkable with vinyl monomers by exposure to an electron beam which is produced in a three step reaction where (1) a siloxane having two or more hydroxy or alkoxy functional groups per molecule is reacted with a C_5 to C_{12} monohydroxy acrylate, i.e., the monohydroxy ester of a C_2 to C_8 diol and acrylic or methacrylic acid, (2) the siloxane-acrylate product of the first reaction step is reacted with a mixture of vinyl monomers at least one constituent monomer of which is selected from glycidyl acrylate and glycidyl methacrylate, and (3) the resultant siloxane-acrylate-vinyl monomer copolymer is reacted with a C_5 to C_{12} monohydroxy acrylate, i.e., the monohydroxy ester of a C_2 to C_8 diol and acrylic or methacrylic acid.

Example 1: An acrylic-siloxane paint binder resin is prepared as described below, in which all parts given are in parts by weight. In the first step (1) 206 parts of an acyclic polysiloxane having molecular weight in the range of 700 to 800 with an average of 3-4 methoxy functional groups per molecule, 40 parts hydroxyethyl acrylate, 0.4 part tetraisopropyl titanate and 0.1 part hydroquinone are reacted together.

The reactants are heated to 100°C. and then to 100° to 150°C. over a 2-hour period. 11 parts by weight methanol distillate are removed via a Barrett distillation receiver. In the second step reaction (2), siloxane-acrylate is reacted with vinyl monomers. Two solutions are prepared. Solution A consists of 100 parts by weight siloxane-acrylate of step (1),

43 parts by weight ethyl acrylate, 42 parts by weight methyl methacrylate, 23 parts by weight glycidyl methacrylate, 5 parts by weight azobisisobutyronitrile and 490 parts by weight xylene. Solution B consists of 0.1 parts by weight hydroquinone, 13 parts by weight methacrylic acid and 0.5 parts by weight tetraethyl ammonium chloride.

Solution A is added to refluxing xylene, evenly, over a 2-hour period. The resultant solution is allowed to reflux for 6 hours. The product has a nonvolatile content of 29%. Solution B is added to the refluxing solution and heating is continued until the acid content drops to 0.053 milliequivalents per gram, (77% reaction). The siloxane-comprising product has a content of 0.5 units alpha-beta olefinic unsaturation per 1,000 units molecular weight.

In the third step reaction (3), siloxane-acrylate resin is reacted with hydroxy acrylate; 350 parts by weight of product of (2) and 17 parts by weight hydroxyethyl acrylate are used. The reactants are heated to 130°C. and three parts by weight distillate are removed in a Barrett receiver. The xylene is then removed by vacuum distillation (10 mm. Hg, 110°C.) and the resin dissolved in 80 parts by weight of methyl methacrylate.

A film of the solution is applied to a metal substrate to an average depth of about 1 mil (0.001 inch) and exposed to an 8 Mrad dose of energy from a 270 kilovolt, 25 milliampere electron beam in a nitrogen atmosphere. A hard, solvent resistant film results. This film has a pencil hardness of F and withstands 13 rubs with a cloth soaked in methyl ethyl ketone.

Example 2: The procedure of Example 1 is repeated except that monobutylether of ethylene is substituted for the hydroxyethyl acrylate in the third step reaction. The film resulting from an 8 Mrad dose of energy from a 270 kilovolt, 25 milliampere electron beam in nitrogen atmosphere is very soft and has poor solvent resistance, i.e., pencil hardness of 2B and three rubs with a cloth soaked in methyl ethyl ketone, thus emphasizing the value of using a hydroxy acrylate in the third step reaction.

Example 3: The procedure of Example 1 is repeated except for the differences that the electron beam has average energy of 325 kilovolts and films having average thickness of 0.2, 0.5, 1.5, 2.5 and 4 mils are irradiated.

Example 4: The procedure of Example 1 is repeated except for the differences that the substrate is wood and the vinyl monomer component of the paint binder solution is a mixture of two molar parts methyl methacrylate and one molar part styrene.

Example 5: The procedure of Example 1 is repeated except for the differences that the substrate is a synthetic polymeric solid, i.e., acrylonitrile-butadiene-styrene copolymer, and the vinyl monomer component of the paint binder solution consists of two molar parts methyl methacrylate, two molar parts ethyl acrylate, one molar part vinyl toluene and one-half molar part 2-ethylhexyl acrylate.

Example 6: The procedure of Example 1 is repeated except for the differences that the substrate is glass, the vinyl monomer component of the paint binder solution consists of one molar part butyl acrylate, one molar part ethyl acrylate and two molar parts styrene, and the paint binder solution is pigmented with particulate titanium dioxide.

In related work, J.D. Nordstrom and C.B. Zelek; U.S. Patent 3,650,811; March 21, 1972; assigned to Ford Motor Company describe an acrylic-siloxane paint binder resin that is cross-linkable with vinyl monomers by exposure to an electron beam which is produced in a three step reaction where (1) a siloxane having two or more hydroxy or alkoxy functional groups per molecule is reacted with a C_5 to C_{12} monohydroxy acrylate, i.e., the monohydroxy ester of a C_2 to C_8 diol and acrylic or methacrylic acid, (2) the siloxane-acrylate product of the first reaction step is reacted with a mixture of vinyl monomers having as a constituent monomer an acrylate selected from glycidyl acrylate and glycidyl methacrylate, and (3) the resultant siloxane-acrylate-vinyl monomer copolymer is reacted with acrylic or methacrylic acid.

Example: An acrylic-siloxane paint binder resin is prepared in the following manner: The first reaction step, i.e., the reaction of siloxane with hydroxy acrylate, is the same as the first step described in Example 1 of U.S. Patent 3,650,812 above. The second step reaction, i.e., siloxane-acrylate with vinyl monomers, is as follows: 100 parts by weight siloxane-acrylate prepared above, 30 parts by weight ethyl acrylate, 32 parts by weight methyl methacrylate, 46 parts by weight glycidyl methacrylate, 5 parts by weight azobisisobutyronitrile and 490 parts by weight xylene are reacted together. The reactants and catalysts are added to refluxing xylene, over a 2-hour period and then allowed to reflux for 6 hours.

The third step reaction is as follows. To the refluxing solution of xylene and the product of the second step reaction is added 26 parts by weight methacrylic acid, 0.1 part by weight hydroquinone, and 0.5 parts by weight tetraethyl ammonium chloride. Heating is continued until the acid content drops to about 0.053 millequivalents per gram. The resin is recovered from the xylene and dissolved in 100 parts by weight methyl methacrylate. The resin is cross-linked in the manner described in Example 1 of U.S. Patent 3,650,812.

Siloxane Reacted with Hydroxy Functional Acrylic Copolymer

In a process described by J.D. Nordstrom and C.B. Zelek; U.S. Patent 3,650,813; March 21, 1972; assigned to the Ford Motor Company the acrylic-siloxane paint binder resins are prepared in a two-step reaction where (1) a siloxane having two or more hydroxy or alkoxy functional groups per molecule is reacted with a hydroxy functional copolymer of acrylic monomers, and (2) the siloxane-comprising reaction product of the first reaction step is reacted with a C_5 to C_{12} monohydroxy acrylate, e.g., the monoester of a C_2 to C_8 diol and acrylic or methacrylic acid.

In a preferred case, the relative proportions of the hydroxylated acrylic resin and the hydroxy or alkoxy functional siloxane are adjusted to yield as nearly as possible a product where one functional group of the siloxane is reacted with a hydroxyl group of a resin molecule leaving the remainder of the functional groups of the siloxane to react with hydroxy acrylate in the second step reaction. Since the preferred siloxanes contain two to five functional groups per molecule, the siloxane molecule of the preferred final product resin would have attached one molecule of the acrylic resin and one to four acrylate molecules. However, perfect control is not to be expected and thus the siloxane molecules in the paint binder solution will have attached on average about 0.5 to 1.5 molecules of the acrylic resin reactant.

Thus, the mol ratio of acrylic resin to siloxane in this reaction step is advantageously in the range of 0.5 to 1.5:1. Sufficient hydroxy acrylate is used in the second step to react with at least a substantial portion, preferably all, of the remaining functional groups of the siloxane. Conversely, the hydroxylated acrylic resin may have a plurality of reactive hydroxyl groups per molecule, advantageously 1 to 10, and thus a number of siloxane molecules may be attached to one molecule of the acrylic resin reactant.

The acrylate reactant of the second reaction step may be the sole means of providing alpha-beta olefinic unsaturation or the unsaturation it provides may exist alongside alpha-beta unsaturation in the acrylic resin previously reacted with the siloxane. Suitable acrylates include 2-hydroxy acrylate or methacrylate, 2-hydroxypropyl acrylate or methacrylate, 2-hydroxybutyl acrylate or methacrylate, 2-hydroxyoctyl acrylate or methacrylate, etc.

The flexibility of the cured paint film formed from the acrylic-siloxane resin thus produced can be varied significantly by varying the molecular weight of the acrylic copolymer and/or the number of acrylic copolymer molecules per siloxane molecule in the acrylic-siloxane resin.

The acrylic-siloxane resin is mixed with C_5 to C_{12} vinyl monomers to form a paint binder solution which is applied by conventional means, e.g., spraying, roll coating, etc., to a substrate and polymerized with ionizing radiation, preferably that of an electron beam having average energy in the range of about 100,000 to 500,000, preferably about 150,000 to 350,000 electron volts.

Example 1: An acrylic-siloxane resin is prepared in the following manner:

(1) An acrylic resin, Resin I, is prepared from 80 mol percent ethyl acrylate, 10 mol percent hydroxyethyl acrylate and 10 mol percent glycidyl methacrylate. The copolymerization is effected by dropwise addition of the monomers, with 1.5 weight percent azobisisobutyronitrile, to an equivalent weight of refluxing xylene. When the reaction is complete, acrylic acid in equimolar ratio with the glycidyl methacrylate is added and reacted with the resin.

(2) Resin I is reacted with siloxane:

Reactants	Parts by Weight
Resin I	75 (0.07 mol hydroxyl)
Hydroxy functional cyclic siloxane*	100 (0.07 mol hydroxyl)
Xylene	120
Hydroquinone	0.1

*A commercially available Dow Corning Z-6018 hydroxy functional, cyclic, polysiloxane.

The reactants are heated at reflux (142°C.) for one hour. The by-product, water, is removed by azeotropic distillation. The product, Resin II, is a slightly hazy, viscous liquid.

(3) Resin II is reacted with hydroxy acrylate:

Reactants	Parts by Weight
Resin II	209
Hydroxyethyl acrylate	11.5

The reactants are heated at reflux for 2 hours. Water of condensation is removed by azeotropic distillation. 65 parts by weight of the resultant acrylic-siloxane resin with 35 parts by weight methyl methacrylate and a film of the solution is drawn down on a steel panel. A continuous nontacky film is formed when this coating is exposed to a 275 kilovolt, 25 milliampere electron beam.

Example 2:   An acrylic-siloxane resin is prepared by the methods of Example 1 with the differences that functionally equivalent amounts of a different acrylic resin and a methoxy functional siloxane are substituted for the acrylic resin and the siloxane employed in Example 1. A catalytic amount (0.4 part by weight) tetraisopropyl titanate is employed as a catalyst.

The acrylic resin employed in this example is prepared by copolymerizing a mixture of monomers whose composition is 40 mol percent methyl methacrylate, 50 mol percent ethyl acrylate and 10 mol percent hydroxyethyl acrylate. The copolymerization is effected by dropwise addition of the monomers with 1.5 weight percent azobisisobutyronitrile to an equivalent weight of refluxing xylene.

The siloxane employed in this example is an acyclic polysiloxane having an average molecular weight in the range of 700 to 800 with an average of 3 to 4 methoxy functional groups per molecule. The resultant acrylic-siloxane resin (65 parts by weight) and methyl methacrylate (35 parts by weight) are mixed to form a paint binder solution. A film of this solution is drawn down on a steel panel and cured to a tack-free state by exposure to 8 Mrads of energy from a 275 kilovolt, 25 milliampere electron beam in a nitrogen atmosphere.

Example 3:   The procedure of Example 2 is repeated except for the difference that the hydroxy acrylate employed in the second reaction step is 2-hydroxyethyl methacrylate.

Example 4:   The procedure of Example 2 is repeated except for the difference that the hydroxy acrylate employed in the second reaction step is 2-hydroxybutyl acrylate.

Example 5:   The procedure of Example 2 is repeated except for the difference that the hydroxy acrylate employed in the second reaction step is 2-hydroxybutyl methacrylate.

Siloxane-Urethane Resins

O.B. Johnson; U.S. Patent 3,585,065; June 15, 1971; assigned to Ford Motor Company describes a radiation-curable, film-forming paint binder comprising in combination (a) an alpha-beta olefinically unsaturated diurethane and (b) an alpha-beta olefinically unsaturated siloxane. The former is formed by reacting an organic diisocyanate with a hydroxy ester of an alpha-beta olefinically unsaturated carboxylic acid. The latter is formed by reacting such an ester with a siloxane having at least two functional groups selected from hydroxyl groups and hydrocarbonoxy groups.

The preferred hydroxyl bearing esters are monohydroxy esters of acrylic or methacrylic acid. Cinnamates and crotonates may also be used.

Example 1: A radiation-curable paint is prepared from the following components in the manner set forth.

(a) Preparation of the Siloxane-Unsaturated Ester Component

Reactants	Parts by Weight
Methoxy functional acyclic siloxane*	178
Hydroxyethyl methacrylate	118
Tetraisopropyl titanate	0.32
Hydroquinone	0.06

*A commercially available methoxylated partial hydrolyzate of monophenyl and phenylmethyl silanes (largely condensed dimethyltriphenyltrimethoxy-trisiloxane) and having the following typical properties:

Average molecular weight	750-850
Average number of silicon atoms per molecule	5-6
Average number of methoxy groups per molecule	3-4

The siloxane, the methacrylate monomer and hydroquinone polymerization inhibitor are heated to 100°C. in a flask fitted with a Barrett type distillation receiver. The titanate catalyst is added and the temperature is raised to 150°C. over a 3 hour period during which time methanol is removed by distillation. The cooled reaction product has a viscosity of 0.6 stoke at 25°C.

(b) Preparation of the Diisocyanate-Unsaturated Ester Component

Reactants	Parts by Weight
2-Hydroxyethyl methacrylate	44.08
Tolylene diisocyanate monomer mixture*	27.00

*80% 2,4-tolylene diisocyanate; 20% 2,6-tolylene diisocyanate.

The diisocyanate is added dropwise to the methacrylate while stirring in a nitrogen gas atmosphere. A rate of addition is maintained so that the exotherm does not exceed 32°C. Stirring is continued for an hour after addition is completed, and the mixture allowed to stand for 16 hours.

(c) Preparation of the Paint Binder Solution

Copolymerizable Components	Parts by Weight
Siloxane-unsaturated ester product of (a)	25
Diisocyanate-unsaturated ester product of (a)	75

The polymerizable reactants in solution with xylene are applied to a metal substrate to an average depth of about 1 mil. The xylene is flashed off and the coating is cross-linked upon the surface of the substrate by exposing the coating to an electron beam. The conditions of irradiation are as follows:

Electron beam potential, kv.	295
Electron beam current, ma.	25
Dose, Mrads	15
Atmosphere	N_2

Example 2:   The procedure of Example 1 is repeated using 75 parts by weight of the siloxane-unsaturated ester product and 25 parts by weight of the diisocyanate-unsaturated ester product.

Example 3:   The procedure of Example 1 is repeated using 50 parts by weight of the siloxane-unsaturated ester product, 25 parts by weight of the diisocyanate-unsaturated ester product, and 50 parts by weight of an equimolar mixture of styrene and methyl methacrylate while omitting the xylene. The film is applied to metal, wood, and synthetic polymeric substrates, i.e., ABS (acrylonitrile-butadiene-styrene) copolymer. The film is applied to an average depth of about 1.2 mils and is cured by exposing it to an electron beam using the conditions of Example 1 and continuing the irradiation until a tack-free film is formed.

Example 4:   The procedure of Example 1 is repeated using 25 parts by weight of the siloxane-unsaturated ester product, 50 parts by weight of the diisocyanate-unsaturated ester product, and 50 parts by weight of a vinyl monomer mixture of alpha-methyl styrene, ethyl acrylate and butyl methacrylate while omitting the xylene. The film is applied to metal, wood and synthetic polymeric substrates, i.e., ABS copolymer. The film is applied to an average depth of about 1.2 mils and is cured by exposing it to an electron beam using the conditions of Example 1 and continuing the irradiation until a tack-free film is formed.

Example 5:   The procedure of Example 1 is repeated using 50 parts by weight of the siloxane-unsaturated ester product, 50 parts by weight of a vinyl monomer mixture of methyl methacrylate, butyl acrylate and 2-ethyl hexyl acrylate while omitting the xylene. The film is applied to the same substrates as in the preceding example to an average depth of about 0.8 mil and is cured by exposing it to an electron beam using the conditions of Example 1 and continuing the irradiation until a tack-free film is formed.

Example 6: The procedure of Example 1 is repeated using 50 parts by weight of the siloxane-unsaturated ester product, 50 parts by weight of the diisocyanate-unsaturated ester product, 50 parts by weight of methyl methacrylate, and 50 parts by weight of an alpha-beta olefinically unsaturated polyester resin while omitting the xylene. The film-forming solution is applied to the previously mentioned substrates to an average depth of about 1.5 mils and cross-linked by exposing the film to an electron beam using the conditions of Example 1 and continuing the irradiation until a tack-free film is formed.

The alpha-beta olefinically unsaturated resin used in this example is prepared as follows:

Starting Materials	Parts by Weight
Maleic anhydride	147
Phthalic anhydride	429
Neopentyl glycol	503

All of the reactants are charged to a four-neck flask fitted with a stirrer, a thermometer, a nitrogen inlet tube, and a ten-inch Vigreux column topped with a Barrett trap for removing the water of condensation. The reactants are slowly heated to 165°C. at which time the first water of condensation distills off. Nitrogen is bubbled through the reactants throughout the reaction. The reaction temperature rises as water is continually removed until a maximum temperature of 225°C. is attained. The column is then removed from the system, 3 weight percent xylene is added to aid azeotropic water removal and heating is continued until the acid number reaches 30. The product is cooled to 100°C. and 0.03 weight percent hydroquinone inhibitor is added and the polymer diluted to 80% nonvolatile content with styrene.

Acryloxypivalyl Acryloxypivalate and Polyesters

In a process described by E.E. Parker, R.M. Christenson and R. Dowbenko; U.S. Patent 3,655,823; April 11, 1972; assigned to PPG Industries, Inc. mixtures of acryloxypivalyl acryloxypivalate and analogous compounds with polyesters are copolymerized by actinic light, free-radical catalysis or radiation curing. The resulting copolymer is a hard, mar resistant, and relatively flexible material.

Example A: A polyester was prepared as follows. A vessel was charged with 882 grams of maleic anhydride, 1,314 grams of adipic acid, and 1,966 grams of neopentyl glycol. The reactants were heated at temperatures from room temperature to 210°C. for 5 1/2 hours, and the polyester had an acid number of 39.8 and a Gardner-Holdt viscosity of C when measured in a 60% solution of ethylene glycol monoethyl ether.

The polyester was solubilized in a vinyl monomer by cooling 65 parts of of the above polyester to 150°C. and adding 0.1 part of trimethylbenzyl-ammonium chloride, and 0.01 part of methylhydroquinone. The mixture was then cooled to 100°C. and 30 parts of 2-ethylhexyl acrylate were added. The resulting polyester–vinyl monomer mixture was allowed to cool at room temperature.

Example B: A polyester was prepared as follows. A vessel was charged with 1,234.8 grams of maleic anhydride, 788.4 grams of adipic acid, and 1,928.16 grams of neopentyl glycol. The reactants were heated at temperatures from room temperature to 210°C. for 7 hours, and the polyester had an acid number of 33 and a Gardner-Holdt viscosity of F when measured in a 60% solution of ethylene glycol monoethyl ether.

The polyester was solubilized in a vinyl monomer by cooling 65 parts of the above polyester to 150°C. and adding 0.1 part of trimethylbenzyl-ammonium chloride, and 0.01 part of methyl hydroquinone. The mixture was then cooled to 100°C. and 30 parts of 2-ethylhexyl acrylate were added. The resulting polyester–vinyl monomer mixture was allowed to cool to room temperature.

Example 1: A copolymer having outstanding physical properties was prepared in the following manner. A vessel was charged with a mixture of 95 parts of the polyester mixture of Example B and 5 parts of acryloxy-pivalyl acryloxypivalate having the formula:

$$CH_2{=}C{-}COOCH_2\overset{\overset{\displaystyle CH_3}{|}}{C}CH_2{-}O{-}CO\overset{\overset{\displaystyle CH_3}{|}}{C}{-}CH_2OOC{-}C{=}CH_2$$
$$\underset{H}{|} \quad \underset{CH_3}{|} \quad \underset{CH_3}{|} \quad \underset{H}{|}$$

0.1 part of methyl hydroquinonine and 0.17 part of trimethylbenzyl-ammonium chloride. The above mixture of acrylic monomer, polyester and 2-ethylhexyl acrylate is then copolymerized by subjecting it to an electron beam source. The mixture is subjected to electron beam impingement at an accelerating voltage of 400 kilovolts and a tube current of 16 milliamps. The total dosage received is 10 megarads. The resulting copolymer is clear, flexible, and has excellent mar resistance and stain resistance.

Example 2:   A copolymer having outstanding physical properties was prepared in the following manner. A vessel was charged with a mixture of 95 parts of the polyester-vinyl monomer mixture of Example B and 5 parts of acryloxypivalyl acryloxypivalate, 0.01 part of methylhydroquinone, and 0.17 part of trimethylbenzylammonium chloride.

The above mixture of acrylic monomer and polyester was then copolymerized by subjecting it to an electron beam source. The mixture was subjected to electron beam impingement at an accelerating voltage of 400 kilovolts and a tube current of 16 milliamps. The total dosage received was 10 megarads. The resulting copolymer was clear, flexible, and had excellent mar resistance and stain resistance.

Examples 3 through 6:   A number of copolymers of acryloxypivalyl acryloxypivalate and various polyesters were prepared and the flexibility of the resulting copolymers after radiation of 5 and 10 megarads was tested by the reverse impact test. The flexibility of the polyester extended acrylic compounds were compared to that of the acrylic compound alone.

The reverse impact test is a test of the flexibility and the impact resistance of the coating, and is measured by dropping various weights from various heights onto a 1/2" diameter steel ball positioned on the coating until a failure in the coating arises. The reverse impact is measured by dropping the weight on the ball on the reverse side of the coating. The test results are given in terms of inch pounds, which is a product of the weight in pounds and the distance in inches in which the weight was dropped.

The following table lists the results where the control is the acryloxypivalyl acryloxypivalate compound alone. All of the coatings were applied to aluminum panels and subjected to electron beam impingement at an accelerating voltage of 375 kv. and a tube current of 14 ma. and in a nitrogen atmosphere.

Example	Maleic acid (moles)	Adipic acid (moles)	Hexahydro-phthalic anhydride (moles)	Neopentyl glycol (moles)	1,6-hexane diol (moles)	Percent acryloxy-pivalyl acryloxy-pivalate	Reverse impact (inch lbs.) 5 mr.	10 mr.
3	3	7		10.6		50	26	34
4	4	6		10.6		50	42	38
5	2		8		10.3	50	50	48
6	6		4		10.5	75	10	4
Control						100	<5	<5

The above tests show the radical increase in reverse impact when the acrylic compound is extended with a polyester.

Crosslinked PVC Film Using Acryloxypivalate Monomer

In a process described by E.A. Hahn; U.S. Patent 3,676,192; July 11, 1972; assigned to PPG Industries, Inc. coating compositions comprising polyvinyl chloride or polyvinylidene fluoride are prepared by mixing a solution of the polyvinyl chloride or polyvinylidene fluoride in an active solvent with an acrylic monomer and subjecting the mixture to ionizing irradiation. The cured films are hard, stain resistant, heat resistant, and mar resistant.

Example 1: A coating composition was formed by adding 10 parts of acryloxypivalyl acryloxypivalate to 50 parts of a 20% by weight solution of polyvinylidene fluoride (Kynar 500) in dimethylacetamide and mixing. A glass substrate was coated with 3 mils of this coating composition and air dried. The coating was then subjected to electron beam impingement at an accelerating potential of 400 kv. and a tube current of 14 ma. The coating received a total dosage of 5 megarads. The resulting coating was crosslinked and found to have excellent mar resistance and stain resistance.

The heat resistance of the above composition was further tested by irradiating with a total dosage of 5 megarads to form a film and taping the film on an aluminum panel and inserting in an oven at 600°F. for a short time. The film was unaffected by the heat resistance test. This was compared to a film formed from polyvinylidene fluoride without the acrylic monomer and irradiation, to the extent of 5 megarads, a film formed from polyvinylidene fluoride along with 5 megarads of irradiation, and the same composition of polyvinylidene fluoride and acrylic monomer without irradiation. The latter three films decomposed when inserted in the oven at 600°F. and melted and turned dark brown.

Example 2: A coating composition was formed by adding 15 parts of acryloxypivalyl acryloxypivalate to 25 parts of a 20% by weight solution of polyvinyl chloride (QYNV) in a 90% dimethylacetamide, 10% methyl ethyl ketone blend and mixing. A glass substrate was coated with 3 mils of this coating composition and air dried. The coating was then subjected to electron beam impingement at an accelerating potential of 400 kv. and a tube current of 14 milliamps. The coating received a total dosage of 5 megarads. The resulting coating was crosslinked and found to have excellent mar resistance and stain resistance. The heat resistance test of Example 1 was repeated with films of polyvinyl chloride and acrylic monomer as above which was irradiated with 5 megarads, films of polyvinyl chloride alone with and without irradiation, and a film of polyvinyl chloride and acrylic monomer without irradiation. The latter three films were decomposed by the treatment while the film formed from the method of this process remained unaffected by the treatment.

Activated Pendant Vinyl Groups in Acrylic Polymers

A process described by G.F. D'Alelio; U.S. Patent 3,654,251; April 4, 1972 relates to crosslinkable polymers having activated vinyl or vinylidene pendant groups extending from the linear polymer chain. Particularly it relates to a specific type of polymer having (1) acrylic or methacrylic ester moieties:

$$CH_2{=}\overset{\overset{\displaystyle R}{|}}{C}{-}COO{-}$$

as well as (2) R_dCOO- ester moieties extending as branches from a linear polymer molecule, where

$$CH_2{=}\overset{\overset{\displaystyle R}{|}}{C}{-}COO{-}$$

is the residue of the acid

$$CH_2{=}\overset{\overset{\displaystyle R}{|}}{C}{-}COOH$$

in which R represents hydrogen and methyl, and R_dCOO- is the residue of an unsaturated, fatty acid containing 16 to 21 carbon atoms consisting of a terminal CH_3 group and at least one and no more than four $-CH{=}CH-$ groups and the remainder consists of $-CH_2-$ groups. The polymers are readily crosslinked by radical, anionic and cationic initiators and effectively, also, by irradiation and by air drying. Related polymers containing only the acrylic or methacrylic ester moieties do not air dry and those containing only R_dCOOH ester groups are not sensitive to irradiation.

Thus, it has been found that crosslinking of various polymeric compositions containing both the

$$CH_2{=}\overset{\overset{\displaystyle R}{|}}{C}{-}COO{-}$$

and R_dCOO- ester linkages can be effected with economical radiation doses by the use of various polymers of this process. It is believed that this is due to the R_d structure in the R_dCOO- ester group in which the $-CH{=}CH-$ react with the oxygen in the air, preventing the oxygen from reacting, during irradiation, with the radiation-sensitive

$$CH_2{=}\overset{\overset{\displaystyle R}{|}}{C}{-}COO{-} \quad \text{groups.}$$

The polymers are conveniently prepared by the reaction of the mixed

$$CH_2=\overset{\overset{\displaystyle R}{|}}{C}-CO-O-OCR_4$$

anhydrides with a polymer having at least one molar percent in the linear chain of a repeating unit having the formula:

$$-CH_2C(R)-$$
$$\overset{|}{\underset{|}{C}}=O$$
$$\overset{|}{O}CH_2CH-\!\!\!-CH_2$$
$$\diagdown\!\!O\diagup$$

where R is methyl or hydrogen as defined above. There are at least two such repeating units per polymer molecule. Preferably the polymer has at least 5 molar percent of the repeating units and optimally 10 to 40%. However, polymers having as high as 100 molar percent can be used, but there is no particular added advantage in these higher percentages since they are economically less desirable.

This reaction can be represented as follows, using the homopolymer for illustrative purposes:

$$-CH_2C(R)- \qquad \overset{R}{\underset{}{CH_2{=}\overset{|}{C}{-}CO}} \qquad -CH_2C(R)-$$

It is believed that the acryloxy group in the repeating unit is attached to the terminal carbon atom of the glycidyl group upon opening of the oxirane ring, and it is believed that the structure is predominantly of this type. However, it is considered equivalent in this process to the arrangement where the acryloxy group is attached to the secondary carbon atom since it is believed that the product actually consists of a mixture of such structures. The reaction is preferably performed in the presence of catalytic amounts of tertiary amine such as triethylamine, tributylamine, pyridine, 1,3,5-tri(dimethylaminomethyl)phenol. The following examples illustrate the process.

Example 1: (a) In a suitable apparatus equipped with a stirrer, reflux condenser, an inert gas inlet, heating mantle and thermostatic control, are placed 450 parts of glycidyl acrylate and 550 parts of methyl ethyl ketone. The apparatus is first swept with nitrogen and a nitrogen atmosphere is maintained above the reaction mass. To the above solution is added 5.0 parts of azobisisobutyronitrile, and the temperature raised to and maintained at 75° to 80°C. for a period of two hours. A clear, viscous solution of homopolymer in quantitative yield is obtained which has the repeating unit structure:

1(a)

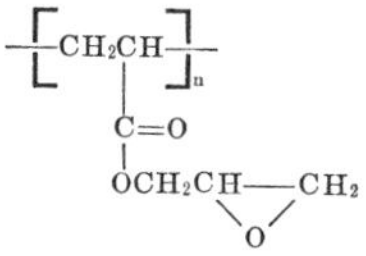

$$\begin{array}{c} -\left[\mathrm{CH_2CH}\right]_n- \\ | \\ \mathrm{C{=}O} \\ | \\ \mathrm{OCH_2CH{-}CH_2} \\ \diagdown\mathrm{O}\diagup \end{array}$$

(b) The solution of the product 1(a) is added slowly with rapid stirring to 2,500 parts of hexane; a fine precipitate forms which is removed by filtration, and is dried in a vacuum oven at 25°C. for 24 hours. There is thereby isolated the solid polymer having an epoxy number corresponding to 97% of structure 1(a).

(c) A mixture of 127 parts of solid polymer 1(a) and 282.5 parts of oleic acid are mixed in a reaction flask and heated under nitrogen at 180°C. for about 30 minutes or until a test sample placed on a glass plate is not unhomogeneous but remains clear when cooled to 30° to 40°C. This gives the crosslinkable polymer having the repeating unit structure:

1(c)

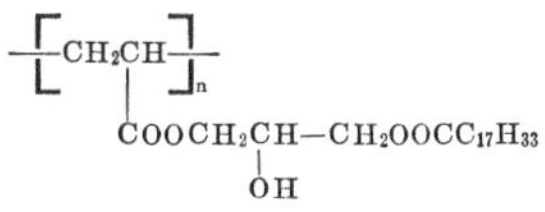

$$\begin{array}{c} -\left[\mathrm{CH_2CH}\right]_n- \\ | \\ \mathrm{COOCH_2CH{-}CH_2OOCC_{17}H_{33}} \\ | \\ \mathrm{OH} \end{array}$$

(d) To a 35% solution of 1(c) in toluene containing 60 parts of trimethylamine is slowly added at 25°C. a solution of 60 parts of acryloyl chloride in 125 parts of toluene during the course of 1 hour. The solid trimethylamine hydrochloride is removed by filtration and the toluene solution is washed with distilled water and dried over anhydrous magnesium sulfate yielding a solution of the polymer having the repeating structure:

1(d)

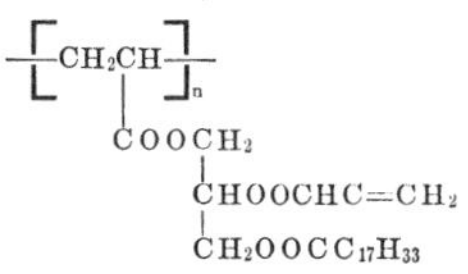

$$\begin{array}{c} -\left[\mathrm{CH_2CH}\right]_n- \\ | \\ \mathrm{COOCH_2} \\ | \\ \mathrm{CHOOCHC{=}CH_2} \\ | \\ \mathrm{CH_2OOCC_{17}H_{33}} \end{array}$$

(e) A sample of the solution 1(d), to which is added 0.05% of a commercial metallic naphthenate drier, is cast on an aluminum sheet and allowed to air dry overnight. The resulting film is insoluble in toluene, acetone and hexane.

(f) A sample of solution 1(d), to which is added 0.25% of cumene hydro-peroxide is cast on a glass plate and placed in an oven at 75°C. At the end of 3 hours, an insoluble, infusible film is obtained. When the hydro-peroxide is omitted from the formulation, curing to the insoluble form occurs when the film is heated at 100° to 118°C. for 30 to 60 minutes.

(g) A sample of solution 1(d) is cast on a glass plate and the solvent re-moved in an oven at 50°C. The resulting soluble film is then exposed, in the presence of air, to the beam of a 1 mev Van de Graaff accelerator and the sample becomes crosslinked and infusible at a dosage of 3.7 to 4.2 megarads; the surface of the film is not tacky. In contrast, when the polymer containing the repeating structure

$$-(CH_2-CH)_n-$$
$$\underset{COOCH_2}{\overset{|}{\underset{CHOH}{\overset{|}{CH_2OOCHC=CH_2}}}}$$

is irradiated in the same manner, a crosslinked polymer is also obtained but the surface is tacky as a result of oxygen inhibition.

(h) The procedure of Example 1(d) above is repeated using 94 parts of methacrylic chloride instead of acrylic chloride and there is obtained the polymer

$$-(CH_2CH)_n-$$
$$\underset{COOCH_2}{\overset{|}{\underset{CHOOCC(CH_3)=CH_2}{\overset{|}{CH_2OOCC_{17}H_{33}}}}}$$

which also yields insoluble films when crosslinked by the procedures of Examples 1(e), 1(f) and 1(g).

Example 2: The procedure of Example 1(c) is repeated 4 times using instead of oleyl chloride equivalent weights of the acid chlorides of linoleic acid, linolenic acid, dehydrated castor oil acid and the mixed fatty acids of tung oil. The four resulting hydroxy esters are converted to the acrylic esters by the procedure of Example 1(d) and are readily cross-linked by the procedures of Examples 1(e), 1(f) and 1(g)

Example 3: The procedure of Example 1(c) is repeated in the presence of 0.5% of tertiary butyl catechol using instead of oleic acid 336 parts of acryloleyl anhydride, $CH_2=CHCOOOCC_{17}H_{33}$, and there is obtained a polymer which crosslinks readily with radical initiators, metallic dryer, ultraviolet light and ionizing radiation, and has the formula shown on the following page.

$$-\!\!\left(\!\text{CH}_2\text{CH}\!\right)_{\!n}\!\!-$$

$$\begin{array}{c}\text{COOCCH}_2\\[2pt]\left[\begin{array}{c}\text{CH}-\\[6pt]\text{CH}_2-\end{array}\right]\!\!\begin{array}{c}-\text{OOCCH}=\text{CH}_2\\-\text{OOCC}_{17}\text{H}_{33}\end{array}\end{array}$$

Additional work with modified glycidyl methacrylate polymers is described by G.F. D'Alelio in U.S. Patent 3,654,240; April 4, 1972.

Ethylene–Acrylic Acid Copolymers and Melamine

V J. Gor and E.H. Manuel; U S. Patent 3,647,520; March 7, 1972; assigned to Continental Can Company, Inc. describe a process for coating articles where a thin film containing a mixture of a copolymer of ethylene and an ethylenically unsaturated carboxylic acid and a melamine composition is applied to the surface of the article and the film mixture crosslinked to form a hardened coating by irradiation with a beam of high-energy electrons.

In the manufacture of containers from metal sheet, a protective organic coating is applied to the side of the metal sheet which is to form the interior of the container. The materials which are employed for coating the metal sheet are generally heat curable, resinous materials which are applied in the form of a solution or dispersion in a volatile solvent. The wet-coated metal surface is passed through an oven in which hot air is circulated to evaporate the solvent and to cure the coating material to the required hardness. As this form of coating application is relatively slow, the art is continually seeking more rapid methods of increasing the speed of coating application.

Among the various methods which have been proposed to increase the speeds at which metal sheet can be coated is to apply a layer of suitable thickness of an ethylenically unsaturated monomer or mixture of monomers to the metal sheet surface whereupon the monomer layer is activated to interact and form a hard, infusible polymer coating by exposing the layer to a source of high-energy electrons. The activation action is carried out at ambient temperatures, and since the action of the high-energy electrons is extremely rapid, the monomer layer can be rapidly polymerized to a cured coating of the required hardness in a continuous flow movement across the source of irradiation.

Although a wide variety of ethylenically unsaturated monomeric compounds have been proposed for use in forming polymeric coatings using high-energy electron curing procedures, these compounds, when exposed to a source of high-energy electron radiation do not always form hard, adherent solvent-resistant coatings of the type which are required for container coatings.

In the process there is provided a rapid method of preparing hard, adherent, polymeric coatings on articles and particularly metal substrates where there is applied to the article substrate a thin film containing a mixture of a melamine composition and a copolymer of ethylene and an ethylenically unsaturated carboxylic acid and the coated side of the substrate is exposed to a source of high-energy electrons to effect crosslinking of the copolymer and to convert the mixture into a continuous, hard coating on the substrate.

The ethylene-ethylenically unsaturated carboxylic acid copolymers employed in the process generally contain about 75 to 90 weight percent ethylene and about 10 to 25 weight percent of the ethylenically unsaturated carboxylic acid and preferably about 75 to 85 weight percent ethylene and 15 to 25 weight percent of the carboxylic acid. The molecular weight of the ethylene-carboxylic acid copolymers as defined in terms of melt index (ASTM D1238-58T) is generally in the range of 100 to 500 g./10 min. and more particularly in the range of 150 to 400 g./10 min.

The ethylenically unsaturated carboxylic acid component of the copolymer is an α,β-ethylenically unsaturated carboxylic acid having from 3 to 8 carbon atoms. Examples of such acids are acrylic acid, methacrylic acid, ethacrylic acid, itaconic acid, maleic acid, fumaric acid, monoesters of dicarboxylic acids, such as methyl hydrogen maleate, methyl hydrogen fumarate, ethyl hydrogen fumarate, and maleic anhydride. Although maleic anhydride is not a carboxylic acid in that it has no hydrogen attached to the carboxyl groups, it can be considered as acid for the purposes of the process because of its chemical reactivity being that of an acid. Similarly other α,β-monoethylenically unsaturated anhydrides of carboxylic acids can be employed.

To effect the coating of an article substrate in accordance with the process, a thin film of the aqueous dispersion of the ethylene-carboxylic acid copolymer and melamine composition is applied to the substrate surface and is heated to about 110° to 250°C. for about 10 to 30 seconds to effect removal of the aqueous dispersing medium and is irradiated with a beam of high-energy electrons to cure the mixture to a hard, infusible coating.

Example: Aqueous colloidal dispersions were prepared having the following composition ranges: ethylene-acrylic acid copolymer (E/AA — acrylic acid content 20 weight percent, melt index 300 decigrams/min. ASTM D-1238-8T), 25 parts by weight; melamine, 5 to 15 parts by weight; NH_4OH (28% solution), 5 parts by weight, and water, 70 parts by weight.

The dispersion compositions were applied to 5 x 3 inch cold-rolled chromium treated steel plates at a thickness of 0.1 mil. The coated plates were left at room temperature for 5 minutes to allow evaporation of the water dispersant. The dried plates were then passed under the window of an electron accelerator and the dried coating irradiated under the following conditions:

Voltage	300,000 volts
Current	30 milliamperes
Conveyor speed	53 feet/minute
Atmosphere	Air
Dosage	4.2 megarads
Distance of coating surface from window	8 inches

The irradiated films were found to be polymerized, hard, adherent coatings. The coated plates were then rubbed with a cloth saturated with methyl ethyl ketone (MEK) to determine whether the coating softens or can be removed by the rubbing. The number of rubs with MEK which are required to soften or remove the coating is a direct indication of the degree to which the coating material has been cured or crosslinked. The results of the MEK test are recorded in the table below.

The procedure of the example was repeated with the exception that Cymel 301, a commercially available hexamethylol ether of hexamethylol melamine was substituted for melamine in the dispersion coating. The results of the MEK rub test on dispersion coatings containing Cymel 301 are also recorded in the table below.

For purposes of comparison, the procedure of the example was repeated with the exception that amine compositions other than melamine-containing materials were substituted for melamine in the dispersion coatings that were crosslinked using high-energy electrons. The results of the MEK test on these comparative coating materials are also recorded in the following table (designated by the symbol "C").

From the table it will be readily apparent that the electron beam irradiation of coating compositions comprised of mixtures of ethylene-acrylic acid copolymers and melamine compounds such as melamine and alkylated melamine-formaldehyde condensates produces coatings which are cured to a more substantial degree than ethylene-acrylic acid copolymer-containing coating compositions which contain amine compounds closely related to melamine.

Test No.	Crosslinking Agent (CLA)	Concentration of CLA in Coating Formulation, parts	MEK* Rubs After Radiation
1	Melamine	5	5
2	Melamine	10	10
3	Melamine	15	6
4	Cymel 301	5	6
5	Cymel 301	10	12
6	Cymel 301	15	8
C_1	Benzoquanamine	5	1
C_2	Benzoquanamine	10	1
C_3	Benzoquanamine	15	1
C_4	Butylated urea-formaldehyde resin**	5	1
C_5	Butylated urea-formaldehyde resin**	10	1
C_6	Butylated urea-formaldehyde resin**	15	1
C_7	None	--	1
C_8	Cymel 301***	98	†

*All coatings could be removed with 1 MEK Rub before panel was
　　irradiated.
**Uformite F-140 commercial available butylated urea-formaldehyde
　　resin.
***No ethylene-acrylic acid copolymer in coating formulation.
†Tacky, did not harden under electron beam radiation.

Ethylene-Acrylic Acid Copolymer and Polyvalent Metal Salt

A. Ravve and J.T. Khamis; U.S Patent 3,671,295; June 20, 1972;
assigned to Continental Can Company, Inc. describe a process for coating
articles where a film of an aqueous colloidal dispersion of a water-
dispersible carboxyl acid-containing olefin polymer and a salt of a polyvalent metal is applied to the surface of the article to be coated, the
aqueous dispersant removed, and the polymer crosslinked to form a hardened
coating using high-energy ionizing radiation. The process is illustrated by
the following examples.

Example 1: 100 grams of an ethylene-acrylic acid copolymer having a
melt index of 300 decig./min. (ASTM D1238-58T) containing 20 weight
percent acrylic acid was dispersed in 400 grams of water heated to 85°C.
To this mixture was added 16.8 grams of 28% ammonium hydroxide and the
mixture heated at 95° to 98°C. for 1 hour to prepare a dispersion of the

copolymer. The dispersion was cooled to room temperature and 7.4 grams of ZnCO₃ was added.

The zinc ion-containing copolymer dispersion was applied to a series of 5 x 3 inch cold-rolled chromium treated steel plates at a film thickness of 0.0001 inch The coated plates were heated to 195°C. for 2 minutes to remove the water and then were irradiated in a nitrogen atmosphere with an electron beam at 300 million electron volts (mev) to dosages ranging from 1.75 to 6.25 megarads. At each dosage, the irradiated polymer films were found to be crosslinked, hard, adherent coatings. The coatings were found to contain 3.2% by weight zinc ion. When heated to 180°F., the coatings were found to be tack free, indicating that they were sufficiently crosslinked.

By way of contrast, coatings on steel plates coated in a manner similar to that of the example with the exception that ZnCO₃ was not present in the dispersion applied to the plate were found to be tacky after the coated plate was heated to 180°F. indicating that the coatings were not adequately crosslinked.

By way of further contrast, coatings on steel plates coated in a manner similar to that of the example with the exception that the ethylene-acrylic acid-ZnCO₃ dispersion was only heat cured at 195°C. for 2 minutes, and not exposed to an electron beam, were found to be tacky after heating of the coated plate to 180°F.

Example 2:   The procedure of Example 1 was repeated with the exception that ZnO was substituted for ZnCO₃ in the polymer dispersion. A series of tests was conducted using ZnO at varying concentrations. The radiation dosage employed in this test series was 4 megarads. The irradiated films were found to be hard, insoluble, adherent coatings The coated plates were rubbed with a cloth saturated with methyl ethyl ketone (MEK) to determine whether the coating softens or is removed by the rubbing. The number of rubs with MEK which are required to soften or remove the coating is a direct indication of the degree to which the coating material has been cured or crosslinked.

The results of the MEK test are recorded in the following table. For purposes of comparison, control runs were made in which ZnO was not present or present in amounts less than 1% by weight of the coating composition. The results of the MEK test on these comparative coating materials are also recorded in the table (C₁ and C₂).

For purposes of further comparison, coating materials which were not irradiated, that is, only heated to 195°C. for 2 minutes, were also evaluated.

The results of these comparison runs are also recorded in the table below.

Run No.	Zinc oxide incorporated in coating composition (percent by weight)	MEK resistance, coating composition	
		Not irradiated (No. of rubs)	Irradiated (No. of rubs)
1	1.0	6	19
2	2.0	12	19
3	3.0	12	20
4	4.0		25
C_1	0	2	3
C_2	0.5	4	7

Thiourethane-Urethane Acrylates

In a process described by T.J. Miranda; U.S. Patent 3,600,359; Aug. 17, 1971; assigned to The O'Brien Corporation a radiation curable compound is produced by reacting a polymercapto compound, a stoichiometric excess of a polyisocyanate (determined on the basis of the mercapto groups) and a hydroxy acrylate.

The acrylate esters are acrylate terminated compounds containing a plurality of intermediate urethane groups and a plurality of intermediate thiourethane groups. The radiation curable acrylate esters may be prepared by first preparing, as an intermediate, a thiourethane containing at least one reactive isocyanate group. Such intermediates may be prepared by reacting a polyisocyanate with a polymercapto compound. The reactive isocyanate group of the intermediate may be then reacted with a hydroxyfunctional acrylate monomer to form the radiation curable mercaptate esters. The acrylates contain both thiourethane and normal urethane or oxyurethane groups, in addition to the acrylate terminal groups.

Example 1: To a 3-neck flask equipped with stirrer, condenser, and inert gas outlet was added 78 g. (0.196 mol) of trimethylolpropane tris(mercaptopropionate). Stirring was begun and 130 g. (0.75 mol) of toluene diisocyanate was added over a 45 minute period, keeping the temperature at 25°C. At the end of the addition, the flask was heated to 70°C. for 30 minutes at which time the temperature adjusted to between 35° and 45°C. and 121 g. (1.04 mols) of 2-hydroxyethyl acrylate was added dropwise. Upon completion of the addition of the hydroxyethyl acrylate, 50 g. of ethyl acetate (solvent) was added, the flask heated to 70°C. for 30 minutes and then cooled. The resulting composition was clear and viscous.

The clear viscous vehicle was drawn down on wood, radiated at 10 Mrads and cured to a hard, solvent-free film. To determine whether lower doses could be obtained, the radiation was repeated on another sample and the dose lowered to 0.75 Mrads in air and the coating was cured to a hard solvent-resistant film. Solvent resistance is determined by rubbing an acetone soaked cloth across the coating 50 times. If no solvent attack is noted the coating is considered to pass. By contrast a vinyl organosol prepared by conventional means is completely removed by acetone rubbing. The coating was resistant to 10% alkali (24 hours). Samples were prepared and cured at 1 Mrad on wood and metal and exposed in Florida for 6 months and compared against conventional clear urethane varnish. At the end of 6 months, the coatings had high gloss, very slight chalking and good appearance.

Example 2:   The resin of Example 1 was reduced as follows. 100 parts of the resin was thinned with 20 parts of vinyl pyrrolidone. The clear solution was drawn down (3 mil film) on wood and subjected to 1.5 and 3.0 Mrad doses, respectively, in air. The coatings cured to hard tack-free films which were solvent resistant.

Example 3:   Into the equipment described in Example 1 was charged 137 grams (0.260 mol) of trimethylolpropane tris(thioglycolate). Temperature was maintained at 25°C. and 274.81 g. (1.55 mols) of toluene diisocyanate were added over a 30 minute period after which time the temperature was increased to 70°C. and held for 30 minutes. The temperature was then lowered to 45°C. and 254.5 g. (2.19 mols) of 2-hydroxyethyl acrylate were added over a 1 hour period. At the end of the addition of the hydroxyethyl acrylate, the temperature was increased to 70°C. and held for 30 minutes.

The resulting resin was cooled and then thinned with 66.68 g. of ethyl acetate. The thin, clear, viscous material was drawn down on wood at 3 mils and given a 10 Mrad dose under the electron beam and cured to a hard, tack-free surface. Samples were also drawn down at the same thickness and given 5 and 1 Mrad doses, respectively, in each case cured to a hard, tack-free solvent-resistant film.

Hydroxy-Containing Vinyl Resin-Unsaturated Isocyanate Reactions

A process described by K. Honda, M. Miyazaki, S. Nomura, K. Akiyama and K. Hirose; U.S. Patent 3,694,415; September 26, 1972; assigned to Dainippon Ink & Chemicals, Inc. and The Dainippon Ink Institute of Chemical Research, Japan provides a vinyl-type resin which can easily be air cured under the action of high-energy ionizing radiation.

The process involves preparing a coating composition by reacting (a) a hydroxy-substituted vinyl polymer or copolymer (referred to as the hydroxy polymer) with (b) 0.1 to 0.4 g. equivalents per 1,000 g. of the hydroxy polymer, of an unsaturated isocyanate obtained by reacting an ester of an α,β-unsaturated monocarboxylic acid containing at least one hydroxyl group in the molecule, with a diisocyanate in a molar ratio of 1.4:1 to 0.6:1, and preferably from 1.2:1 to 0.8:1, where the unsaturated isocyanate may conveniently be used in the form of a mixture obtained by the reaction without isolating it. A monomer which is copolymerizable with the resin may be added to the isocyanate. The following examples illustrate the process.

Example 1:   A mixture of 50 parts of methyl methacrylate, 30 parts of butyl methacrylate, 20 parts of 2-hydroxyethyl methacrylate, and 1 part of azobisisobutyronitrile was added dropwise to 50 parts of boiling butyl acetate over a period of 2 hours. After completion of polymerization, 50 parts of methyl methacrylate was added to the polymerization mixture to obtain a hydroxy polymer solution (A-1).

In another reaction vessel, 168 parts (1.0 mol) of hexamethylene diisocyanate was reacted with 100 parts (0.86 mol) of 2-hydroxyethyl acrylate at 50°C. for 3 hours and then diluted with 170 parts of methyl methacrylate to obtain a mixture (I-1) of an isocyanate equivalent of 400 containing an unsaturated isocyanate as a main component.

135 parts of (A-1) was reacted (urethanized) with 10 parts of (I-1) at 60°C. for 5 hours. The resinous solution thus obtained was applied by means of a bar coater to a steel panel to form a coating film of a thickness of 40 microns and the coating film was irradiated at room temperature at a dose of 3 Mrad by means of an electron accelerator of an acceleration voltage of 500 kev. The irradiated coating film exhibited excellent properties. The results of the performance tests are shown in the following table.

Example 2:   A cured coating film of good properties was obtained by applying and curing a resinous solution under the same conditions which solution was prepared in the same manner as in Example 1, except that 5 parts of the unsaturated isocyanate mixture (I-1) in urethane formation was used. The results of the performance tests are shown in the table on the following page.

Example 3:   A monomer mixture consisting of 40 parts of methyl methacrylate, 20 parts of vinyl acetate, 25 parts of butyl acrylate and 15 parts of 2-hydroxyethyl methacrylate was polymerized in the same solution polymerization process as in Example 1, and then diluted with methyl methacrylate. 100 parts of the polymer solution thus obtained was reacted in the same manner as in Example 1 with 4.4 parts of a mixture obtained by reacting in another reaction vessel 188 parts (1.0 mol) of xylylene diisocyanate with 128 parts (1.1 mol) of 3-hydroxyethyl acrylate at 50°C. for 3 hours to an isocyanate equivalent of 355, to obtain a solution of a copolymer resin having active unsaturated bond in side chains. 90 parts of the resinous solution and 10 parts of rutile-type titanium oxide were

milled in a sand mill to obtain a white enamel. The enamel was applied
to a steel panel and cured by the irradiation of electron rays in the manner as in Example 1. The coating film thus obtained exhibited excellent
properties which are shown in the table below.

Example 4:   100 parts of a resinous solution prepared in the same manner
as in Example 1 were mixed with 10 parts of a polyisocyanate resin obtained by reacting trimethylolpropane with tolylene diisocyanate in the
molar ratio of 1:3 at 60° to 70°C., and the mixture was applied and cured
in the manner as in Example 1 to obtain a coating film having excellent
hardness, solvent resistance and other properties. The results are shown
in the table below.

Example No.	1	2	3	4
Gram equivalents of unsaturated isocyanate per 1,000 g. of hydroxy polymer	0.4	0.2	0.3	0.4
Hardness (pencil hardness)	H	F	H	2H
Scratch resistance	○	○	○	○
Gasoline resistance (2 hours immersion)	□	□	□	□
Water resistance (40° C., 24 hours)	□	□	□	□
Acid resistance (5% aq. HCl, 24 hours)	□	□	□	□
Alkali resistance (5% aq. NaOH, 24 hours)	□	□	□	○

NOTE.—□: Not changed at all, ○: Slightly changed, △: Changed to a great extent, X: Completely deteriorated.

RESISTS

Dibutyltin Maleate Sensitizer

A process described by B. Broyde; U.S. Patent 3,594,170; July 20, 1971;
assigned to Western Electric Company, Incorporated relates to additives
to standard negative photoresists which result in increased capture of
electrons per unit flux of incident electrons, and increased reactivity
of the photoresist itself. The process has particular application in the
generation of microminiature circuit patterns by electron beam exposure
of negative photoresists.

A negative photoresist is an organic material which, when exposed to
radiation, undergoes chemical reactions of the type referred to as cross-
linking, which reactions result in insolubilizing the exposed photoresist.
The crosslinking reactions are of the type that can be initiated either by
light or by electrons. Because it is possible to generate electron beams of
substantial energy but only 0.1μ or smaller diameter, their use in the
generation of extremely small circuit patterns is preferred to the use of
light.

Electron beams also have a much better resolution capability than is possible when using an optical mask and light exposure, and they have a
much greater depth of focus. The exposure of a conventional positive

photoresist involves solubilization of the exposed areas, and the chemical reactions involved are of the scission of degradation type, which also require absorption of light or electrons. Because this type of photoresist requires higher flux densities for proper exposure than negative photoresists require, electron beams are not widely employed in this service. Materials that have been successfully used as electron-sensitive positive photoresists are discussed by Haller et al, IBM Journal, May 1968, pp. 251-256.

The most common negative photoresists in current use are Kodak Photoresist (KPR) and Kodak Thin Film Resist (KTFR) The KPR composition is based on the dimerization of polyvinyl cinnamate, while KTFR is based on the crosslinking of polymerized isoprene dimers. Other members of the KPR group are KPR II and KPR III, and KOR (Kodak Ortho Resist). Another product, KMER (Kodak Metal Etch Resist), belongs to the KTFR group. The process will be described with primary reference to use of KPR and KTFR.

The crosslinking and insolubilization of resists is a complex phenomenon, but is believed to be describable, broadly, as follows. A KPR-type resist has the following general formula.

(1)

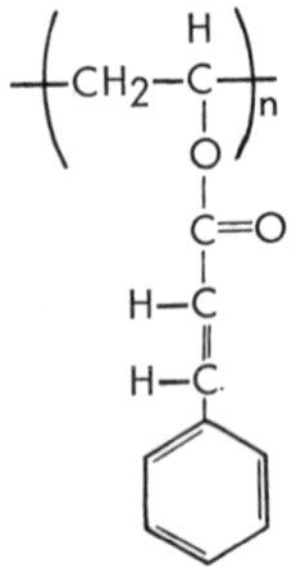

$$\left(CH_2-\overset{\overset{\displaystyle H}{|}}{\underset{\underset{\displaystyle \begin{matrix} C=O \\ | \\ H-C \\ \parallel \\ H-C \end{matrix}}{|}}{C}}\right)_n$$

The number average molecular weight (NAMW) is 180,000-230,000, and the weight average molecular weight (WAMW) is 315,000-350,000. Upon exposure to light or electron energy, a diradical is formed:

(2)

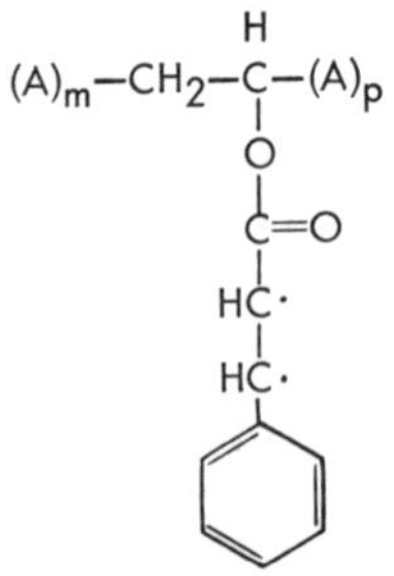

$$(A)_m-CH_2-\overset{\overset{\displaystyle H}{|}}{\underset{\underset{\underset{\displaystyle HC\cdot}{\displaystyle |}}{\underset{\displaystyle HC\cdot}{|}}}{C}}-(A)_p$$

A is the KPR monomer [structure (1) where n = 1)]

The diradical then reacts with another diradical to form a 4-member ring:

(3)

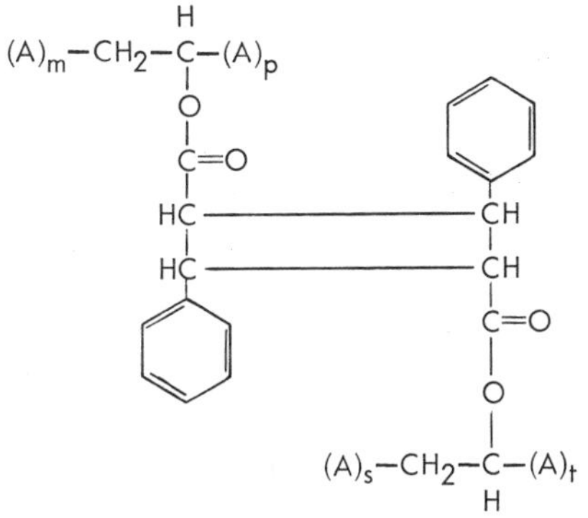

Further excitation and dimerization leads to an insoluble product; no free radicals participate in these reactions. The KTFR-type resists can be characterized as follows.

(4)

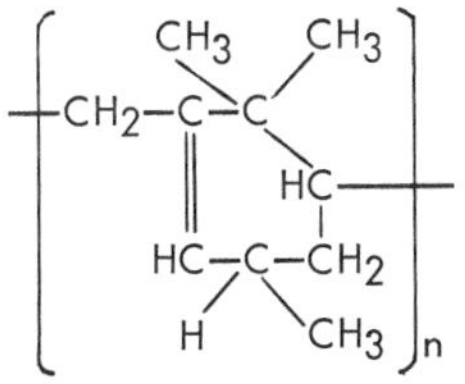

These materials have a NAMW of $65,000 \pm 5,000$ and a WAMW of about $100,000$, and are insolubilized by free radical reactions. Thus, radiation produces a diradical:

(5)

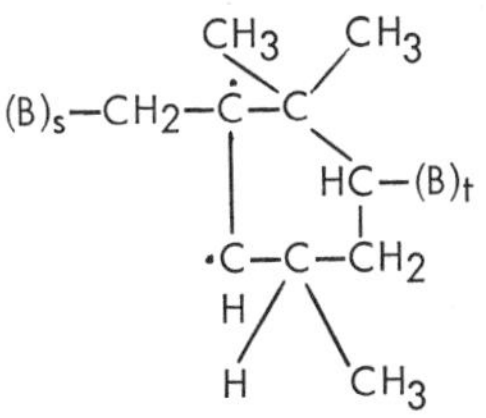

where B is the monomer of (4). The diradical reacts with other molecules until the free radical terminates. For good resolution, additives are believed to be incorporated to keep the chain short. In all of the above structural formulae, the subscripts (n, m, p, s, t) refer to integers which are determinative of molecular weight. While KPR and KTFR are insolubilized by different mechanisms, both result in crosslinked systems.

The procedures for generating a microminiature pattern circuit by electron bombardment of a photoresist are well established, and are summarized briefly below. The substrate is typically an oxidized silicon wafer or a chromium coated glass plate. The photoresist is dissolved in a suitable solvent and applied to the substrate, which may then be spun at a high speed to leave an even film of the photoresist, having a controlled thickness, on the substrate surface.

Alternatively, the photoresist solvent solution may be sprayed on. In either case, most of the solvent evaporates immediately. The photoresist coated substrate is then dried or baked briefly to drive off any remaining solvent and to improve adhesion. The coated substrate is then placed in a vacuum chamber and, when the vacuum has been established, it is radiated in the desired pattern and with an appropriate dosage. The coated and radiated substrate is then placed in a developer, which is a solvent for the soluble portion of the resist, to dissolve and remove the unexposed portions. It is again dried or baked. The desired pattern area on the substrate is now free of any covering film, and etching, plating or oxidizing follows. After this step, the remaining resist is stripped off. There are a variety of limitations imposed upon the radiation step.

Briefly, the amount of radiation must fully expose the photoresist all the way down to the substrate, or else the developed photoresist will float off when the underlying undeveloped photoresist is dissolved in the developer. On the other hand, too much radiation will cause stripping problems and even polymer degradation. The amount of radiation necessary to form an insoluble photoresist is a function of the molecular weight of the material, and the gross amount of radiation. The efficiency of the crosslinking reactions is related to the accelerating potential of the electrons, penetration range (also a function of potential) and other factors.

For instance, it has been determined that the maximum film thickness that can be developed by 5 kv electrons is about 6500 A., and by 10 kv. electrons is about 2μ. On the other hand, photoresists should initially be at least 6000 A. thick to avoid pinhole problems (a 6000 A. film will shrink to about 4000 A. when developed). Other limitations which must be considered are electron scatter within the film and back-scatter from the substrate, though these are of lesser order.

Other workers have carried out extensive studies on the above limitations, particularly with respect to the sensitivity and resolution capability of standard resists. This work is referenced below for background information.

Thornley et al, "Electron Beam Exposure of Photoresists,"
 J. Electrochem. Soc., Vol. 112, No. 11, November 1965,
 pp. 1151-1153.
Broers, "Combined Electron and Ion Beam Process for

Microelectronics," Microelectronics and Reliability, Vol. 4,
1965, pp. 103-104.

Kayaya et al, "Measurement of Spot Size and Current Density
Distribution of Electron Probes by Using Electron Beam Exposure
of Kodak Photoresist Films," Zeit. f. Licht-und Elektronioptik,
Vol. 25, No. 5, 1967, p. 31.

Matta, "High Resolution Electron Beam Exposure of Photoresists,"
Electrochemical Technology, Vol. 5, Nos. 7-8, July-August
1967, pp. 382-385.

None of these workers have made any effort to alter conventional photoresist compositions, although it is significant to note that Thornley et al
appreciated the problems which they pose: "For serial exposures, such
as may be required in printed circuit generators, the maximum exposure
rates are limited by the sensitivities of presently available resists."
(Thornley et al, op. cit., p. 1151).

While many workers who have studied electron beam development of resists
to generate small patterns have worked only with the available resists,
workers in the field of photolithography, where photoresists were first employed, have proposed literally thousands of compounds as photopolymerization initiators, catalyzers and sensitizers. The end in view was generally to increase the sensitivity or resolution of the photoresist to light
of a particular wavelength. This work is not readily summarized, but the
following U.S. patents are considered representative: U.S. Patents
2,816,091; 2,831,768; 2,861,057; 3,168,404; 3,178,283; 3,257,664;
and 3,331,761.

In essence, this process comprises the addition to a photoresist solvent
solution, in small amounts, of solvent-soluble organometallic compounds
having a heavy metal moiety selected from groups III, IV or V of the
Periodic Table. The concentration of the organometallic compound in
the solvent resist solution is below 5% and is generally within the range
of about 0.1 to 2% (all percentages used are weight percent). Concentrations above 2% will not generally be soluble in the solvent and, if the
concentration exceeds the solubility limit, resolution will be lost. In
the dried (solvent free) photoresist film, the organometallic concentration
will be generally less than 50%.

If one knows the average molecular weight of the photoresist film and the
electron accelerating potential, and makes certain assumptions regarding
electron penetration, scatter and energy transfer, the gel dose of energy
can be calculated from theory. (The gel dose is the electron flux necessary
to record an image in the film surface, i.e., the minimum dose to cause
insolubility.) Experimental results are in fair agreement with such theory.

Similar calculations, which take into account the presence of a heavy metal ion in the film, show that the gel dose should be lower. This is not entirely unexpected, since heavy metals are noted for their ability to stop or capture electrons (explaining the use of lead as radiation shielding). More precisely, the energy loss when an electron collides with a heavy metal is quite high, and energy transfer sites will overlap when a metal is present in the film. This means that insolubilization should be more efficient. For example, a theoretical calculation will show that an equimolar mixture (based on monomer composition) of KTFR and hexaphenyldilead, which is about a 1% solution, should absorb about 15% more energy than KTFR alone, reducing the gel dose by a corresponding amount. However, in this instance the theory fails to predict experimental results; in fact, such a mixture reduces the gel dose by about half.

It is not known whether the theory is defective in some unknown manner or whether the additive participates in the crosslinking reaction in some undefined way. Presumably, the theory is defective in not taking the unknown additive participation into account. Following is a representative list of compounds which are effective in carrying out the process:

Group III: Cyclopentadienylthallium.
Group IV: Hexaphenyldilead, dibutyltin maleate and tetra-
cyclohexyltin.
Group V: Triphenylarsine, triphenylstibene and triphenylbismuth.

While the decrease in gel dosage provided by these additives is in itself significant, it is not a necessary conclusion that the improvement will be of the same magnitude when the flux density necessary to expose a resist to a thickness of 3000 or 4000 A. is considered. More surprisingly, the magnitude of improvement is increased under these conditions. The following examples will illustrate this, but it is first necessary to delineate standard procedures and dosages for comparison purposes.

Both KPR and KTFR are supplied dissolved in a solvent. With the latter composition, a KTFR thinner may also be employed; this acts merely to reduce viscosity and produce a thinner film. The solvent system used for KPR is 86 to 87% chlorobenzene and 13 to 14% cyclohexanone. The KTFR solvent system is 12% ethylbenzene, 82% mixed xylenes and 6% methylcellosolve. Both systems also contain a sensitizer; in KTFR this is believed to be 2,6-bis(p-azidobenzilidene)-4-methylcyclohexanone. The KTFR thinner is primarily mixed xylenes.

To establish a basis for comparison, tests were first made with KPR and KTFR without any additives. KPR was applied to a chromium coated glass plate. This was then spun so that the resulting coating, after baking at

150°C. for 10 minutes, was 6000 A. thick. The coated plate was then placed in a vacuum chamber and radiated with electrons accelerated at 15 kv. The plate was developed with KPR developer and baked at 150°C. for 10 minutes. The following results were obtained.

(a) Flux needed to record an image (gel dose) =
1.1×10^{-6} coul./cm.^{2}.
(b) Flux needed to form 3000 A. thick resist layer =
6×10^{-6} coul./cm.^{2}.
(c) Flux needed to form maximum thickness (after develop-ment, 4000 A.) resist = 10×10^{-6} coul./cm.^{2}

KTFR was mixed with KTFR thinner in a 1:3 ratio and the mixture applied to a chromium coated glass plate (or, alternatively, to a silicon slice on-to which a 18,000 A. SiO_2 layer had been grown), and then spun to a thickness of 8000 A. After baking at 150°C. for 10 minutes, the film was 6000 A. thick. The coated plates were then put into a vacuum chamber and radiated with 15 kv. electrons. The plate was developed with KTFR developer and KTFR rinse, and baked at 150°C. for 10 minutes. The following results were found.

(a) Flux needed to record image (gel dose) =
0.9×10^{-6} coul./cm.^{2}.
(b) Flux needed to form 3000 A. film = 4×10^{-6} coul./cm.^{2}.
(c) Flux needed to form maximum (4000 A.) thickness =
7.5×10^{-6} coul./cm.^{2}.

Under identical conditions, but with 5 kv. electrons, the dose densities required to expose KTFR films are:

(a) 0.5×10^{-6} coul./cm.^{2}
(b) 0.75×10^{-6} coul./cm.^{2}
(c) 2×10^{-6} coul./cm.^{2}

Example 1: A saturated solution (2%) of dibutyltin maleate in KPR was prepared, and the same procedure was followed as outlined above. Results with 15 kv. electrons were as follows.

(a) 1.0×10^{-6} coul./cm.^{2}
(b) 1.5×10^{-6} coul./cm.^{2}
(c) 2.2×10^{-6} coul./cm.^{2}

It will be noted that with this additive, the improvement in gel dose was only nominal (about 10%), but was substantial for exposure to 3000 and 4000 A.

Figure 5.3a below graphically illustrates the improvement in sensitivity achieved by using dibutyltin maleate as an additive. It also illustrates that dibutyl maleate, by itself, has no effect on sensitivity.

Example 2:   The same tests were carried out on a film prepared from a saturated solution (1%) of hexaphenyldilead in KTFR and thinner. Results with 15 kv. electrons were as follows.

(a) 0.5×10^{-6} coul./cm.^{2}
(b) 1.0×10^{-6} coul./cm.^{2}
(c) 1.5×10^{-6} coul./cm.^{2}

Here, the reduction in gel dosage was 45%, and the reduction in dosage for 4000 A. film was five-fold. This is the most effective additive known. Figure 5.3b illustrates the rather dramatic increase in sensitivity of KTFR when hexaphenyldilead is used as an additive.

FIGURE 5.3:   SENSITIZERS FOR NEGATIVE PHOTORESISTS

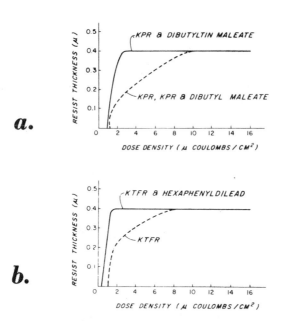

(a)(b) Resist Thickness vs. Flux Density for Exposure of 6000 A. with Dibutyltin Maleate (a) and Hexaphenyldilead (b).

Source: B. Broyde; U.S. Patent 3,594,170; July 20, 1971

Examples 3 through 7: The standard procedures were repeated on five other additives in KTFR and thinner, with 15 kv. electrons, and concentrations and results as set forth below.

Ex. 3: 1% Tetracyclohexyltin
        (a) 0.5×10^{-6} coul./cm.^{2}
        (b) 2×10^{-6} coul./cm.^{2}
        (c) 4×10^{-6} coul./cm.^{2}

Ex. 4: 2% Triphenylbismuth
        (a) 0.85×10^{-6} coul./cm.^{2}
        (b) 1.5×10^{-6} coul./cm.^{2}
        (c) 1.85×10^{-6} coul./cm.^{2}

Ex. 5: 2% Triphenylstibene
        (a) 0.7×10^{-6} coul./cm.^{2}
        (b) 1.85×10^{-6} coul./cm.^{2}
        (c) 2.5×10^{-6} coul./cm.^{2}

Ex. 6: 2% Triphenylarsine
        (a) 0.7×10^{-6} coul./cm.^{2}
        (b) 0.9×10^{-6} coul./cm.^{2}
        (c) 1.85×10^{-6} coul./cm.^{2}

Ex. 7: 1% Cyclopentadienylthallium
        (a) 0.5×10^{-6} coul./cm.^{2}
        (b) 2×10^{-6} coul./cm.^{2}
        (c) 3×10^{-6} coul./cm.^{2}

The magnitude of improvement brought about by each of the additives is readily seen in the table below, where the percent reduction in dose for each of the three levels, as compared to the photoresist without any additives, is set forth.

	Reduction in Dose, Percent		
Example	(a) Gel Dose	(b) 3000 A.	(c) 4000 A.
1	9.1	75	78
2	45	75	80
3	45	50	47
4	5.5	63	75
5	22	54	67
6	22	77	75
7	45	50	60

It will be noted that one of the effects of the organometallic additives is to increase the slope of the plot of resist thickness vs. dose density to near

infinity near the gel point (see Figures 5.3a and 5.3b). By using the minimum dose density needed to achieve the desired thickness, back-scattered
electrons or scattered primary electrons are minimized if not eliminated,
and resolution capability of the resist is correspondingly increased. Under
these conditions, an edge definition of about 300 A. can be expected as
an upper limit. This is significantly better than previously reported
definition.

It will be further noted by comparing the KTFR radiated with 5 and 15 kv.
electrons, that the 5 kv. samples required less energy at all three stages.
It is quite true, in fact, that lower energy electrons act much more
efficiently than higher energy electrons; on the average, about 2.5 times
the number of molecules at each energy transfer point will react at 5 kv.
than will at 15 kv. It would seem appropriate then, to utilize lower
energy electrons, but control of the size of the beam is more difficult at
low energies. If very high potentials are used (+20 kv.) the efficiency
of crosslinking drops too low and back-scatter can become a significant
problem. For these reasons, a 15 kv. accelerating potential is preferred.

Maleic Anhydride Copolymers

A process described by A.S. Deutsch and W.G. Herrick; U.S. Patent
3,594,243; July 20, 1971; assigned to General Aniline & Film Corp.
involves polymeric resist image formation involving the irradiation of a
maleic anhydride polymer coating with corpuscular radiation, e.g., an
electron beam to insolubilize the exposed areas, resist image formation
being effected by solvent treatment of the polymer coating. The following examples illustrate the process.

Example 1: A bimetallic plate comprising copper coated aluminum is
immersed into a dilute nitric acid solution in order to clean the surface.
After rinsing with water and drying, a resist-forming composition comprising a 5% acetone solution of a 1:1 maleic anhydride-methyl vinyl
ether copolymer having a specific viscosity of 2.0 measured at 25°C. as
a 1% solution in methyl ethyl ketone and commercially available under
the trade name designation Gantrez AN 149 is flow coated onto the copper layer to a thickness of 6 microns and allowed to dry.

The coated plate is inserted into an enclosed chamber containing an electron beam gun. The internal pressure of the chamber is reduced to 10^{-6}
mm. by evacuation. The exposure step is effected by subjecting a small
section of the methyl vinyl ether maleic anhydride copolymer layer to
the electron gun accelerated by a potential of 10 kv. to yield an exposure
value of 10^{14} electrons/cm.^{2}. Upon completion of the exposure dosage,
the entire assemblage is immersed into an acetone solution in order to

remove those portions of the coating not subjected to the electron beam. The nonexposed portions of the polymer coating are readily removed by the solvent treatment while those portions of the coating rendered insoluble by electron beam exposure remain totally unaffected by the acetone solvent. The solvent-etched assembly is placed in an oven heated to a temperature of 130°C. for about 1/2 hour. Upon completion of the heat treatment the plate is immersed in an etch solution of the following composition maintained at a temperature of 55°C.

Ammonium persulfate	227 g.
Mercuric chloride solution (1.34 g./100 ml. water)	0.5 ml.
Sulfuric acid (conc.)	15 ml.
Water	1,892 ml.

Those portions of the copper coating which are not protected by the insolubilized resist portions are readily removed by the etch solution in a period of approximately 10 min. However, those portions of copper layer protected by the insolubilized resist portions remain totally unaffected by the etch solution with absolutely no trace of spurious etch solution diffusion being detected. The insolubilized resist areas are removed from the copper surface by immersion in a dilute ammonium hydroxide solution thereby laying bare those portions of the copper layer unaffected by the etching treatment.

Example 2:   Example 1 is repeated except that a copper clad laminate commercially available as Copper Fluoroply laminate sheet is employed in lieu of the copper-aluminum bimetallic plate of Example 1.

Example 3:   A copper laminate identical with that employed in Example 2 is coated with a composition comprising a 2.5% methyl ethyl ketone solution of a 1:1 maleic anhydride-isobutyl vinyl ether copolymer having a specific viscosity of 2.9 measured with a 1% methyl ethyl ketone polymer solution. The coating is applied to thicknesses of 5 microns. A portion of the isobutyl vinyl ether-maleic anhydride copolymer layer is subjected to an electron beam exposure of 10^{14} electrons/cm.^{2} accelerated by 14 kv. Physical removal of the noninsolubilized resist portions is effected in the manner described in Example 1.

The plate element is then treated directly, i.e., absent any intermediate heat treatment with the etch solution having the composition described in Example 1. As it is with the preceding examples, those portions of the copper layer protected by the insolubilized resist portions remain totally unaffected by the copper etch. Moreover, no spurious diffusion of solution is detected. The insolubilized polymer portions are then removed by treatment with a dilute solution of ammonium hydroxide.

Example 4:   This example illustrates the process utilizing silicon wafers
as the support. A type N chemically polished silicon wafer is coated
with a 10% methyl ethyl ketone solution of a 1:1 maleic anhydride-isobutyl vinyl ether copolymer having a specific viscosity of 0.6 as measured
in a 1% solution of methyl ethyl ketone. The element thus coated is
heated for 15 min. at 100°C. and then subjected to an electron beam
exposure of 5×10^{14} electrons/cm.^{2} accelerated by 14 kv. The non-
solubilized parts of the layer are removed by treatment with acetone. The
resist obtained is found to be resistive to a solution comprised of 4 parts
70% nitric acid, 3 parts acetic acid and 3 parts 48% hydrofluoric acid
which etches the silicon.

Example 5:   This example illustrates the process utilizing glass as the
support. Glass plate which has been etched with 6% hydrofluoric acid
is coated with a 10% methyl ethyl ketone solution of a 1:1 maleic anhydride-isobutyl vinyl ether copolymer having a specific viscosity of 0.6
as measured in a 1% solution of methyl ethyl ketone. The coating is subjected to an electron beam exposure of 5×10^{14} electrons/cm.^{2} accelerated by 14 kv. The noninsolubilized parts of the coating areas are then
removed by treatment with methyl ethyl ketone. The resist obtained is
resistant to 25% hydrofluroic acid solution that readily etches glass.

Unsaturated Partial Esters of Maleic Anhydride Copolymers

H.S. Cole, Jr.; U.S. Patent 3,703,402; November 21, 1972; assigned
to General Electric Company has found that the copolymers of maleic
anhydride with at least one copolymerizable monomer, selected from the
group consisting of C_{2-3}-alkenes, vinyl C_{1-20}-alkyl ethers, C_{1-8}-alkyl
esters of acrylic acid and C_{1-8}-alkyl esters of α-(C_{1-8}-alkyl)acrylic
acids, although not very electron-sensitive per se, do become extremely
electron-sensitive if they are esterified with an ethylenically unsaturated
alcohol selected from the group consisting of allyl alcohol, α-(C_{1-8}-alkyl)
allyl alcohols, propargyl alcohol, monoacrylate esters of the alkylene
glycols and the α-(C_{1-8}-alkyl) acrylate esters of the alkylene glycols.

From the above description it is apparent that the C_{2-3}-alkenes are
ethylene and propylene, the acrylate and α-acrylate esters have the
formula:

$$CH_2 {=\!=} C - COOR_a$$
$$|$$
$$R_b$$

allyl alcohol and the α-alkylallyl alcohols have the formula:

$$CH_2 {=\!=} C - CH_2OH$$
$$|$$
$$R_b$$

the acrylate and α-alkylacrylate monoesters of the alkylene glycols have the formula:

$$CH_2{=}C{-}COO{-}R_c{-}OH$$
$$|$$
$$R_b$$

and the alkyl vinyl ethers have the formula $CH_2{=}CH{-}O{-}R_d$ where R_a is C_{1-8}-alkyl, R_b is hydrogen or C_{1-8}-alkyl, R_c is C_{2-8}-alkylene and R_d is C_{1-20}-alkyl.

In general, since this work has shown that electron sensitivity comes from the unsaturated ester groups, it is preferable to have as many anhydride groups present in the initial copolymer so that its esterification produces the maximum number of ester groups. Thus, it has been found that the anhydride copolymer should contain a minimum of 25 mol percent anhydride groups with the maximum being 50 mol percent which represents the maximum amount of maleic anhydride that can be incorporated in any copolymer. The following examples illustrate the process.

Example 1: An essentially equimolar copolymer of maleic anhydride and octadecyl vinyl ether is commercially available as a 40% solution in toluene. The allyl half-ester of this polymer was prepared by refluxing a mixture of 100 g. of the solution of this polymer with 100 ml. of allyl alcohol overnight on a steam bath. The reaction mixture was added dropwise over 5 minutes to 2 liters of methanol which caused the half-ester to precipitate as an easily filterable solid. After filtering the semidry powder weighed 120 g. It was dried under vacuum at room temperature to constant weight to give a yield of 44 g. of the allyl half-ester. The IR spectrum of the product showed no evidence of the anhydride group initially present in the polymer.

Example 2: In similar manner the crotyl half-ester of the anhydride copolymer of Example 1 was prepared using 25 g. of the copolymer solution in toluene and 25 ml. of crotyl alcohol (2-buten-1-ol). A yield of 13 g. of the half-ester was obtained whose IR spectrum showed no evidence of the anhydride group.

Example 3: In a similar manner as described in Example 1, the propargyl half-ester of the maleic anhydride copolymer with octadecyl vinyl ether was prepared by refluxing a mixture of the 40% toluene solution of the copolymer, 25 ml. of additional toluene and 25 ml. of propargyl alcohol overnight. The half-ester was isolated by evaporation of the volatile materials under vacuum of a water aspirator with a liquid nitrogen trap in the vacuum line. The product was a waxy powder initially, but became a dry, white powder weighing 9 g. after 2 hours drying in a vacuum oven at 35°C. The yield was low in this case because of initial unsuccessful attempts to isolate the polymer by precipitation techniques. Again, the IR spectrum showed no evidence of the anhydride group. It did show a strong absorption peak at 3.05 microns characteristic of the acetylenic group.

Example 4:   In a similar manner as described in Example 1 the half-ester of 2-hydroxyethylmethacrylate and maleic anhydride copolymer with octadecyl vinyl ether was prepared by reacting an excess of 2-hydroxy-ethylmethacrylate with the maleic anhydride-octadecyl vinyl ether co-polymer solution overnight. The 2-methacryloxyethyl half-ester was precipitated by pouring the reaction mixture into an excess of methanol. It weighed 16 g.

Example 5:   In a similar manner as described in Example 1 the propyl half-ester of the maleic anhydride-octadecyl vinyl ether copolymer was prepared by refluxing a mixture of 25 g. of the copolymer solution and 25 ml. of n-propyl alcohol overnight. The propyl half-ester was pre-cipitated by pouring the reaction mixture into two liters of methanol and drying to constant weight. A yield of 12 g. was obtained.

Example 6:   An essentially equimolar copolymer of maleic anhydride and isobutyl vinyl ether is commercially available as a white powder. The allyl half-ester was prepared by refluxing a solution of 10 g. of this material with 80 g. of allyl alcohol on a steam bath for 18 hours. The half-ester solution in the excess alcohol was diluted with methanol to give about a 7% solution of the half-ester. This solution was used in Example 11.

Example 7:   The maleic anhydride copolymer with methyl vinyl ether in essentially equimolar amounts is commercially available as a white powder. The allyl half-ester was prepared by refluxing a solution of 10 g. of this copolymer in 20 g. allyl alcohol for 2 hours on a steam bath. At the end of this time IR spectrum showed no evidence of anhydride group being present. The solution of the half-ester in excess allyl alcohol was diluted with methanol to give about a 15% solution which was used in Example 11.

Example 8:   The maleic anhydride copolymer with ethylene in essen-tially equimolar amounts is commercially available. The allyl half-ester was prepared by refluxing a solution of 10 g. of this copolymer in 20 g. of allyl alcohol for 2 hours on a steam bath. At the end of this time the IR spectrum showed no evidence of the anhydride group. The half-ester in excess allyl alcohol was diluted with methanol to give about a 15% solution which was used in Example 11.

Example 9: The maleic anhydride copolymer with styrene in essentially equimolar amounts is available commercially. The allyl half-ester was readily prepared by refluxing overnight a solution of 10 g. of the copolymer in 20 g. of allyl alcohol and 20 g. of xylene. The IR spectrum showed that about 70% of the anhydride groups had been esterified to the half-ester. After evaporating the volatiles from the reaction mixture on a steam bath, the half-ester was dissolved with benzene to give about a 15% solution which was used in Example 11.

Example 10: According to De Wilde et al, in J. Polymer Sci. 5, 253 (1949), to prepare a maleic anhydride copolymer with methyl acrylate which is essentially equimolar, the polymerizable mixture of monomers must contain a ratio of about 9 mols of maleic anhydride to 1 mol of methyl acrylate. After flushing a 2-liter 3-neck flask containing 1 liter of benzene for 1 hour with dry nitrogen, 176 g. of maleic anhydride and 17.4 g. of methyl acrylate and 0.5 g. of lauryl peroxide were added and heated at reflux for 3 hours, after which it was cooled and allowed to stand at room temperature overnight. From this reaction mixture, 1.5 g. of maleic anhydride copolymer with methyl acrylate was isolated by precipitation by addition of petroleum ether as described in the reference.

The IR spectrum and the determination of the acid number indicated that essentially an equimolar copolymer had been prepared. It was converted to the allyl half-ester by refluxing a solution of the 1.5 g. prepared above in 20 ml. of allyl alcohol for 5 hours by which time essentially all of the anhydride groups had been converted to the half-ester as shown by the IR spectrum. The solution of the half-ester in excess allyl alcohol was diluted with acetone to give about an 8.5% solution.

Example 11: The electron beam sensitivity of the compositions prepared above was determined by the following general procedure. Commercially available glass plates having a 3000 A. thick chromium layer (optical density of 4) on one of the two major surfaces was used in all cases as a substrate. Solutions of the various half-esters were prepared with the concentration and the solvent being chosen so that on spin coating the chromium layer with the solution, smooth pinhole-free coatings of the half-esters, approximately 0.2 to 0.5 micron, were obtained. After evaporation of the solvent, the coated plates were exposed to an electron beam to give a dot pattern with each dot being exposed to a progressively higher radiation dose than the previous dot, with the range being sufficiently broad to underexpose some dots while overexposing other dots.

The exposed plates were then placed in a stirred solvent for the coating which caused the unexposed areas and the insufficiently exposed dots to dissolve leaving those dots which had been sufficiently crosslinked that

they were no longer soluble in the solvent. By this means it was possible
to determine the minimum radiation dose which produced a dot which
would no longer dissolve in the solvent. The results on the above compositions are shown in the following table.

Half-ester	Solvent	Concentration developer solvent [1]	Minimum dosage required, coul./cm.[2]
Example:			
1	Benzene	10% 80/20 benzene methanol	4×10^{-8}
2	do	5% benzene	1×10^{-6}
3	do	10% 50/50 benzene butyl-acetate.	3×10^{-7}
4	n-Butylacetate	10% n-butylacetate	3×10^{-8}
5	Benzene	10% benzene	$> 10^{-8}$
6	} As prepared	50/50 methanol n-butyl-acetate. {	3×10^{-8}
7			1×10^{-8}
8			2×10^{-7}
9	As prepared	Benzene	2×10^{-6}
10	do	Acetone	3×10^{-7}

[1] Mixtures are on a volume basis.

The results in the above table show clearly that for the compositions to
require a radiation dose no greater than 5×10^{-7} coul./cm.^{2} an unsaturated half-ester is definitely required and that the unsaturation must be
terminal unsaturation, i.e., $CH_2{=}C{<}$ or a terminal acetylenic group, i.e.,
$CH{\equiv}C{-}$. The results further show that an aryl group should not be present
in the composition.

Surface Deformation Imaging Process

S.H. Boyd, Jr.; U.S. Patent 3,674,591; July 4, 1972; assigned to
Stromberg DatagraphiX, Inc. describes a surface deformation electron
beam imaging system which includes the steps of forming a layer of a
highly viscous photopolymer, exposing the photopolymer to an electron
beam in a vacuum, whereby the exposed areas on the photopolymer are
simultaneously polymerized and form a surface deformation image. The
photopolymer layer comprises a material which is capable of polymerization or crosslinking when subjected to electron beam bombardment. The
material could be a monomer, a mixture of monomer and polymer, or a
prepolymer which further polymerizes or crosslinks when irradiated.

Background areas on the photopolymer layer may be fixed by exposure to
suitable radiation, such as ultraviolet light, which completes the polymerization process throughout. Thus, different areas of the layer may be imaged
by the electron beams at different times followed by eventual exposure of
the entire surface to ultraviolet light to fix the material between imaged
areas. No heating or chemical treatment of the layer is required.

While the mechanism by which this imaging process operates is not fully
understood, it is through that it is a combination of photopolymerization
and electrostatic charge repulsion processes operating simultaneously in
those areas bombarded by the electron beam. The polymerization process

proceeds through the formation of free radical monomers in those areas penetrated by the electron beam. The surface deformation appears to be due to changes in the viscosity of the layer in a random manner depending on such things as charge distribution on the layer surface. During the polymerization in the exposed areas, the highly viscous material is capable of migration into the polymerizing areas. The nonuniform reaction results in microscopic cracks in the layer at the surface giving a cracked, folded or wrinkled appearance which scatters light.

The effect of differing voltages on the surface deformation image appears to be related to the penetration of the electron beam into the photopolymerizable layer. At low voltages only the surface, or a portion of the surface, polymerizes. As the beam voltage increases, the layer is penetrated to greater depth, resulting in more uniform polymerization. Since the surface is then less able to flow and deform in response to electrostatic repulsion forces caused by the electrostatic charge pattern created by the electron beam, less surface deformation and smaller cracks occur. Ultimately, at a sufficiently high voltage, a completely homogeneous reaction occurs and the polymer forms with no light scattering surface deformation.

This system has high photographic speed resulting from the fact that one electron can form many radicals. The polymerization process proceeds by a free radical mechanism causing many monomers to react for each radical initially formed. It appears that the characteristics of the surface deformation image are both voltage and current sensitive. It has been found that as the current is increased (with constant voltage) the intensity of the surface deformation image also increases. Also, as the voltage is increased at constant current, the coarseness of the surface deformation decreases until at sufficiently high voltages no surface deformation occurs.

It has been found that a useful combination of photopolymerization and surface deformation imaging occurs over a voltage range of from about 5,000 to 23,000 volts. The surface can be polymerized without surface deformation at voltages above about 30,000 volts. This is useful, for example, to permit fixing of background areas. The useful current density range has been found to be from about 10^{-8} to 10^{-2} amp./cm.^{2}. Depending upon the image characteristics required, it may be desirable to use low voltages with high currents or high voltages with low currents.

The exposure to the electron beam should take place in a vacuum. In general, a vacuum of 10^{-4} torr or better is preferred. Excessive gas in the vacuum chamber decreases the effectiveness of the electron beam and causes scattering of the beam which decreases image resolution. Further, it is very desirable that oxygen be excluded during this step since oxygen tends to inhibit the polymerization reaction. Any conventional electron beam generating system may be used. For example, the imaging material may be scanned in a raster made and the beam current modulated by a conventional television video signal to produce alphanumeric or pictorial images. Alphanumeric information can also be imaged by techniques in which dots or short lines are produced on the imaging material by the electron beam in patterns which form the desired characters. An especially useful character generating technique (such as is described in U.S. Patent 2,735,956) utilizes a character-shaped electron beam.

Any suitable photopolymerizable material may be used in the photo-sensitive layer. Results are often improved by the addition of polymerization initiators and thermal polymerization inhibiters. Best results are generally obtained with photosensitive layers comprising from about 10 to 90 weight percent of an addition polymerizable ethylenically un-saturated compound and from about 90 to 10 weight percent of a thermo-plastic polymer. Often, image quality and process efficiency can be fur-ther improved by the incorporation of up to about 10 weight percent of an addition polymerization initiator and up to about 3 weight percent of a thermal polymerization inhibitor.

Example: The imaging material is prepared by mixing together about 100 parts polyethylene glycol diacrylate, about 1 part phenanthrene-quinone and about 100 parts cellulose acetate butyrate in sufficient acetone to give a solution containing about 10% solids. This solution is coated onto a polyethylene terephthalate substrate to a wet thickness of about 1.5 mils. After drying for 2 hours at a temperature of about 50°C., the layer has a thickness of about 0.5 mil and is tacky to the touch.

The resulting composite sheet is placed on a grounded aluminum plate in a demountable cathode ray tube which is then closed and evacuated, to a vacuum below about 10^{-4} torr. An electron beam, operated at a voltage of about 10 kv. and a current density of about 10^{-4} amps./cm.^{2} is caused to scan the sheet in a conventional manner to produce an image. As the electron beam exposure proceeds, the exposed areas are seen to form micro-scopic cracks or wrinkles, giving these areas a frosted appearance. The sheet is then removed from the tube and exposed to ultraviolet radiation from a 150 watt short arc mercury lamp for about 1 minute at a lamp-to-sheet distance of about 18" in a nitrogen atmosphere. The sheet is found to be no longer tacky and to have an excellent light scattering image on a clear, handleable background.

PLASTICS AND TEXTILES

PLASTICS

Fiber-Reinforced Polymeric Composites

J.H. Lemelson; U.S. Patent 3,676,249; July 11, 1972 describes a method for irradiating material on a continuous basis to improve or predeterminately change its physical and chemical characteristics. In one form,
polymeric material is continuously formed by extrusion and is continuously combined with or sealed to a second material to form an integral unitary structure after which the unitary structure is driven and guided in a
predetermined path through a field of intense radiation which is operative to change the physical characteristics of one or more of the components of the composite material such as by effecting the cross linking of
the polymer.

In another form, a material such as a polymer is continuously fed or formed
into a sheet or other suitable shape and a plurality of reinforcing elements
such as glass filaments are encapsulated within the polymer or otherwise
secured on a continuous basis. The resulting composite filament-reinforced
member is treated to effect the cross linking of the polymer to improve its
physical characteristics and to improve the bond between the polymer and
the reinforcing filaments.

Figure 6.1a illustrates an apparatus for continuously forming and operating on a laminate made of two or more materials such as a base sheet
and a reinforcing component. In particular, the apparatus of Figure 6.1a
may be utilized for fabricating filament reinforced sheet or tape such as
tape used in strapping or for winding purposes.

FIGURE 6.1:   PRODUCTION OF FIBER REINFORCED POLYMERIC COMPOSITES

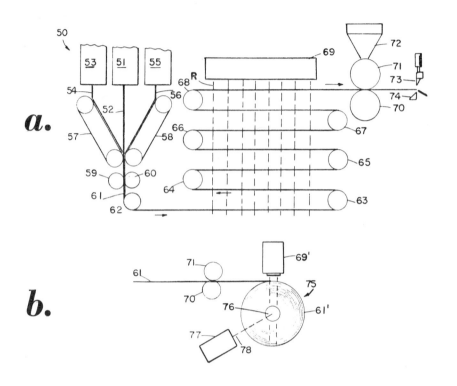

(a)   Side View of Apparatus for Continuous Processing
(b)   Side View of Modified Form of Apparatus

Source: J.H. Lemelson; U.S. Patent 3,676,249; July 11, 1972

The apparatus (50) includes a first supply means (51) for a reinforcing material (52) which is shown disposed between two other supply means (53) and (55) each of which continuously supply respective sheet materials (54) and (56) which are fed downwardly therefrom towards the downwardly feeding reinforcing material (52). The supply means (51), (53), and (55) may each comprise a respective extruder for continuously extrusion forming their materials or one or more of the devices may comprise means for otherwise providing its material such as a spirally wound coil formation thereof and suitable means for feeding and guiding the material. Reinforcing material (52) may comprise a single sheet or strip of a plurality of filaments or wires of glass which are fed in parallel array downwardly and of a width such as to cover substantially the width of the sheet members (54) and (56). Endless belt conveyors (57) and (58) respectively guide sheets (54) and (56) into abutment with the reinforcing member (52) and, in certain instances, with each other so as to encapsulate member (52).

Power driven rolls (59) and (60) operate to receive and compress the sheet members (54) and (56) against the central element of filaments (52) in a manner to completely encapsulate same between the sheets and to weld the sheets together as they are fed. The composite formation (61) may comprise a single sheet or ribbon of polymeric material which is internally reinforced with a plurality of filaments, whiskers or wires extending through the central portion thereof in a direction parallel to the longitudinal axis of member (61). The elongated formation (61) is power driven back and forth around a plurality of rolls (63) to (68) to cause the formation to loop back and forth a number of times in alignment with a device (69) for generating high energy radiation as described and directing same through the looped formation to intersect different portions of member (61) as it travels back and forth between the rollers.

Radiation generating means (69) may comprise a Van de Graaff generator, an atomic pile or other suitable source of atomic fission or an electric glow discharge means operating at high frequency and high voltage glow discharge directly in alignment with one or a plurality of the loops of the composite material (61). Radiation from the high intensity radiation generator or source (69) is of such a characteristic and is operative to irradiate a sufficient area or areas of the composite material (61) during its travel through the field thereof such that a desired and predetermined degree of cross linking of the cross linkable portion of the composite material is obtained.

In other words, by providing a source or sources of radiation of predetermined intensity, locating the radiation source or sources so as to irradiate a predetermined effective length of the continuously fed composite material, supplying the components (52), (54) and (56) at such a rate of flow that the composite formation (61) travels through the radiation field at a rate to effect the exposure of any unit area thereof to a predetermined quantity of radiation dosage such that predetermined changes or degree of cross linking occur in the composite material by the time it has been completely irradiated.

The end effect may be such as to convert, for example, a thermoplastic polymer such as polyethylene comprising sheet members (54) and (56) from a relatively soft material having a low melting point to a cross linked material of substantially greater rigidity, strength and higher melting point. The reinforcing material (52) which is fed between sheet members (54)

and (56) may or may not be also improved in physical and chemical characteristics by the action of the intense radiation. If it comprises glass filaments or fibers fed between sheets (54) and (56) and encapsulated by compression of the sheets to completely surround the filaments and become wel-
ded together, then the radiation may be such as to improve the bond not only
between sheets (54) and (56) but also between the material of the sheets
and the filaments (52) so as to provide a substantially improved end product.

Also illustrated in Figure 6.1a are means for coating an adhesive on at least
one surface of the composite sheet member (61) which comprises a pair of
rolls (70) and (71) one of which is power rotated and operative to receive
the irradiated sheet (61) and apply suitable pressure-sensitive adhesive to,
for example, the upper surface thereof from a supply reservoir (72) of the
adhesive. The member (61) may be slitted into separate filaments which are
immediately coiled into rolls for dispensing as filament reinforced adhesive
tape or may be further processed or coiled before slitting. Notations (73)
and (74) refer to cooperating cutting blades which are predeterminately operated to cut predetermined lengths of sheet (61) from the main sheet.

It is noted that the apparatus (50) of Figure 6.1a which includes the continuous supply means or extrusion heads (51), (53) and (55) for continuously
fabricating an elongated composite member such as a sheet which is internally reinforced with a plurality of filaments or netting, may be provided
per se or in combination with a similar array of extrusion heads at the head
of a packaging machine to supply one or more reinforcing sheets of material
to define the walls of containers or bags which are continuously formed.

It is also noted that the laminating means illustrated in Figure 6.1a may be
modified whereby a single sheet of thermoplastic polymer such as sheet (54)
is continuously formed and fed downwardly as described into abutment with
a plurality of reinforcing filaments or netting such as (52) which are also
continuously fed downwardly and both formations are thereafter compressed
together by the bite of a plurality of rolls or belts which are operative to
force the filaments or netting into the surface of the extrusion softened sheet
to form an integral assembly and bond between the two prior to the irradiating or container formation.

Figure 6.1b illustrates means for irradiating a sheet of material such as the
composite material (61) produced as in Figure 6.1a. A suitable sheet of any
material to be predeterminately irradiated to effect, for example, cross
linking of one or more components thereof is fed to a core member or drum
(76) upon which the sheet is wound for storage thereafter prior to dispensing
same. The member (61) may be a strap, tape, ribbon or band with or without an adhesive coating applied as described.

In Figure 6.1b irradiation processing of the material (61) is effected as it is wound onto its core or drum (76) by means of a suitable winding means (not shown) which is preferably operative at constant speed. A source (69') of intense radiation such as a Van de Graaff generator, quantity of radio-active material, cathode ray tube or other suitable radiation generating means, is disposed to direct radiation of predetermined intensity against and through the outer layer or ply of the winding material through the subsequent turns of the coil formation (61') during the entire winding operation.

In one form, the radiation emitted by the generator (69') is generated at a constant intensity and the desired degree of cross-linking is attained by subjecting the winding coil formations to a radiation dosage of such an intensity as to provide the entire length of the winding material in the desired physical condition by the time winding is completed or shortly after. The shielding effected by each layer of the winding material (61) for previously wound turns may be such that constant radiation dosage is directed against the entire length of material so wound.

However, for those situations where the wound portions of the coil formation (61') which are closer to the core or center thereof are subjected to substantially higher dosages of radiation which may be of such a nature as to degrade same in order to provide sufficient radiation against the outer turns of the coil formation to effect a predetermined change in the characteristics of the material, one or both of two radiation variables may be predetermi-nately changed during a winding cycle so as to expose the complete length of wound material to substantially the same dosage or to reduce the amount of dosage to which the first wound portions of material are exposed so as not to degrade same.

To effect such a process, the intensity of radiation emitted by (69') may be varied during a winding cycle. The direction of the beam may be shifted during a winding cycle with respect to one or more turns to control dosage or the coil may be shifted with respect thereto during a winding cycle to provide changes in the physical characteristics of the entire length of material being wound.

As an example of the radiation dosage required to effect cross-linking of a suitable polymeric material applicable to improving material which is fabricated and processed by means of the type described, it is noted that a Van de Graaff electron accelerator capable of generating beam energy having an output of 2,000,000 volts at a power output of 500 watts may be utilized in locations with its output being a foot or less from the surface of the plastic materials described to effect suitable cross-linking of such plastics as polyethylene, polypropylene and polyethers. Polyethylene, for example, having a melt index of 1.8 and a molecular weight of about

20,000 may be improved in its adhesion and heat sealing characteristics
by exposure with high particle energy generated by a Van de Graaff generator of the type defined above for periods of one minute or more and the
exposure may be effected by means of a single source of radiation disposed
and operative as illustrated in the drawings to simultaneously irradiate different portions of loops of the material wound or guided back and forth
through the field of radiation.

Exposure to the direct beam of radiation of such a generator or a corona
discharge device may also be operative to affect the surface of the sheet
material moving through the field in such a manner as to increase its ability
to retain and adhere an adhesive such as pressure-sensitive adhesive applied
to the sheet material after being so processed. Exposure of polyethylene,
for example, to electrical energy in the order of 10,000 to 30,000 watt
seconds per square foot for a period of 10 to 30 seconds will substantially
increase the adhesion of polyethylene and polyvinyl chloride to each other
or to reinforcing material such as glass provided as filaments or fibers for
reinforcing purposes as described.

Polyester Lamination Process

R.P. Hall; U.S. Patent 3,658,620; April 25, 1972; assigned to SCM
Corporation describes a process for preparing a laminable sheet from a
substantially catalyst-free system containing a polymerizable organic
unsaturated resin susceptible to free-radical catalysis comprising assembling
a film of the resin in contacting substantially coplanar relation with a membrane, and then exposing the resulting assembly while overlying a substrate
to high energy radiation.

Many thermosetting resins, such as those typified by thermosetting, unsaturated polyester resins, exhibit air-inhibited curing at their air-con-
tacting surfaces. Such surfaces are softer than the interiors of the resins
and are therefore more easily scratched and marred. Obviously, these
qualities are undesirable, especially when such a resin is to be used for
coating purposes. Several techniques have been suggested to overcome
air-inhibition in the curing of resins. For example, U.S. Patent No.
3,210,441 is based on the discovery that the presence of esterified resi-
dues of monohydroxy acetals in polyester resins of particular formulation
are free of air-inhibition.

Within relatively recent years, the polymerization of resinuous materials
by electron radiation has increasingly become of interest. However, the
use of this technique has encountered the same difficulty with many thermosetting resins, namely, air-inhibition at the resin-air interface. During
penetration by high energy radiation, the resinuous material undergoes an

"ionizing effect" which induces chemical reactions including polymerization; note U.S. Patent 2,863,812. Radiation, such as a beam of electrons, has not been found to have any appreciable ionization effect at the exposed surface of irradiated material. The desired ionization effect is obtained only after penetration of the resinous material. It would be advantageous to use high energy radiation to cure completely such air-retarding polymerization, without requiring chemical modification of the resin or additional radiation apparatus.

Previous attempts have been directed to modifying the radiated energy so as to obtain an ionization effect after relatively short distances of penetration. For example, in U.S. Patent 2,863,812, electrons pass through an electrically conductive shield before impinging upon the material to be radiated. This technique, of course, increases and complicates the type of apparatus used for the radiation.

In this process, a film of the resinous system is assembled with a membrane in contacting, substantially coplanar relation, and then the assembly is exposed to high energy radiation while the assembly overlies a substrate. The side of the assembly shielded from the atmosphere is cured to a tack-free, mar-resistant condition, while the side of the assembly open to the atmosphere remains relatively tacky and mar-susceptible. This may be due to the chemical combination of oxygen or other elements (from the air) which render the open surface incapable of cure to a mar-resistant state. The resulting laminable sheet can then be laminated in a separate operation to a suitable substrate or cooperating lamina, using the relatively tacky side of the assembly to face and contact the substrate.

When the process includes lamination as well, for example on a continuous basis, the described assembly of resin film and membrane is passed successively through at least two treating zones. The objective of the first zone treatment is to impart a tack-free, mar-resistant surface to a shielded side of the assembly, as before, and also to impart mass integrity to the assembly, so that it may be treated as a self-supporting sheet, although portions of the resin of the assembly may still be capable of further cure.

In this regard, the membrane serves as a reinforcing member in addition to other uses. The objective of the second zone treatment is to complete all possible cure of the resin and, preferably simultaneously, laminate the assembly to another lamina. The following examples illustrate the process.

Example 1:   A thermosetting polyester resin was prepared by reacting equal molar portions of 1,3-propylene glycol and maleic anhydride. Water was removed until the resin had an acid number of 35. An amount of 70 parts of the cooled reaction product was then mixed with 30 parts of styrene monomer, all by weight.

Referring to Figure 6.2, a supply (10) of the resulting polyester resin mix was periodically dumped onto a slowly rotating drum (11) having a chrome plated surface to minimize adherence with the mix. A doctor knife (12) smoothed the mix to a film form (13) which is enlarged in the figure for purposes of illustration. A paper membrane of about 65 pounds grade, advanced from a roll (15) to be caught in the resin (10), such that the membrane passed beneath the doctor knife (12), in the embodiment illustrated, substantially equidistant between the faces of the film (13).

The membrane (14) might, if desired, be mechanically advanced as by power-driven rollers. Normally, however, once the continuous process starts, the advance of the film (13) with the membrane (14) embedded therein is sufficient as illustrated to pull the trailing portion of the membrane from its roll (15) at a proper speed.

An electron accelerator (16) of standard construction bombarded the film (13) with a radiation of 20 megarads as it passed at a rate of about 20 feet per minute. In general, the radiation strength of the gun (16) and the speed of rotation of the drum (11) are synchronized to cure at least enough of the film that it has sufficient mass integrity to be stripped from the drum (11) as by a knife edge (17) without rupturing; and also to provide a tack-free, hard undersurface to the film (13) as previously described. If high energy radiation had not been used for this step, the drum (11) could have been internally heated as by steam; or the gun (16) could have been replaced by an infrared lamp, an oil or gas fired burner, or the like.

A continuous conveyor (18) caught the film and membrane assembly (13) atop a plywood substrate (22) as the two laminae together passed between compressing rollers (19) and (24). The plywood substrate (22) has an adhesive coating (23) of a polymerizable, unsaturated polyester resin. This resin was prepared by partially reacting equal molar amounts of diethylene glycol and maleic anhydride with removal of water until an acid number of (35) resulted. An amount of 70 parts of the resulting condensation product was dissolved in 30 parts of a styrene monomer all by weight.

The film (13) and plywood (22) together passed beneath a second accelerator gun (20) while riding the conveyor. This accelerator gun would have been necessary if no high energy radiation had been used adjacent the drum (11) as described. Otherwise the use of the accelerator (20) is optional, and it could be replaced by more conventional heat sources.

In any event, it is emphasized that the top side of the film (13) at the

FIGURE 6.2:   IRRADIATION LAMINATION PROCESS FOR AIR-INHIBITED
POLYMERS

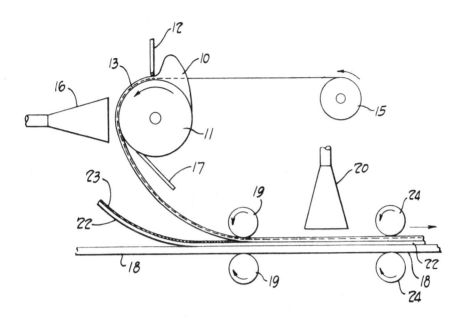

Source: R.P. Hall; U.S. Patent 3,658,620; April 25, 1972

second treating zone was the underside of the film when it overlay the
drum (11). Subsequent exposure to radiation from the electron accelerator
(20) not only completed any possible further cure of the film but also cured
the adhesive coating (23) on the plywood (22) as well and chemically bonded the film (13) to the plywood to form a laminate. This laminate was
subsequently lifted from the conveyor (18) and conventionally cut to size.

Example 2: An unsaturated polyester resin was prepared by reacting 696
grams of ethylene glycol and 2,128 grams of propylene glycol with 3,098
grams of isophthalic acid and 2,249 grams of maleic anhydride until esterification was substantially complete, as indicated by an acid number of about
15 to 20. The resulting polyester was then diluted with 2,249 g. of toluene.

A procedure was carried out with this resin mix like the procedure of Example 1, except that after the initial radiation exposure on drum (11), the
laminable sheet was cut to size and the resulting sheets stored for a period

of time, while remaining readily laminable to a suitable substrate with or without the use of intervening adhesives. The impregnated laminable sheet of this example at this juncture contained from about 60 percent to about 70 percent by weight resin when intended for forming decorative "overlays." Sheets having 70 percent or more resin by weight are used in making very hard, high glossy, clear laminates.

The stored resin-impregnated sheets were laminable, as a batch operation in a match die or flat press, over a wide range of pressures (100 to 600 psig), temperatures (about 240° to 300°F.), and time cycles (about 2 to 10 minutes) with lower temperatures requiring longer time cycles. Since the impregnation resin was thermosetting, the laminates can be stacked hot out of the press.

A typical cycle for laminating a resin impregnated sheet was 6 minutes at 200 psig and 260°F., although laminates could have been produced in cycles of 4 minutes at 275°F. and 200 psig to 7 minutes at 260°F. and 170 pounds per square inch gauge. Such laminated articles are attractive in appearance and have durable stain and impact resistance, excellent color retention, and chemical resistance.

In related work R.P. Hall; U.S. Patent 3,644,161; February 22, 1972; assigned to SCM Corporation describes a process where a film or coat of an incompletely cured, normally air-inhibited, thermosetting resin is formed over a substrate and then exposed to high-energy radiation from a suitable, conventional source to cure at least a depthwise portion of the film that is contiguous to the substrate and remote from the source. Normally, the radiation exposure cures the entire film except for the upper, exposed surface which remains tacky and mar-susceptible.

This may be due to the chemical combination of oxygen or other elements (from the air) which render such upper surface incapable of cure to a mar-resistant state. In any event, the depthwise cure is at least sufficient to impart mass integrity to the film. The film is then turned over usually, but not necessarily, on the same substrate to present topside the nontacky, mar-resistant surface of the film. The film may then be heated at temperatures to complete any possible further cure, especially when it is desired to laminate the film to a substrate. Preferably, however, the film is again exposed to high-energy radiation. This step also may be used to effect lamination and without the need for externally applied heat.

Foamed Polyethylene

A process for the production of an expanded polyethylene product having a density greater than 1 lb. per cubic ft. and not more than 5 lbs. per cubic foot is described by A. Cooper; U.S. Patent 3,640,915; February 8, 1972; assigned to Expanded Rubber & Plastics Ltd., England. The process comprises raising the melting point by submitting the polymer to a cross-linking treatment equivalent to that produced by an irradiation dose of between 2 to 6 megarads or a suitable organic peroxide content of 0.1 to 0.3 part by weight per 100 parts by weight of polymer and thereafter heating the cross-linked polymer to a temperature above the softening temperature in an atmosphere of nitrogen at a pressure of 2,000 to 12,000 lbs. per sq. in. to impregnate the polymer with nitrogen and expanding the nitrogen impregnated polymer by releasing the pressure.

The cross-linking may be effected by an ionizing radiation dose of e.g. 4 megarads, or alternatively the cross-linking may be effected by intimately blending the polymer with e.g. 0.25 part by weight of organic peroxide per 100 parts by weight of polymer and heating the blend to a temperature above the temperature at which the peroxide dissociates into free radicals. The polyethylene polymer preferably has a density of 0.90 to 0.94 g./cc prior to the modification treatment.

Polyethylene Foam

W.A. Patterson, S.N. Weissman and H.G. Schirmer; U.S. Patent 3,592,785; July 13, 1971; assigned to W.R. Grace & Co. have found that improved foamable polyethylene bodies can be prepared by (a) forming a homogeneous mixture of 100 parts by weight of a normally solid polyethylene material and from 1 to 15 parts by weight of a normally solid, heat decomposable, organic foaming agent having a decomposition temperature at least 10°C. above the melting point of the polyethylene material while (b) maintaining the mixture below the decomposition temperature of the foaming agent; (c) subjecting the mixture to high energy ionizing radiation of a total dosage of 10 to 50 megarads and sufficient to provide a percent gel of 30 to 80% in the irradiated material.

The foaming agent is a normally solid heat-decomposable organic foaming agent which has a decomposition temperature at least about 10° centigrade above the melting point of the polyethylene material. Several commercially available materials having high decomposition temperatures permitting universal utilization in the process are azobisformamide (Celogen AZ, also called azodicarbonamide), N,N'-dinitrosopentamethylene tetramine (Unicel NDX), p,p'-oxybis-(benzenesulfonyl-semicarbazide) (Celogen BH), trihydrazino-symtriazine (THT, German Patent 1,001,488), bis-benzenesulfonylhydrazide (BBSH) and barium azodicarboxylate (Expandex 177).

Foaming agents with lower decomposition temperatures and hence useful only with low melting temperature polyethylene materials include p,p'-oxybis-(benzene-sulfonyl hydrazide) (Celogen), azobis (isobutyronitrile) (Genitron

AZDN), and benzene-1,3-disulfonyl hydrazide (Porofor B-13). The use of
azobisformamide is highly preferred because of its ease of incorporation;
controllable blowing action; colorless, non-toxic residue and other desirable properties.

The process provides foamable polyethylene bodies which upon foaming have
extremely small, discrete uniformly distributed cells. The average cell size
is usually less than about 0.003 inch (0.075 mm.) whereas maximum cell
size is 0.008 inch (0.20 mm.) or less. Foams with cell sizes of this order
of magnitude have not previously been producible with the facility and e-
conomy of operation that is inherent in this process.

The following examples are presented to illustrate the process. In each of
the examples irradiation was accomplished using a General Electric one
million volt resonant transformer unit until the indicated dosage measured
in megarads (mr) was received. Gel percentage of the irradiated samples
were determined by using the following procedure.

Specimens of irradiated samples weighing between 0.46 and 0.50 g. were
weighed to 0.1 mg. and to an accuracy of ±0.05 mg. Specimens were cut
into smaller pieces, approximately 1 square cm. in size, and transferred to
a 22 x 80 mm. single-thickness Whatman extraction thimble which had been
reduced in length by cutting approximately 10 mm. off of the top. As a
precautionary measure to insure against loss of sample in transfer, the thimbles were weighed before and after the samples were added. The samples
were then extracted over a 20 hour period in an apparatus designed for
ASTM D-147 using toluene (analytical reagent grade) as the solvent.

Upon completion of the extraction, the thimbles were removed and a
visual inspection made of the gel. If the gel was found to be in a cohe-
sive form and capable of total removal with forceps, it was transferred
directly to aluminum weighing cups and dried under reduced pressure in a
vacuum oven at 55° to 60°C. for a period of no less than 48 hours.

If the gel could not be removed without fear of loss to the thimble, then
the hot extract was analyzed. It was transferred to evaporating dishes
which had previously been weighed to 0.1 mg. The flasks from which the
solution had been transferred were washed twice with 20 to 25 ml. of hot
toluene and the washings added to the solution   Toluene was partially
evaporated while cooling under a hood in air prior to drying under the same
conditions as was the gel.

Materials balances of 99.6% to 101.1% obtained prior to the investigation
and from spot checks during the investigation justified the use of sol weight
for some samples and the gel weight for others. Gel content was determined

directly from the gel whenever possible, as it was found to be a much more convenient procedure. Determination of cell size in the foamed products was made by microscopic examination. Melt indices of the polyethylene materials were determined in accordance with the procedure of ASTM-1258-521.

Example 1: One-hundred parts of a commercially available, powdered ethylene-butene copolymer (about 95% ethylene) having a density of 0.950 g. per cc and a melt index of 9.0 were dry-blended with 3 parts of commercially available azobisformamide blowing agent (Celogen-AZ) in a Patterson-Kelley twin shell blender for 15 minutes to produce a homogeneous mixture. A portion of the dry blend was compression molded at temperatures below the decomposition temperature of the Celogen AZ into a 9 x 9 x 0.075 inch plaque in a commercial hydraulic press.

The compression molded plaque was irradiated at room temperature until it had received a dosage of 16 mr. A circular portion (about 4 inches in diameter) was die cut from the center of the plaque and retained between a pair of circular clamps. The clamped disc was then preheated at a temperature of about 280°F. for about 2 to 3 minutes after which it was immediately transferred to a high temperature (59°F.) foaming oven. Foaming occurred within about 30 seconds in the high temperature oven.

The sample was kept in the foaming oven for a few seconds to permit substantially complete decomposition of the blowing agent, and was then immediately removed from the oven. Quite surprisingly it was noted that during foaming, the disc expanded in a substantial amount in three dimensions, i.e., there was an increase of 40% or more in the diameter of the disc as well as in thickness. Because of the restraint on diametral expansion due to the circular clamps, the disc expanded into a foamed dome-like bubble. After removal from the oven, the foam was air cooled in the ambient atmosphere.

Samples were cut from the center of the dome for density measurements and examination of foam structure. The sample had density of 0.47 g. per cc calculated from a determination of its volume and weight. The surfaces of the foam were smooth and glossy. It was observed by microscopic examination that the average size of cells in the foam was less than 0.001 inch with a maximum cell size of 0.002 inch. The cells were substantially spherical in shape and very uniformly distributed. The foam structure was essentially a closed cell structure; i.e., virtually every cell was discrete from other cells.

Expandable Thermoplastic Resinous Products

L.C. Rubens and W.B. Walsh; U.S. Patent 3,579,472; May 18, 1971; assigned to The Dow Chemical Company describe a method for the fabrication of a foamable structure by polymerizing a monomer capable of providing an expandable thermoplastic resin in the presence of a blowing agent or expanding agent at a temperature of from about 0° to about 80°C. under atmospheric pressure while the monomeric material and the expanding agent are restrained in a mold having the form of the desired end product.

The thermoplastic resinous materials which can be prepared in accordance with the process are those which are prepared by polymerizing monomers selected from the group consisting of monochlorostyrenes, dichlorostyrenes, methylmethacrylate, and mixtures containing up to about 30 weight percent styrene. Optionally for a maximum degree of replication, minor quantities of a difunctional copolymerizable monomer, such as divinylbenzene, may be incorporated in quantities of from about 0.035% to about 1% by weight and preferably from about 0.04% to about 0.25% by weight.

The percentages being based on the weight of the polymerizable monomers exclusive of the cross-linking agent, such as pentaerythritol tetramethacrylate, diethylene glycol dimethacrylate, polypropylene glycol 150 dimethacrylate, trimethylol propane trimethacrylate, polyethylene glycol 400 dimethacrylate.

Example 1: A mixture of the following components was prepared. 89.91% orthochlorostyrene, 0.045% divinylbenzene, 0.045% ethylvinylbenzene, 10.0% trichloromonofluoromethane (all percentages by weight). A quantity of this mixture was placed in a rectangular bag prepared by sealing two 1 mil thick polyvinyl fluoride films together at their edges. The bag in the form of a flat packet was placed between two sheets of a magnesium alloy having a thickness of 50 mils. The polyvinyl fluoride bag and contents formed an assembly that was compressed to a thickness of about 1/10 of an inch.

The assembled unit was subjected to gamma radiation from a cobalt-60 source at a dose rate of 100,000 roentgens per hour for a period of 24 hours at a temperature of 68°F. The total radiation dose was 2.4 megarads. On removing the magnesium alloy sheets and the polyvinyl fluoride film, a clear bubble-free sheet of hard polymeric material was obtained. The sheet was cut into squares which measured 1/2 inch on the side and samples were placed in an air oven at a temperature of 128°C. and other samples in an air oven having a temperature of 142°C. Foam volume against time is observed and the results are set forth in the following table.

TABLE 1

No.	Oven temp. (° C.)	Time (minutes)	$\frac{V_{foam}}{V_{solid}}$	Foam density lb./ft.³
1	128	4	5.8	10.8
2	128	6	8.2	7.6
3	128	10	12	5.2
4	128	20	15.2	4.1
5	128	60	20	3.1
6	128	120	25.3	2.5
7	128	390	42.5	1.5
8	128	1,200	76.8	0.81
1A	142	4	5.8	10.8
2A	142	6	12.8	4.9
3A	142	10	17.3	3.6
4A	142	20	22.2	2.8
5A	142	60	38.5	1.6
6A	142	120	60.4	1.03
7A	142	290	87.8	0.71

Thus, depending upon the foaming time and temperature chosen, a wide variety of foam densities can be obtained. In each case, the expanded piece is an expanded replica of the original. The excellent stability of the foam against shrinkage at the high foaming temperature is evident. The resultant foam had fine uniform cells and the piece expanded into the replica of the original, that is, the foam of the unfoamed samples was unchanged. Only the dimensions were altered.

Example 2:   In a manner similar to Example 1, the following composition was polymerized and foamed. 89.91% orthochlorostyrene, 0.045% divinylbenzene, 0.045% ethylvinylbenzene, and 10.0% isopentane. All percentages are weight percentages. The results are set forth in the following table.

TABLE 2:   FOAMING OF COMPOSITION

No.	Oven temp. (° C.)	Time (minutes)	$\frac{V_{foam}}{V_{solid}}$	Foam density lb./ft.³
1	128	4	16.2	3.86
2	128	6	19	3.29
3	128	10	23.3	2.68
4	128	20	29.1	2.14
5	128	60	36.4	1.71
6	128	120	52	1.2
7	128	1,440	130	0.48
8	128	2,520	181	0.34
1A	142	4	16.5	3.78
2A	142	6	23.3	2.69
3A	142	10	28.1	2.22
4A	142	20	41.4	1.51
5A	142	60	65.1	0.959
6A	142	1,320	260	0.240

The foaming rate of the above composition was more rapid than that of Example 1. The foamed particles had fine, uniform cells and the expanded bodies were enlarged replicas of the unfoamed samples. Similar results were obtained when the procedure was repeated utilizing neohexane, neopentane, and tetrafluorodichloroethane as blowing agents.

Example 3:   A casting resin syrup was prepared by dissolving 30 g. of poly-
orthochlorostyrene in 70 g. of orthochlorostyrene. The resulting solution had
a viscosity of 6,300 cp. at 25°C. To this casting syrup was added blowing
agents and cross-linking agents as set forth in the following table. The fluid
mixtures were placed into polytetrafluoroethylene molds which had cavity
dimensions of 0.5 inch in diameter and 0.375 inch in height. The open top
of the mold was covered with a 1 mil thick polyvinyl fluoride film and the
mixtures were polymerized for 24 hours at 25°C. At the end of this period
hard resin cylinders were removed from the molds and heated for 15 min. in
an air oven at 140°C. The results are set forth in the following table.

TABLE 3:   PREPARATION AND FOAMING OF CAST POLYCHLORO-
STYRENE RESINS

No.	o-Chloro-styrene, wt. percent	DVB-EVB mixture, [1] wt. percent	CFCl₃, wt. percent	Celogen, [2] wt. percent	IPPC, [3] wt. percent	Foaming behavior of cured sample in 15 minutes at 140° C.	
						Foam vol./ init. vol.	Avg. cell size (mm.)
1	90.9	0.1	8	0	1	13.8	1.5
2	90.85	0.1	8	0.05	1	25.9	0.3
3	90.8	0.1	8	0.1	1	30.4	0.25
4	90.6	0.1	8	0.3	1	20.1	0.2
5	90.4	0.1	8	0.5	1	19.25	0.2
6	89.9	0.1	8	1.0	1	16.78	0.2
7	87.9	0.1	10	1.0	1	23.88	0.2
8	85.9	0.1	12	1.0	1	28.1	0.2
9	83.9	0.1	14	1.0	1	30.4	0.2
10	81.9	0.1	16	1.0	1	36.0	0.2

[1] 50 weight percent divinylbenzene, 50 weight percent ethylvinylbenzene.
[2] Oxy-bis(benzene sulfonyl hydrazide).
[3] Diisopropyl peroxydicarbonate.

Inclusion of a low concentration of the N₂ releasing compound actually
results in both a smaller cell size and greater expansion. 0.1% of the
N₂ releasing compound can only generate a maximum of about 0.21 cc of
gas per g. of polymer. In each case, the foamed article was an expanded
replica of the original cylinder.

Example 4:   A mold was prepared having a cavity in the form of a generally rectangular plate, the inner surfaces of the mold were covered with a
1 mil thick polyvinyl fluoride film. The mold was used to prepare a plurality of sheets by filling with a polymerizable mixture containing a blowing
agent, maintaining the mold at a temperature of about 86°F. and exposing
the mold to a source of high energy ionizing radiation from a cobalt-60
source to provide gamma radiation at a dose rate of 10^5 rads per hour until
the polymerizable mixture within the mold had been subjected to a total
radiation dose of 1.6 megarads. Subsequently, the polymerized samples

were removed from the mold and the results are set forth in Table 4.

TABLE 4

No.	Composition Methyl methacrylate (wt. percent)	Blowing agent Wt. percent	Type	Appearance after polymerization
1	90	10	n-Pentane.	Hard opaque white resin.
2	90	10	Isopentane.	Do.
3	90	10	Cyclopentane.	Hard clear resin.
4	90	10	n-Hexane.	Hard opaque white resin.
5	90	10	Isohexane.	Do.
6	90	10	Neohexane.	Do.
7	90	10	CFCl$_3$.	Hard clear resin.
8	90	10	C$_2$F$_3$Cl$_3$.	Do.

The polymerized sheets in Table 4 were then heated in a hot air oven and the foam volume and time observed. The results are set forth in Table 5, wherein the sample numbers designate the samples prepared in Table 4.

TABLE 5

No.	Blowing agent Wt. percent	Type	Oven temp. (° C.)	V_F/V_S, after indicated heating time of— 5 min.	15 min.	30 min.	60 min.
1	10	n-Pentane.	130	1	6.7	---------	13
1	10	----do----------	140	4	8	13.9	22
2	10	Isopentane.	130	1	3.6	5.6	14
2	10	----do----------	140	3.6	7.6	15	18
4	10	n-Hexane.	130	---------	16.8	19.1	25.6
5	10	Isohexane.	130	---------	15.8	20	25.4
6	10	Neohexane.	130	---------	13.5	21	21.8
7	10	CFCl$_3$.	130	---------	1	1	1.5

All of the samples subjected to the hot air oven produced a fine celled foam having a very smooth surface and were enlarged replicas of the unfoamed sheet.

Multicellular Product Based on Acrylonitrile

N. Sagane, I. Kuwazuru and I. Kaetsu; U.S. Patent 3,673,129; June 27, 1972; assigned to Sekisui Kagaku Kogyo K.K., Japan describe a process for preparing multicellular products which comprises mixing urea and/or a urea derivative, and an acid with a monomeric mixture comprising 60% to 95% by weight of acrylonitrile and 40% to 5% by weight of a vinyl monomer copolymerizable with acrylonitrile, polymerizing the mixture with the use of catalyst and/or by irradiation, and thereafter heating the resulting polymer at 100° to 250°C. The multicellular products produced by the above process, have cells with an average diameter of less than 1.0 mm. which are substantially uniformly dispersed in the polymer and which product has a specific gravity of less than 0.3 g/cm^3.

Example 1: One hundred parts of acrylonitrile, 50 parts of methyl methacrylate, 15 parts of urea, 31 parts of benzoic acid and 24 parts of monochloroacetic acid were mixed and stirred to obtain a uniform transparent solution. After polymerizing this solution by radiating 5.9 x 10^6 roentgens of a gamma ray (cobalt 60) having a strength of 4.7 x 10^4 roentgens/hour, the so obtained polymer was heated at 180°C. for 15 minutes, as a result 20-time expanded yellow colored multicellular product was obtained.

This multicellular product had an average cell diameter of 0.89 mm. and a specific gravity of 0.06. When this multicellular product was used for a long period of time at 160°C., it did not shrink, being excellent in heat resistance.

Example 2: One hundred parts of acrylonitrile, 50 parts of methyl methacrylate, 15 parts of urea and 33 parts of glacial acetic acid were mixed to obtain a uniform transparent solution. After polymerizing this solution by irradiation a dose of 5.9 x 106 roentgens of a gamma ray having a dose rate of 4.7 x 10^4 roentgens/hour, and heating the so obtained polymer lump in a constant temperature oven at 180°C. for 15 minutes, a 25-time expanded light yellow multicellular product was obtained. This multicellular product had an average cell diameter of 1.0 mm. and a specific gravity of 0.048.

Example 3: One hundred parts of acrylonitrile, 40 parts of methyl methacrylate, 5 parts of urea, 17.2 parts of a 95% sulfuric acid were mixed to obtain a transparent solution. To this solution, 0.5% by weight based on this solution each of benzoyl peroxide and N,N'-dimethylaniline was added and polymerization was carried out at 35°C. for 12 hours in a polymerization vessel having an internal capacity of 20 x 240 x 300 mm. to obtain a polymer of a size of 15 x 240 x 300 mm.

Thereafter, when this copolymer was heated in a constant temperature oven at 190°C. for 15 minutes, a light yellow fine foamed multicellular product having a specific gravity of 0.037, a foaming magnification of 32 times, and an average cell diameter of 0.01 mm. was obtained. This multicellular product had excellent mechanical strengths of a bending strength of 10 kg./cm.^{2}, a tensile strength of 13 kg./cm.^{2} and a 50% compressed strength of 8 kg./cm.^{2}, at the same time, exhibiting a high heat-resistance of enduring a long use at 160°C.

Example 4:   100 parts of acrylonitrile, 40 parts of methyl methacrylate,

5 parts of urea, 27 parts of trichloroacetic acid, 5 parts of a 85% phosphoric acid and 4 parts of water were mixed to obtain a transparent solution. To this solution, 1% by weight based on this solution each of tertiary butyl-hydroperoxide and N, N'-dimethylaniline was added, polymerization was carried out at 40°C. for 16 hours and at 50°C. for 10 hours, and the obtained polymer was foamed by heating at 190°C. to obtain a 35-time expanded light yellow multicellular product having a fine cell structure. This multicellular product had an average cell diameter of 1 mm. and a specific gravity of 0.034.

Heat Recoverable Thermoplastics

A process described by P.M. Cook; U.S. Patent 3,597,372; August 3, 1971; assigned to Raychem Corporation involves the production of elastomeric articles having heat-activated dimensional memory characteristics.

A series of important commercial products have been developed over the past few years, based upon the property of plastic memory. Two different techniques are used for the production of such so-called dimensionally heat-unstable or perhaps more properly "heat-recoverable" thermoplastic materials, i.e., products which change their size and shape upon the application of heat without the necessity for the application of external forces. The first technique is that of imparting a considerable amount of built-in stresses during fabrication, followed by a cold temperature quench to hold the molecules in the stressed condition.

Upon careful heating, this fabricated product will tend to reform or recover to the original configuration. However, upon slight overheating, or upon heating too long, such thermoplastic materials will melt and relax to a new size and shape. More recently, a series of cross-linked thermoplastic products have been fabricated wherein the memory characteristic of the plastic is obtained by a 3-dimensional network rather than built-in stresses in a 2-dimensional system. For example, a cross-linked polyethylene can be heated to above the crystalline melting temperature, at which point it behaves as an elastomer where the application of a force will lead to a deformation directly proportional to that force.

If, while the cross-linked polyethylene is in the elastomeric state, a force is applied to cause deformation proportional to the force, and this is followed by a reduction in temperature, crystallization will take place which will maintain the cross-linked polyethylene in its deformed condition. Upon the subsequent application of heat sufficient to remelt the crystals (in the absence of a deforming force), the material will rapidly recover to the exact size and shape in which it has been cross-linked. However, such materials, being of crystalline thermoplastic nature, will exhibit normal thermoplastic properties while in the crystalline state, and will act as elastomers only at the elevated temperatures wherein the crystals are melted. A very large variety of applications of commercial importance can be envisioned for elastomeric articles which are heat-recoverable, i.e., having the properties of changing shape and/or size upon the application of heat but not external forces, but exhibiting essentially the elastomeric property of elastic deformation under stress.

According to this process, it has been found that elastomer products can be made having the properties of elastic deformation substantially equal to true elastomers, and at the same time having the properties of changing shape and/or size merely upon the application of heat and recovering to the original vulcanized or cross-linked shape and size.

The sequence of steps utilized in carrying out the process is as follows. (1) Intimately mixing the thermoplastic or resinous material with the elastomer or rubber gum in the uncured state, or mixing the plasticizer and cross-linkable thermoplastic material. The usual fillers, extenders, curing agents, accelerators and the like are included at this stage if desired, depending upon the desired properties of the final article.

(2) The composition formed in step (1) is then fabricated into an article of predetermined configuration and cross-linked or vulcanized. This may be accomplished by a chemical cross-linking technique where a vulcanizing or cross-linking agent is added, the subsequent application of heat and/or pressure bringing about the desired cure. Alternatively, the cross-linking may be brought about by exposure of the article to high energy radiation such as from accelerated electrons, x-rays, gamma-rays, alpha particles, beta particles, neutrons, etc., without the necessity for the addition of cross-linking or vulcanizing agents. Further, the cross-linking or vulcanization can be accomplished by a combination of these two techniques.

The degree of chemical cross-linking or the radiation dosage are sufficient to produce at least the minimum high temperature modulus of elasticity of 10 psi referred to hereinabove. Generally, the minimum radiation dosage is of the order of 2×10^6 rads.

(3) The cross-linked and cured article is then heated to a temperature sufficiently high to soften the thermoplastic component, i.e., above the melting or softening point or range, and while the material is maintained at that temperature an external force or forces are applied to change the size and/or shape of the article to a more convenient configuration for later application and use.

(4) The deformed article is cooled or quenched while still under the external

deforming stresses, whereupon the article will retain the deformed shape upon the release of the external stresses. The article is now in the heat-recoverable state but may be left for an indefinite period of time at room temperature without danger of its recovering back to its original size and shape.
The following methods of sample preparation and test procedures were utilized in carrying out the examples.

Mixing Technique: The method of incorporating the resin into the elastomer is important. If the resin is not dispersed thoroughly and completely throughout the elastomer, the properties of the compound are impaired.
The preferred method used in the examples is as follows. A quantity of the
resin is placed on a 2-roll mixer operating at a temperature sufficient to
soften or melt the resin.

The resin is milled until completely softened and then an equal amount of
elastomer is slowly added to the resin. Mixing is continued until an homogeneous composition is secured. This is removed from the rolls and cooled.
The balance of the mixing is done by placing a requisite amount of mixture on a cold 2-roll mill and adding the additional elastomer along with
antioxidants, accelerators, fillers, plasticizers, etc., as required.

Molding Technique: Molded slabs were prepared using a 6" x 6" x 0.062"
rubber mold as described in ASTM D-15. The time-temperature cycles were
varied depending upon the particular rubber-memory plastic system.

Irradiation Technique: Samples were irradiated using a General Electric
resonant transformer operating effectively at 850 kv. and 5 milliamps. The
samples were cross-fired to insure uniform irradiation through the sample,
and the irradiation dose was predetermined by Faraday cage measurements.
A well-grounded thermocouple in contact with the sample served to measure
the temperature of the sample.

Tensile Strength, Young's Modulus and Elongation: All tests were run at
room temperature using the variable speed automatic recording Instron
tester. All samples were tested at a crosshead speed of $20 + 1$ inch per
minute. The tensile strength and elongation were determined in accordance with ASTM D-412. Young's modulus was determined in accordance
with ASTM D-638.

Determination of Modulus of Elasticity and Ultimate Strength: The basic
technique for determining these values has been described by Black, R.M.,
The Electrical Manufacturer, October 1957. For this investigation, a
similar apparatus was used, consisting of a vertical glass tube with a glass
jacket (similar to a Liebig condenser). The jacketed space was filled with
boiling cyclohexanone and this kept the interior tube at 150°C.

Strips of the cross-linked compound were prepared (0.062" x 0.125"x 6"). Bench marks 1" apart were stamped in the middle portion of the sample. This strip was placed in the center tube and fastened securely at the top. Stress was applied to the sample by hanging weights on the bottom of the sample. Strain was measured by noting the increase in distance between the bench marks, the measurement being made at equilibrium after each addition of weight. The weights were increased until the sample broke. From the stress-strain data obtained, a modulus chart was prepared. The slope of the line was determined as the M100 figure, or stress necessary to effect a strain of 100%. The breaking force was recorded as the Ultimate Strength or U.S.

Importance of Modulus of Elasticity, Ultimate Strength Relationship: For memory devices the modified elastomer must have certain physical proper-ties. For heat shrinkable devices, the M100-U.S. ratio is a vital one. For the best use of such devices, either tubing or molded items, they must be capable of a significant amount of stretching or expansion without split-ting at elevated temperatures. The M100 figure expresses the stress neces-sary to effect a stretch of 100% at a temperature above the resin softening point. The Ultimate Strength (U.S.) is the stress at the breaking point. The elongation at the breaking point expressed in percent is:

$$100 \times \frac{U.S.}{M100}$$

Thus the larger the ratio, the more a compound can be stretched without danger of breaking.

Description of Plastic Memory Test: Strips of cross-linked compound, 1/8" x 0.062" x 6" were marked in the middle with 2 parallel ink impres-sions 1" apart. The strip was heated 1 minute in a 150°C. glycerine bath, stretched until the 1" lines were 3 inches apart (200% stretch) where pos-sible, removed from the hot bath and plunged into cold water. Five minutes after the cooling, the distance between the marks was measured as the ex-tended length. This distance was expressed as percent increase in length over the original 1" and recorded as the "Memory". Twenty-four hours later the extended strip was placed in the 150°C. glycerine bath for 1 min-ute and allowed to freely retract or shrink. It was then cooled and the dis-tance measured between the marks as the retracted distance. The "Retrac-tion", calculated as a percentage as follows, was recorded:

$$\text{Retraction} = \frac{\text{extended length} - \text{retracted length}}{\text{extended length} - 1} \times 100$$

Example 1:   In carrying out this example several specimens comprising polychloroprene elastomer and polyvinyl chloride were made and tested. As
indicated by the data set forth below, as little as 10% of polyvinyl chloride
based on the total of the polyvinyl chloride and elastomer, imparts memory
properties to the article in accordance with the process. By increasing the
amount of polyvinyl chloride the memory properties are progressively improved, excellent properties being obtained with as much as 40% polyvinyl chloride without sacrifice in tensile strength and with substantial retention of the elastomeric properties of the elastomer. The following compositions or mixes were press cured for 10 minutes at 350°F.

Compositions

Specimen	1	2	3	4
Geon 101 EP	10	20	30	40
Neoprene W	90	80	70	60
JZF	2	2	2	2
Stearic acid	1	1	1	1
Vancide 51Z	1	1	1	1
Maglite D	4	4	4	4
Sterling V	15	15	15	15
Flexol TOF	9.2	8.4	7.6	6.8
ZnO	5	5	5	5

Tensile and Elongation

	Average		
	P.s.i.	Percent elongation	Young's modulus (p.s.i.)
Specimen:			
1	1,151	533	113
2	1,146	457	178
3	1,094	407	360
4	917	193	656

Elastic Modulus and Ultimate Strength at 150°C.

	M100 (p.s.i.)	U.S. (p.s.i.)
Specimen:		
1	79	194
2	70	187
3	68	218
4	63	133

Memory Characteristics

	1	2	3	4
Memory, percent	40	80	130	200
Retraction, percent	100	100	100	100

Example 2: This example illustrates that a wide variety of thermoplastics, resins and waxes can be incorporated in polychloroprene compositions to produce articles within the scope of this process. The basic formulation was as follows. Compounds were prepared as described in mixing procedure.

Effect of Various Memory Plastics in Heat Shrinkable Neoprene Rubber

	Parts
Memory ingredient	35
Neoprene W	65
Stearic acid	1
MgO	4
Sterling V	15
Vancide 51Z	1
Flexol TOF	5
ZnO	5
JZF	2

Test specimen slabs made up using the memory ingredient indicated were press cured for 20 minutes at 340°F. and tested with the following results:

	Specimen No.												
	1	2	3	4	5	6	7	8	9	10	11	12	13
	Memory Ingredient												
	Silicone R4281	Cumar S	Atlac 382	Epon 1031	Halowax 2141	Acrowax B	Styron 666	Geon 101 EP	Tenite 812	Alathon 34	6001 Poly-ethylene	Marbon 8000	Hi-Fax 1400
Tensile strength (psi)	926	959	1,062	736	1,157	384	1,071	1,479	770	826	985	1,206	1,144
Elongation (%)	480	540	440	180	460	400	420	317	200	230	252	518	338
M100 (psi)	58	76	73	123	36	35	56	109	32	45	70	38	60
U.S. (psi)	288	121	251	222	218	56	182	390	100	120	200	130	190
Memory (%)	180	190	200	150	70	190	200	200	125	100	100	100	150
Retraction (%)	100	94	100	93	100	100	95	100	86	100	100	82	100

Example 3: This example illustrates that the conventional polychloroprene rubber accelerators function as such in the compositions of the process. In carrying out this example the following compositions were press cured for 10 minutes at 350°F.

Specimen No.	1	2	3	4	5	6
Geon 101 EP	40	40	40	40	40	40
Neoprene W	60	60	60	60	60	60
JZF	1	1	1	1	1	1
Maglite D	4	4	4	4	4	4
Stearic acid	1	1	1	1	1	1
Flexol TOF	5	5	5	5	5	5
Sterling V	15	15	15	15	15	15
ZnO	5	5	5	5	5	5
Ottacide P			1	1	1	1
Methyl zimate			1			0.5
Zetax				1		0.5
Thiate A					1	
Vancide 51Z		1				

Test results for the articles thus formed were as follows:

Specimen No.	1	2	3	4	5	6
Tensile strength (p.s.i.)	1,220	1,911	1,806	1,448	2,410	1,709
Elongation (percent)	253	230	273	297	203	320
M100 (p.s.i.)	18	26	62	60	142	58
U.S. (p.s.i.)	157	220	295	302	272	246
Memory (percent)	200	200	200	200	200	200
Retraction (percent)	100	100	100	100	100	100

Example 4:   To illustrate that by the inclusion of a suitable low-temperature plasticizer, heat-recoverable polychloroprene articles can be produced having low temperature flexibilities equivalent to those of conventional polychloroprene articles, and in fact having adequate flexibility at -55°C. the following compositions were prepared and press cured for 10 minutes at 350°F.

Specimen No.	1	2	3
Geon 101 E P	35	35	35
Neoprene W	65	65	64
JZF	1	1	1
Stearic acid	1	1	1
MgO	4	4	4
ZnO	5	5	5
Sterling V	15	15	15
Vancide 51Z	1	1	1
Flexol TO F	2.5	5	10

Cold Bend tests were performed in accordance with MIL-R-6855, as follows: The test specimens were strips 5 1/2" x 1/4" x 0.062". The bending devices consisted of 2 parallel jaws 2 1/2" apart. The ends of the strips were inserted into the jaws for a distance of 3/4" and fastened firmly with the middle portion of the strip forming a loop between the jaws. This assembly was conditioned in dry air at the indicated temperature for 5 hours. While still at this low temperature, the jaws were moved rapidly from the 2 1/2" to a 1" separation. Failure was denoted by cracking of the sample. The results were as follows:

Specimen No.	1	2	3
-40° C	OK	OK	OK
-45° C	(¹)	OK	OK
-50° C		OK	OK
-55° C		(¹)	OK
Tensile strength (p.s.i.)	1,421	1,409	1,521
Elongation (percent)	230	250	260
M100 (p.s.i.)	71	75	77
U.S. (p.s.i.)	276	308	299
Memory (percent)	200	190	170
Retraction (percent)	100	97	94

¹ Broke.

Example 5:   A number of heat-recoverable polychloroprene articles were prepared by press-curing the following compositions for 20 minutes at 340°F. to illustrate that the properties of the articles can be changed by varying the amounts and type of compounding ingredients:

Mix No.	1	2	3	4	5	6
Geon 101 EP	50	50	50	50	50	50
Neoprene W	50	50	50	50	50	50
Stearic acid	1	1	1	1	1	1
JZF	2	2	2	2	2	2
Vancide 51Z	1	1	1	1	1	1
Flexol TOF	3	3	3	3	3	3
MgO	4	4	4	4	4	4
ZnO	5	5	2½	2½	5	2½
DOTG			.5			
Sulfur				.25		
Sterling V		15	15	15		
Thermax					5	5

Following are the results of the physical tests:

Specimen No.	1	2	3	4	5	6
Tensile strength (p.s.i.)	827	1,088	978	1,282	846	810
Elongation (percent)	133	103	113	80	133	140
M100 (p.s.i.)	26	66	58	140	31	31
U.S. (p.s.i.)	101	173	162	224	123	111
Memory (percent)	120	160	140	200	170	140
Retraction (percent)	96	100	100	100	97	92

Example 6:   A heat-shrinkable elastomeric tubing was produced by extruding the composition set forth below in a 1 1/2" Davis Standard extruder with a conventional thermoplastic screw. The tubing had a 0.250 i.d. and a 0.031" wall thickness. The tubing was vulcanized in a cylindrical steam vulcanizer at a pressure of 70 psi for 2 hours (Sample No. 1) and 3 hours (Sample No. 2).

	Percent
Geon 101 EP vinyl resin	29.0
Neoprene W	43.0
JZF	1.5
Flexol TOF	7.5
Maglite D	3.0
Stearic acid	1.0
Vancide 51Z	1.0
Sterling S.O.	10.0
Zinc oxide	4.0

To produce the heat-shrinkable tubing, the samples were expanded using a hot glycerine bath and a heated mandrel to 100% and 200%. The tubing was then cooled while over the mandrel in cold water and removed. The tubing showed an initial shrinkage of 10 to 15% and thereupon remained expanded until it was heated above the softening range of the polyvinyl chloride whereupon it shrank to its original dimension. Test results were as illustrated on the following page.

Elastic Modulus and Ultimate Strength at 150°C.

Sample No.	M100 (psi)	U.S. (psi)
1	29	192
2	38	181

Tensile Strength and Elongation

Sample No.	Average (psi)	Percent Elongation
1	1,077	250
2	1,007	260

Polyurethane Insulating and Burn Resistant Compositions

G.A. Kuhar; U.S. Patent 3,699,023; October 17, 1972; assigned to The Goodyear Tire & Rubber Company describes a method for preparing a high temperature insulator comprising the steps of preparing a reaction mixture of a polyester polyol, an organo-peroxide, an organic polyisocyanate and a filler having a resistance to burning; and setting and curing the reaction mixture at a temperature no greater than about 250°F. for about 10 to 20 hours followed by subjecting the cured reaction mixture to a treatment of at least 300°F. for at least 10 hours or an ionizing radiation treatment of about 5 to about 30 megarads.

It has been found that by subjecting a cured polyurethane composition containing a refractive filler to a post-cure treatment of at least about 10 hours at a temperature in excess of 300°F. the post-cured composition has improved insulating properties in addition to greater resistance to burn. Alternately the post-cure treatment may be effected by subjecting the cured polyurethane composition to an ionization radiation sufficient to give a dosage of at least about 5 megarads. Usually for economic reasons the ionizing radiation treatment will not exceed about 30 megarads. The following examples illustrate the process.

Example 1: An unsaturated polyester was formed by the condensation of adipic acid with a mixture consisting of 85 mol percent of propylene glycol and 15 mol percent of glycerol allyl ether. This polyester had a molecular weight of about 2,000, a reactive number of 60 and an acid number of less than 5. This unsaturated polyester (600 parts) was mixed with Spodumene (180 parts), sodium borate (180 parts) and 6 parts dicumyl peroxide. This mixture was mixed at 90°C. with 0.5 part of phenyl-beta-naphthylamine and a vacuum was maintained upon the system during the mixing for 15 minutes.

This treatment removed any occluded free moisture present in these reactants to give substantially an anhydrous mixture. Then 102 parts of tolidine diisocyanate was added and stirred into the above degassed ingredients at 90° to 102°C. The stirring was continued for 10 minutes before the fluid mixture was poured into molds 4" x 4" x 1/2" and held at 212°F. for 19 hours to set and cure the fluid mixture. The cured material was removed from the mold, and had a good structure free of pores, a Shore A hardness of 70 and good flexibility. When this cast block was subjected to 5000°F. oxyacetylene torch burn test the burn rate was 4.0 mils per second.

A duplicate casting 4" x 4" x 1/2" was given a post-cure for 96 hours at 325°F. under a nitrogen atmosphere and then was subjected to the 5000°F. oxyacetylene torch test. The post-cured sample had a burn rate of 2.6 mils per second.

Example 2:   The recipes set forth in Table 1 were used to make cast insulating compositions. These cast insulating compositions were formed in blocks 3/8" thick, then part of the blocks were subjected to the post-cure treatment in Table 1 before being subjected to the oxyacetylene burning tests. The results of this test are shown in Table 1.

TABLE 1

R108X	366	367	368	369
Polyester*	600	600	600	600
Phenyl-beta-naphthylamine	0.5	0.5	0.5	0.5
Spodumene	180	180	180	180
Sodium borate	180	180	180	180
Tolidine diisocyanate	78	90	102	90
Peroxide 10**	6	6	6	3
Burn Rate in Mils/Second				
Original cure	4.0	3.6	3.6	6.3
Post-cured with 10 MR	3.1	3.2	3.5	4.0
Post-cured 96 hrs. at 325°F. under N$_2$	2.6	2.8	1.8	3.2

  *The same polyester as that used in Example 1.
  **A material reported to contain in excess of 95% dicumyl peroxide
     with the rest being inerts.

Similar results may be obtained by substituting other unsaturated polyesters, such as propylene adipate fumarate for the one of this example.

Example 3:   A polyurethane reaction mixture was made using the recipes indicated in Table 2 and then the resulting liquid reaction mixture was allowed to react and then was cured by subjecting the reaction mixture which contained the peroxide 10 therein to a cure treatment which consisted of letting the reaction mixture stand at room temperature overnight and then heating at 212°F. for 16 hours. Then part of the cured samples were subjected to ionizing radiation in a cobalt 60 radiation cell for sufficient time to give a radiation dosage of the indicated megarads. These samples were subjected to burn rate determination in an arc image furnace at about 5000°F. The burn rate on the original cured sample and on the samples which received the post-cure treatment are listed in Table 2 along with the Shore A hardness of these samples.

TABLE 2

R108X	520	522	524	520	522	524
Polyester [1]	600	600	300			
Mica, 160 mesh	90		45			
Mica, 325 mesh	90	90	45			
Phenyl beta naphthylamine	1.0	0.5	0.5			
Tolidine diisocyanate	144.6	74.4	37.2			
Peroxide 10	6	6	3			
Asbestos	12					
Glycerol allyl ether	35.5					
Cork		50				
Calcium metaborate			45			
	Burn rate			Shore A hardness		
Cured sample	3.3	3.9	4.3	79	83	52
Post cure: [2]						
5.3	2.4	3.3	3.8	89	91	82
10	2.7	3.4	3.5	89	94	84
30	2.5	3.5	2.9	94	97	91
110	2.4	3.3	2.7	93	96	91
150	2.5	3.7	2.7	92	96	91
205	2.4	3.8	2.6	90	96	91

[1] The same polyester as that used in Example I.
[2] The post cure treatment is expressed as megarads.

Example 4:   An unsaturated polyester was prepared by the condensation of adipic acid with a mixture consisting essentially of 85 parts of propylene glycol and 15 parts of glycerol allyl ether to obtain an unsaturated ester having an acid number of less than 2 and a total reactive number of about 40 to 60. This polyester contained one double bond for about every 1,500 units of molecular weight.

This polyester (600 parts) containing 1.0 part of phenyl-beta-naphthylamine was mixed with mica (90 parts 160 mesh and 90 parts 325 mesh), and 180 parts of sodium borate and heated with stirring at 90° to 109°C. under a vacuum for 30 minutes to degas the mixture. Six parts of a commercial dicumyl peroxide was added and stirred into the mixture. This addition was followed by the addition of 102 parts of orthotolidine diisocyanate.

After stirring for 9 minutes, the pasty mixture was poured into aluminum molds 3/8" on a side and covered with an aluminum lid. These molds were held at 100°C. for 20 hours to set and cure the polyurethane. Part of the coatings were cured by treatment with 10 megarads of gamma radiation, others were cured by heating at 325°F. in a nitrogen atmosphere for 120 hours and 576 hours. The 10 megarads sample had an oxyacetylene torch (5000°F.) burn rate of 3.1 mils per second and formed a tough char. While the burn rate for samples cured in a nitrogen atmosphere at 325°F. for 120 hours and 576 hours were 2.0 and 2.6 respectively.

The cured samples which were 4 x 4 x 1/2 inch on the side had a thermo-couple placed 1/4 inch from the top of the sample and a thermocouple placed in the hole drilled in the sample in alignment with the focal point of the arc image. The burn rate was run and the results are reported in Table 3 as the time required for the thermocouple to reach the temperature indicated. These results for the original cured sample are reported in the column headed "Original Cure" and the results for the samples which had received the post-cure treatment for 48 hours at 325°F. are reported in the column headed "Post-Cure".

TABLE 3:   THERMOCOUPLE HEAT TRANSFER TEST

Thermocouple Temp., °F.	Original Cure, Minutes	Post-Cure, Minutes
200	1.13	1.49
250	1.39	1.89
300	1.74	2.27

Heat Recoverable Polyurethane Rubbers

E.C. Stivers; U.S. Patent 3,624,045; November 30, 1971; assigned to Raychem Corporation describes a class of heat-recoverable cross-linked polyurethane rubbers having heat-activated dimensional memory character-istics.

Polyurethane elastomers or rubbers are known materials for use in conveyor belts and tubing. These known polyurethane rubbers are cross-linked and are comparable in many ways to vulcanized natural rubber and possess the physical property of elastic extensibility. Hence, such polyurethane rub-bers will stretch an amount directly proportional to the applied force and will recover upon the removal of the force. Thus, when the force is re-moved, the polyurethane rubber returns to its original cross-linked size and shape. More recently there has been marketed a class of uncross-linked thermoplastic polyurethane rubbers which behave at ordinary, e.g., room temperature as if they are cross-linked, but which behave as a thermoplastic, i.e., melt and flow, at elevated temperatures. Typical of such thermoplastic polyurethane rubbers is one having the following structural formula. These polymers are known as Estane.

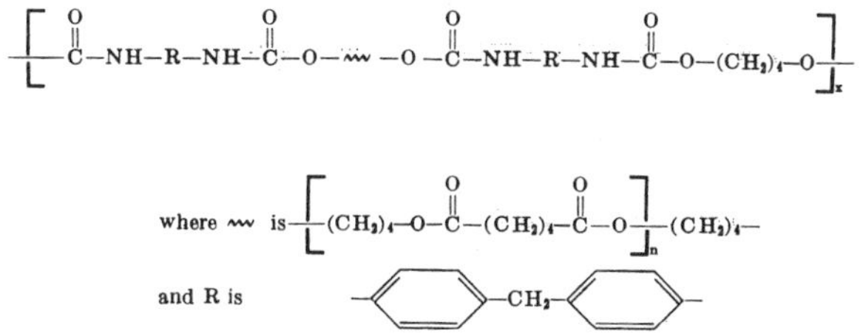

$$\left[\begin{array}{c} O \\ \| \\ -C-NH-R-NH-C-O-\text{\Large$\sim$}\!-O-C-NH-R-NH-C-O-(CH_2)_4-O- \\ \begin{array}{cccc} O & & O & O \end{array} \end{array}\right]_x$$

where $\text{\Large$\sim$}$ is $\left[-(CH_2)_4-O-\overset{O}{\overset{\|}{C}}-(CH_2)_4-\overset{O}{\overset{\|}{C}}-O-\right]_n(CH_2)_4-$

and R is (structure: two phenyl rings joined by $-CH_2-$)

Another class of such thermoplastic polyurethane rubber is called Texin. These isocyanate-terminated polymers are somewhat higher in molecular weight than Estane, the polymer shown above. The class of polymer are prepared from hydroxyl-terminated polyesters, methylene-p-phenylene diisocyanate, and a diol resulting in solid, thermoplastic products. The classical rubberlike behavior of these uncross-linked polymers at moderate temperatures can probably be ascribed to interchain hydrogen bonding which behave as secondary cross-links, e.g.:

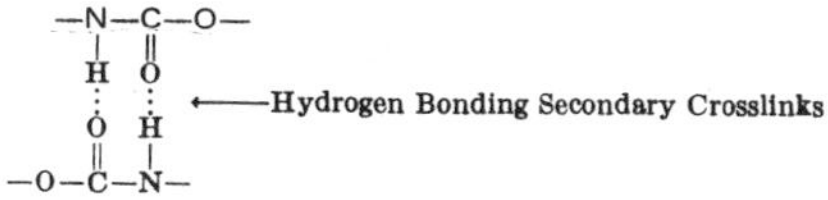

$$\begin{array}{c} -N-C-O- \\ |\ \ \| \\ H\ \ O \\ \vdots\ \ \vdots \\ O\ \ H \\ \|\ \ | \\ -O-C-N- \end{array} \longleftarrow \text{Hydrogen Bonding Secondary Crosslinks}$$

Dimensionally heat-unstable or heat-recoverable thermoplastic polyurethane articles of little utility may be produced by imparting a considerable amount of force or stress to the heated material after initial fabrication, followed by a cold temperature quench to hold the molecules in the stressed, usually elongated condition. Subsequently, upon carefully heating, the fabricated product will tend to recover or reform to the original configuration. However, these polyurethanes have essentially no strength at elevated temperatures and thus can be readily deformed to an undesired shape or form. Generally, the hydrogen bonds in these materials disappear with increasing temperature in a manner which may be estimated by the Arrhenius equation

$v_S/V \cong Ae^{-E_a}/RT$ where v_S/V is the secondary cross-link concentration, A a constant, R the general gas constant, T the absolute temperature, and E_a is the activation energy of bonding. Thus at temperature of the order of 120° to 160°C. these polyurethanes show a decrease in elastic modulus from a room temperature value of 1,800 psi to 5 psi. A primary contribution of this process is the provision of articles prepared from thermoplastic polyurethane rubbers which are heat-recoverable and exhibit the property of elastic deformation of extensibility under stress over a wide range of temperatures, i.e., have strength at elevated temperatures.

The heat recoverable articles of the process are produced by deforming articles comprising cross-linked thermoplastic polyurethanes under conditions such that the article is substantially permanently deformed and retains a deformed configuration until heated to the recovery temperature of the article. This deformation may be most conveniently carried out while the article is maintained at an elevated temperature, e.g., between about 100° and 160°C., depending upon the particular polyurethane used, and then quenching the article at a relatively low temperature, e.g., 25°C.

However, such deformation at elevated temperatures is preferred because, among other things, a lesser degree of deformation is required to produce a given degree of permanent deformation. For example, when identical samples produced according to the process were stretched at room temperature and at 100°C., it was found that the former sample required an expansion of 560% to produce a retained expansion of 100% whereas the latter sample required an expansion of about 145% to produce a retained expansion of 100%. It is, of course, to be understood that the optimum temperature for deformation will vary from composition to composition and will depend upon the composition of the thermoplastic polyurethane, the degree of cross-linking, etc.

Typical heat-recoverable methods by which polyurethane rubber articles may be produced include forming or fabricating into the desired configuration, an article comprising:

(1) A thermoplastic polyurethane of the type having the following general formula:

$$\left[-\overset{\overset{O}{\|}}{C}-NH-R-NH-\overset{\overset{O}{\|}}{C}-O-\text{\textasciitilde}-O-\overset{\overset{O}{\|}}{C}-NH-R-NH-\overset{\overset{O}{\|}}{C}-O-(CH_2)_4-O-\right]_x$$

where $\text{\textasciitilde}$ is $\left[-(CH_2)_4-O-\overset{\overset{O}{\|}}{C}-(CH_2)_4-\overset{\overset{O}{\|}}{C}-O-\right]_n(CH_2)_4-$

and R is $\langle\!\!\!\!\bigcirc\!\!\!\!\rangle-CH_2-\langle\!\!\!\!\bigcirc\!\!\!\!\rangle-$

which has been cross-linked by (a) irradiating such a polymer into which has been incorporated a cross-linking monomer such as N,N'-methylene-bis-acrylamide (MBA), and/or
N,N'-hexamethylene-bis-maleimide, (b) chemically cross-
linked, e.g., with an organic peroxide, alone or together
with a cross-linking monomer such as those mentioned above.

(2) Heating the material to an elevated temperature, e.g., about
100°C., applying an external force to deform the article to
the desired heat-recoverable configuration and then quenching the article at a lower temperature while in the deformed
state.

After release of the external force, the article remains in a deformed configuration, not returning to its original cross-linked configuration, as would
be the case in ordinary rubber. In this state, the article has the property
of heat-recoverability. The following examples illustrate the process.

Example 1:   Various polyurethane compositions known as Estane were mill
mixed on a two-roll laboratory mill at a temperature of 150° to 175°C.,
and the resulting mixtures were pressed into 0.6 mm. slabs. The slabs were
then subjected to various irradiation doses, using a 1 mev General Electric
Resonant Transformer operating at 5 to 6 ma. The modulus of elasticity and
ultimate strength at 160°C. and gel content of these cross-linked polyurethanes were measured and the results are tabulated in Table 1.

As is shown in Table 1, two different cross-linking monomers were incorporated in the thermoplastic polyurethane and irratiated to total doses of
5, 10, 20, and 40 megarads. At 160°C., the polyurethane containing no
cross-linking monomer requires much higher doses to obtain equivalent
modulus than samples containing monomer. It can be seen in the data in
Table 1 that N,N'-hexamethylene-bis-maleimide and N,N'-methylene-
bis-acrylamide are very effective cross-linking monomers.

TABLE 1

Estane No.	Monomer used	Monomer level, pts./100 pts./resin	5 Mrad, 160° C.			10 Mrad, 160° C.			20 Mrad, 160° C.			40 Mrad, 160° C.		
			Percent gel	Mod., p.s.i.	U.S. p.s.i.	Percent gel	Mod., p.s.i.	U.S. p.s.i.	Percent gel	Mod., p.s.i.	U.S. p.s.i.	Percent gel	Mod., p.s.i.	U.S. p.s.i.
5740XI	Nothing		0	5.0	5.9	0	7.4	13	26.5	12	67	50.0	23	53
5740XI	MBA¹	1	0	18	69	46.4	19	61	53.5	30	113	51.7	37	75
		2	53.1	41	129	56.5	75	160	64.5	74	112	72.4	123	122
		4	58.9	81	129	66.2	82	156	72.9	106	136	76.3	161	113
58013 Nat	Nothing		0	7	4	0	≈10	23	32.3	8	62	52.8	30	65
58013 Nat	MBA¹	½	52	7	52	55.3	21	94	64.0	32	68			
		1	57.3	13	88	59.3	50	118	68.0	60	94	65.7	66	72
		2	67.6	23	94	71.5	60	127	76.5	71	142	78.1	115	62
		4	70.5	73	143	76.9	116	149	77.3	137	120	79.1	173	106
58013 Nat	HMBMI²	½	26.0	29	29	33.7	26.5	63	47.8	41	96			
		1	50.6	36	67	55.9	51	85	59.5	48	155			
		2	67.5	58	128	70.0	79	112	70.7	89	113			
5740X2	Nothing		0	0	2.2	0	8	27	43.7	13.4	48	58.7	25	59
5740X2	MBA¹	1	38.4	15	75	51.3	38	98	62.3	44	94	68.8	78	73
		2	43.0	21	101	55.2	47	111	67.5	75	85	72.0	105	74
		4	50.2	39	104	64.3	75	154	73.4	109	135	76.0	164	124
5740X7	Nothing		0	15	15	0	27	18	Trace	28	32	42.7	36	51
5740X7	MBA¹	1	56.5	42	54	60.4	49	60	65.4	44	82	69.6	62	73
		2	56.3	53	77	64.0	45	70	68.8	58	63	73.0	80	70
		4	68.2	81	90	77.0	63	88	78.4	97	82.0	133	104

¹ N,N'-methylene bis-acrylamide.
² N,N'-hexamethylene bis-maleimide.

Example 2: Several of the cross-linked samples of Example 1 were heat aged at temperatures of 175° and 200°C. to determine the effect of cross-linked density on the heat aging properties. The results of these tabulated in Table 2. The data in Table 2 shows that the cross-linked thermoplastic polyurethanes have vastly improved high temperature dimensional stability over the uncross-linked materials. The cross-linked polyurethane rubbers are, therefore, generally suitable over a wide range of temperatures.

TABLE 2

Estane type	Monomer/ level (pt./100 pts.)	Dose, mr.	Heat aging observations	
			200° C.	175° C.
5740X1	None	0	Flowed in less than 1 hour	Flowed in less than 1 hour,
5740X1	do	20	....do	Slightly elongated at 1 hour, otherwise O K after 700 hours.
5740X1	do	40	....do	Elongated some after 52 hours, otherwise O K after 700 hours.
58013 Nat	do	0	....do	Flowed in less than 1 hour.
58013 Nat	do	20	....do	Do.
58013 Nat	do	40	....do	Elongated after 1 hour, flowed after 5 hours.
5740X1	MBA/2	5	Elongated in less than 1 hour, flowed in 4 hours.	
5740X1	MBA/2	20	Elongated some between 20-25 hours, otherwise O K after 87 hours.	O K after 700 hours.
5740X1	MBA/2	40	Elongated some at 41 hours, otherwise O K after 87 hours. [1]	Do.
58013 Nat	MBA/2	20	Flowed in less than 1 hour	Elongated after 5 hours, flowed at 13 hours.
58013 Nat	MBA/2	40	....do	Melted at 13 hours..
5740X1	MBA/4	5	Elongated in 1 hour, otherwise O K after 36 hours.	
5740X1	MBA/4	20	O K after 87 hours [1]	O K after 700 hours.
5740X1	MBA/4	40	....do [1]	Do.
58013 Nat	MBA/4	20	Elongated at 1 hour, flowed between 1 and 2 hours.	Elongated at 13 hours, otherwise O K after 700 hours. [2]
58013 Nat	MBA/4	40	Elongated at 1 hour, otherwise O K after 87 hours. [1]	Elongated at 37 hours, otherwise O K after 700 hours. [2]
5740X2	MBA/4	40	Elongated at 18 hours, otherwise O K after 36 hours.	
5740X7	None	5	Flowed in less than 1 hour	
5740X7	MBA/2	5	Elongated in less than 1 hour, otherwise O K after 36 hours.	
5740X7	MBA/4	5	....do	

[1] Became brittle somewhere between 87 and 231 hours.
[2] Became "puffy" and "porous" at 350-380 hours.

Example 3: Several samples of the cross-linked polyurethanes of Example 1 were tested for their ability to retain "locked-in" elongation. This was done by heating strip 6" x 1/4" x 0.025", which had 1" bench marks centrally located on the strips, in a glycerine bath maintained at various temperatures, stretching the samples while at temperature, and, while in the stretched condition, quenching in a cold water bath. To test the heat recoverable characteristics of these materials, stretched samples were kept at room temperature for several days and subsequently placed in a glycerine bath at different temperatures. The recovered dimension was then measured. A comparison of uncross-linked and cross-linked polyurethane known as Estane 5740X1 is shown in Table 3.

As is shown in Table 3, the polyurethane materials of this process are able to retain the elongation to a high degree and yet upon the final recovery, by heating in a 160°C. glycerin bath, the original length is subsequently regained. As shown in Table 3, when the same experiment is performed on the thermoplastic polyurethane, which has not been cross-linked a low degree of recovery is obtained at a temperature below the flow temperature of this polymer. It is not practicable to obtain a high degree of recovery since the material flows at temperature necessary to obtain a high degree of recovery.

TABLE 3

Expansion bath temp., ° C.	Stretched elongation, percent (stretched at bath temp.)	Elongation (percent) after quenching in water bath (1 hr. after quick release at 25° C.)	Final recovery (percent) increase in length over original length at— 110° C.	160° C.
4 pts. MBA/100 pts. Estane 5740X1 at 20 Mrad				
25	550	100	....	2
60	175	100	....	3
80	160	100	....	5
100	....	200	20	5
100	150	100	....	2
120	160	100	....	5
140	175	100	....	12
Estane 5740X1 (not crosslinked)				
160	200	100	....	10
100	....	200	40	Flowed
160	Flowed	....		

Example 4:   To demonstrate the chemical cross-linking of thermoplastic polyurethanes, two different polyurethanes known as Estane were mill mixed at a temperature of 90° to 120°C. with two different organic peroxides and with and without a cross-linking monomer. Following the mill mixing, slabs were cured in a mold at 170°C. for either 8 or 10 minutes, depending upon the specific peroxide used. The modulus of elasticity at 160°C. and percent gel were measured as described previously. Results are tabulated in Table 4.

TABLE 4

Estane type	MBA [1] level (pts./100 pts.)	Peroxide level (1 pt./100 pts.)	Percent gel	Modulus of elasticity and ultimate strength at 160° C. Mod. (p.s.i.)	US (p.s.i.)
5740X2	0	Varox [2]	54.8	55	44
5740X2	4	.....do.....	85.9	208	137
58013	0	Di Cup [3]	0	0	0.5
58013	4	.....do.....	39.3	80	43

[1] N,N'-methylene bis-acrylamide.
[2] Varox, 2, 5-dimethylhexane-2, 5-di-t-butyl diperoxide (the pure liquid), samples cured at 170° C. for 10 minutes.
[3] Di Cup, dicumyl peroxide—samples cured at 170° C. for 8 minutes.

In general, it can be seen that the peroxides cross-link these thermoplastic polyurethanes and higher elastic modulus values are obtained when cross-linking monomers are used.

Bioriented Vinyl Fluoride Polymer Webs

G.N. Foster and W. Sacks; U.S. Patent 3,594,458; July 20, 1971;
assigned to Union Carbide Corporation describe a process for producing
bioriented vinyl fluoride polymer web containing at least 50% by weight
vinyl fluoride which comprises irradiating the web to a total dosage of 0.5
to 100 megareps and thereafter heating the web to a temperature in the
range of 60°C. below the melting point of the web composition to about
10°C. above and biorienting the web at such temperature.

Thus, by the method, formerly difficult to orient vinyl fluoride polymer
webs, upon suitable irradiation undergo considerable cross-linking and can
be easily bioriented to form a strong uniform tubular or nontubular web including film. Web, in tubular form, is typically extruded as a tube up to
35 mils thick or more, cooled by a fluid such as air or water, to a set temperature, therefor, irradiated, heated to an orientation temperature therefor, and then biaxially stretched several diameters to a clear tubular web
or a clear glossy film as thin as 1 mil or less.

Polyvinyl fluoride webs, including films oriented by the process, display
high weathering resistance, high clarity, high toughness and strength, including tensile strength of from 15,000 to 25,000 psi or more for oriented
film of 0.5 to 4 mils as well as high impact strength, high gloss, clarity
and transparency. Line draw and fibrillation are reduced to a minimum
and often eliminated. The webs further have high solvent resistance, showing high resistance to aliphatic, aromatic, chlorinated and ketonic solvents
even at boiling temperatures. The following example is illustrative of this
process.

Example: A stabilized, unirradiated, vinyl fluoride copolymer of 10.2%
by weight ethylene and 0.77 reduced solution viscosity in cyclohexanone
at a concentration of 0.2 g./dl., was melt extruded at 193°C. into blown
film. The unoriented film has a tensile strength of 5,230 psi/5,045 psi
(MD/TD) and an elongation of 573%/801%. An extruded primary tube,
15 mils in thickness, was irradiated at 10 megareps by passing the tube in
proximity with a Van de Graaff Electron Accelerator, model AD with a
rating of 2 mev. and 500 watts.

The irradiated tube was heated to a temperature near its melting point,
about 160°C. and then was easily stretched by inflating the primary tube
with air to yield a biaxially oriented 1.0 mil film. The irradiated biaxially oriented film had a tensile strength of 8,050/17,530 psi (MD/TD)
and an elongation of 320%/82%, respectively. The unirradiated polymer could not be successfully biaxially oriented either at or below its
melting point.

Chlorinated Polyethylene–Acrylonitrile–Styrene Polymers

A. Takahashi, H. Kojima, M. Ogawa, H. Osuka and S. Kobayashi; U.S. Patent 3,673,279; June 27, 1972; assigned to Showa Denka Kabushiki Kaisha, Japan describe a thermoplastic resin having high impact strength and a process for producing the thermoplastic resin. The thermoplastic resin is obtained by polymerizing a homogeneous mixture comprising (a) chlorinated polyethylene; (b) acrylonitrile; and (c) styrene. The following examples illustrate the process.

Example 1: An ampoule is charged with 5 parts of chlorinated polyethylene having a degree of chlorination 34%, 15 parts of acrylonitrile, 70 parts styrene and 20 parts chloroform. The monomers and the solvent are frozen, and then the system is evaporated for 5 minutes at 10^{-2} mm./Hg following which the system is closed and the contents melted. After repeating this operation twice, the ampoule is melt-sealed. Then, placing the ampoule in a 60°C. bath, it is irradiated with a 4.8×10^5 rad dose of gamma rays of cobalt-60 at the dose rate of 10^4 rad per hour.

The contents are then poured into methanol and the polymer is precipitated, following which it is isolated from the unreacted substances and the solvent and is then dried at 50°C. under reduced pressure. The conversion of the monomeric mixture of the acrylonitrile and styrene was 63.3%, indicating that this is a three-component resin of chlorinated polyethylene, acrylonitrile and styrene in which the content of the chlorinated polyethylene is 30.6%. After putting this product through mixing rolls, it is formed into test specimen using a heated press.

The specimen is then submitted to a tensile test in accordance with ASTMD-638-61T. The specimen had a tensile strength of 3.4 kg./mm.^{2} and an elongation at break of 150% at a pulling speed of 50 mm./min. This molded product was practically colorless. The burning speed as determined by the flammability test prescribed by ASTMD-635-56 was 144 seconds. A similar test conducted on "Cycolac" (an ABS polymer) resulted in a time of 64 seconds.

Example 2: Except that the chlorinated polyethylene, acrylonitrile and styrene are used in the amounts of 2, 8 and 2 parts respectively, and the dose used is 2.4×10^5 rad, otherwise the experiment is carried out as in Example 1. The conversion of the monomeric mixture being 45.4%, the resulting product is a three-component resin of chlorinated polyethylene, acrylonitrile and styrene in which the content of the chlorinated polyethylene is 31%. The molded product was colorless.

Example 3: Except that the chlorinated polyethylene, acrylonitrile and styrene are used in the amounts of 5, 5 and 15 parts respectively, a solvent is not used, 25°C. is used as the reaction temperature, and as the dose is used 1.2×10^6 rad at the dose rate of 6×10^4 rad/hr., otherwise the experiment is carried out as in Example 1. The conversion of the monomeric mixture of acrylonitrile and styrene was 95%. Hence, this product is a three-component resin of chlorinated polyethylene, acrylonitrile and styrene in which the content of the chlorinated polyethylene is 20.8%. A test specimen obtained by molding this resin as in Example 1 had a tensile strength of 5.5 kg./mm.^{2} at a pulling speed of 500 mm./min. This molded article had a slight yellowish tinge.

Honeycomb Construction

A process described by C.L. Kehr, W.R. Wszolek and C.B. Lundsager; U.S. Patent 3,660,217; May 2, 1972; assigned to W.R. Grace & Co. relates to a method for laminating superimposed layers of the same or different material by adhering the layers with a curable liquid adhesive composition comprising a polyene and a polythiol which, on exposure to ionizing radiation or a free radical generating agent, cures to a solid adhesive under ambient conditions. The adhesive can be used in bonding structural cellular material such as honeycomb cores and facings.

Honeycomb core may be produced by applying to opposite sides of an elongated web or sheet of material the polyene/polythiol adhesive in alternately spaced or staggered lines. By lapping the sheet back and forth over itself, superimposed layers of the material with the adhesive lines thereon are built up into a stack of unexpanded honeycomb. The unexpanded honeycomb is then exposed to ionizing radiation to cure the adhesive and bond the layers in the stack together.

The stack can be sliced, preferably transversely, through the lines of adhesive to form one or several blocks or slices of unexpanded honeycomb. The unexpanded honeycomb can be expanded by conventional means such as applying uniform expansion forces to opposite end webs of the block or stack at a plurality of substantially equidistantly spaced sites throughout the lengths of each end web of the material. The expanded cores can then be adhered to facings using a different adhesive system or, if appropriate, the same adhesive composition cured by ionizing radiation.

The dose rate for the irradiation operable to cure the adhesive is in the range of 0.00001 to 10.0 megarads/second. Radiation dosages of less than a megarad up to 20 megarads or more for electrons are operable, preferably 0.1 to 5 megarads energy absorbed are employed. For gamma-rays or x-rays, radiation dosages in the range 0.0001 to 5.0 megarads energy absorbed are operable. The irradiation step can be performed at temperatures ranging from below room temperature up to 150°C. or more with the temperatures above the polymer melting points in most cases affording greater cross-linking rates and efficiency.

When using ionizing radiation, the depth of penetration is dependent upon the density of the material to be penetrated. When the ionizing irradiation is in the form of electrons, 0.5 to 12 million electron volts (mev) are usually employed. Where gamma-rays or x-rays are employed, a range of 0.01 to 5.0 million electron volts is used. The above range of voltage will allow penetration of aluminum in the range of 1 to 1,200 mils, of titanium in the range of 1 to 800 mils and of plastics in the range of 1 to 4,000 mils or more. Additionally if the plastic is in the form of precorrugated foam, e.g., Styrofoam, with densities as low as 0.5 lbs./ft. far greater depths of penetration are operable.

Figure 6.3a illustrates one method of producing the article shown in Figure 6.3. The reference numeral (10) indicates a continuous web of sheet material, such as metal foil, e.g., aluminum, paper, sheet plastic, foam plastic, resin impregnated textile or glass fabric, or the like being fed from a continuous roll (11) over a guide roll (12) and through an adhesive application station indicated generally at (13). More particularly, station (13) can comprise a rotogravure type applicating cylinder (14) mounted so that its lower periphery is submerged in a bath of the liquid polyene/polythiol adhesive (15).

The circumference of the cylinder (14) is provided with etched lines (16) and (17) which define the pattern of the adhesive lines being applied to surface (18) of the continuous web as the latter moves between the gravure cylinder (14) and a conventional backup roller (19). In Figure 6.3a the etched lines (16) extend halfway around the periphery of drum (14) along parallel axis; while etched lines (17) extend the other half the circumference of the drum along parallel axis thereby forming staggered lines of adhesive on each layer in the stack.

Thus, rotation of the drum through 180° will cause adhesive line (16) to be applied to surface (18) of the web whereas rotation of the drum through its other half cycle will cause adhesive line (17) to be imprinted thereon. This causes the pattern of the etched lines (20) and (21) to be affixed to the web in staggered relationship. The continuous web with the adhesive is then cut into rectangular sheets of equal dimension by a conventional suitable cutter mechanisms indicated at (22). Specifically, the web material is sheared along spaced transverse cut lines (23a) and (23b), which coincide with the line of juncture between adhesive lines (20) and (21) as the lines appear repetitively along the length of the web.

The result of the above process is to provide rectangular sheets of the web material of equal dimension such as indicated at (24a to d) the sheets being provided with parallel adhesive bonding lines (20) and (21) in staggered relationship to each other. The sheets are superimposed in a stack (25) touching one another along the spaced parallel adhesive bonding lines (20) and (21), the bonding lines uniting each sheet with the bonding lines touching each sheet with the first adjacent sheet being in staggered relationship to the bonding lines uniting the sheet with a second adjacent sheet. The stack of material (25) is placed on a conveyor belt (26) and passed through compressive means, e.g., nip rolls (27) maintain the layers in pressed uniformly stacked relationship and to exclude air from between the layers.

The stack (25) on exiting the nip rolls (27) is passed under ionizing irradation source (28) whereat the stack across its entire width is subjected to high energy particle irradiation or gamma-rays or x-rays penetrable to the entire depth of the stack. Such irradiation causes curing of the polyene/polythiol adhesive composition almost instantaneously, i.e., 0.5 to 5 seconds.

The thus cured stack (25) may be expanded or opened out into cellular pattern, or the stack may be stored or sheared along parallel cut lines (C_1) or (C_2) to form smaller sections or slices of the laminated material which are capable of being expanded or opened out into cellular structure resembling the honeycomb material shown in Figure 6.3c. Various conventional methods of expanding the laminated material are known in the art and can be employed in the process such as the method described in U.S. Patent 2,674,295.

In Figure 6.3d (50) is a roll of sheet material e.g., paper, plastic, metal foil or the like which is fed through a printing apparatus (51) where the adhesive is applied in a line pattern running lengthwise on the sheet material (52). The printed sheet is wound on a large drum (53) in helical fashion with the adhesive layer towards the center of the drum. An electron beam generator (54) irradiates the adhesive through the layer of sheet metal. Such beams can penetrate several layers of material and the curing of the adhesive may be completed on the first pass or after several passes around the drum, depending on the intensity of the beam and the speed of the drum.

This process is continued until sufficient number of layers are built up on the drum. In order to give the staggering of the adhesive lines on successive layers, which is necessary for expanding into a honeycomb, the printing roller (55) is moved axially by one-half of the distance between the adhesive lines, once per turn of the drum (53). This axial movement is done by mechanical or other means, not shown, and is controlled by a trigger

(56) on the drum. To avoid smearing of the adhesive the backup roll (57) can be lifted momentarily while the displacement of the print roll takes place. The result will be that a small part of the lay-up of sheet material on the drum at the same position on the drum will not be adhesive coated. When sufficient layers have been built up on the roll the winding is stopped after cutting the sheet before the printing station. The rotation must be continued long enough to assume complete cure of the topmost adhesive layer in the lay-up.

The complete lay-up is moved from the drum axially, and is subsequently cut at that point of its circumference where the discontinuity of all the adhesive lines is located. The cylindrical lay-up is then flattened out by conventional means such as a sheet rolling mill. In order to make this possible, without excessive distortion of the individual layers in the lay-up, the diameter of the drum (53) must be large relative to the total thickness of the lay-up. By large is meant 10 times or more the thickness of the lay-up, preferably 20 times or more. The flattened lay-up is then sheeted as previously described to provide slices for expansion into honeycomb.

FIGURE 6.3: HONEYCOMB CONSTRUCTION

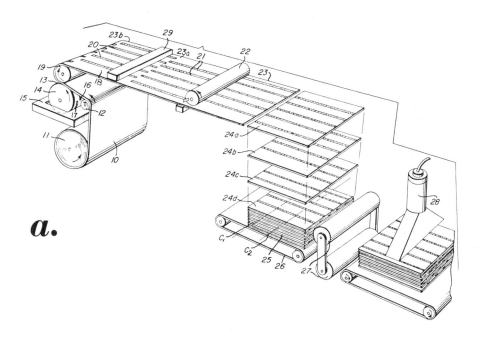

a.

Apparatus for Electron Beam Curing   (continued)

FIGURE 6.3:   (continued)

Cured Stack of Laminated Material   Side View of Expanded Section

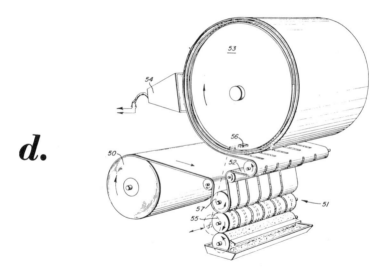

Continuous Process Equipment for Making Honeycomb Slab

(continued)

FIGURE 6.3:   (continued)

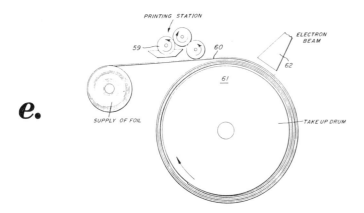

e.

Alternate Form of Apparatus

Source: C.L. Kehr, W.R. Wszolek and C.B. Lundsager; U.S. Patent
          3,660,217; May 2, 1972

In Figure 6.3e, the printing from printing station (59) is applied to the
side of the sheet material (60) on the side away from the center of the take
up drum (61). Thus, by passing under the electron beam (62) the adhesive
is cured without the shielding of the layer of sheet material on top. The
speed must be great enough to give partial curing only, so that sufficient
tack of the adhesive is retained to form a good bond when covered with the
next layer of sheet material.

As before, the curing continues to completion on successive turns of the
drum, and similar means to those described with Figure 6.3d are used to
give staggered printing of the adhesive. This process, by allowing more
intensive cure initially has the advantage of operating at greater speed
and because of the partial cure of the adhesive before it becomes part of
the lay-up, will give a narrower glue line which can be desirable.

Carbon Fiber Composites — Thermal Neutrons

R. Prescott; U.S. Patent 3,671,285; June 20, 1972; assigned to Great
Lakes Carbon Corporation has found that in composites, preimpregnated
tapes, and sized-carbon fiber, prepared by coating carbon fibers with a
high cross section metal or metalloid and resin and then curing the resin
while irradiating the fiber with a source of thermal neutrons, improves

bonding between resin and fiber. Composite materials, for use in the aero-space industry, are well-known to the art. Such materials comprise a resinous binder, as for example a polymerized epoxide and a filler, as for example asbestos, glass fibers, or carbon fibers.

Of the above named fillers, carbon fibers have received attention due to their high corrosion and temperature resistance, low density, high tensile strength and high modulus of elasticity. Uses for such carbon-fiber rein-forced composites include aerospace structural components, rocket motor casings, deep submergence vehicles, and ablative materials for heat shields on reentry vehicles.

The incorporation of carbon or graphite particles in resin bases in amounts of up to 60% by volume will impart a heat-conducting property but not an electrical conductivity to the component. Litant, in U.S. Patent 3,406,126, describes the addition of carbon yarn in as little as 0.05% by volume to the resinous matrix to impart electrical conductivity to the re-sulting composite. Such composites can be prepared from polyesters, poly-vinyl chloride, polyepoxides, or the resins, and carbonized rayon, acrylic, or like fibers.

High modulus composites usually have low shear strengths parallel to the direction of the fibers of about 3,000 to 4,000 psi. These low shear strengths are probably due to poor bonding between the carbon fibers and the matrix. Attempts to improve this bonding, particularly between rayon-based carbon fiber fillers and an epoxy matrix have been partially success-ful, but have resulted in a degradation of the ultimate tensile strength of the fiber and also of the fabricated composite.

Improved bonding has been accomplished by plating the fiber with various metals, as for example tantalum, with metal carbides, as for example whiskers of silicon carbide, and with nitrides. More recently, carbon fibers have been treated with various oxidizing agents in order to etch the surface of the fiber. Such oxidizing agents have included air, ozone, concentrated nitric acid, and chromic-sulfuric acid. In most cases the oxidative treatment of rayon-based carbon fibers resulted in a decrease in ultimate tensile strength of the fiber and of the fiber-resin composite.

The primary structural properties of fiber-resin composites improve as car-bon fiber content is increased up to about 65 volume percent then decreases as the fiber content exceeds that aforementioned figure. The preferred range of carbon fiber content is about 45 to 65 volume percent of fiber in the fabricated composite. This process involves coating a carbon fiber with a metal, metalloid, or compound thereof possessing a high cross section for the capture of thermal neutrons, then sizing the coated fiber with resin and curing the resin while irradiating the fiber with thermal neutrons. Alternatively, the metalloid, metal, or compound coated fiber can be used to prepare a composite material, or a preimpregnated tape, consisting of the fiber and a resin matrix. The preimpregnated tape is cured to the B-stage or the composite is completely cured in the presence of a source of thermal neutron irradiation. Absorption of thermal neutrons by the metal or metalloid causes release of energetic particles which ionize the molecules at the interface improving bonding of carbon and matrix. The following examples illustrate the process.

Example 1:   A 5 g. sample of carbon fibers is coated with a monomolecular layer of boron by vapor deposition at a temperature of about 1300°C. utilizing resistance heating technique such as described in U.S. Patent 3,409,467. The coated fiber is then formed into a composite by aligning the strands in a parallel manner and coating the whole with 10 ml. of an epoxy matrix resin containing a catalytic amount of meta-phenylenediamine. Pressure, 300 psig, is applied to remove the excess resin and compress the fibers. The compressed tape is subjected to a curing temperature of about 100°C. for 2 hours and a bombardment of 10^{12} neutrons/cm.^{2}/sec. from a source of thermal neutrons.

Example 2:   An aqueous solution of 5% by weight of boric acid is prepared. Five grams of the carbon fiber is immersed in the solution, then allowed to air dry. The dried fiber is treated as above to prepare a composite and subjected to cure at 100°C. for 2 hours in the presence of 10^{12} neutrons/cm.^{2}/sec. from a source of thermal neutrons.

Example 3:   A 5 g. sample of carbon fibers is immersed in a 5% aqueous solution of sodium tetraborate, removed, and allowed to air dry. The coated fiber is then immersed in an epoxy resin containing a catalytic amount of m-phenylenediamine. The fiber is removed and heated to 100°C. for 2 hours in the presence of a thermal neutron source of 10^{12} neutrons/cm.^{2}/sec.

TEXTILES

Perfluoro n-Alkyl-Vinyl Ethers for Water Repellancy

W. Loffler and M. Rieber; U.S. Patent 3,617,355; November 2, 1971; assigned to Farbwerke Hoechst AG, Germany have found that excellent water and oil repellent finishes are obtained on synthetic fibrous materials and foils by grafting onto the substrate perfluoro n-alkyl-vinyl ethers in the presence of trifluoroacetic acid and/or methylene chloride and under the action of an ionizing irradiation and/or with the aid of radical forming

catalysts. The finishes obtained according to the process have the substantial advantage of being extraordinarily stable to mechanical wear and
to cleansing operations. As perfluoro n-alkyl-vinyl ether compounds,
there are used compounds of the general formula $F_3C-(CF_2)_x-O-CH=CH_2$,
in which x stands for an integer from 0 to 17, preferably 2 to 8.

It was found that the course of the grafting reaction is considerably influenced by auxiliary agents or solvents present during the grafting reaction.
In this regard it was found that the grafting reaction is influenced in a particularly favorable manner by the use of trifluoroacetic acid and/or methylene chloride as auxiliary agent.

The mixture consisting of the monomers and auxiliary agents or of their
solution or emulsion may be applied in any desired manner onto the synthetic fibrous materials or any foils, for example by immersion, padding,
spraying or similar processes. In addition, there is the possibility of applying the monomers in the gaseous state onto the goods which have already
been wetted with the auxiliary agents and then to carry out the grafting
reaction. The following examples illustrate the process.

Example 1:   25 g. of a polyethylene terephthalate taffeta having a weight
of about 55 g./m^2 were wetted with a mixture of 8 g. of heptafluoropropyl-
vinyl ether and 2 g. of trifluoroacetic acid and then irradiated in a closed
vessel, under exclusion of oxygen, with a dose of 4 Mrad of gamma-rays
(Co 60). The dose rate was 2.1×10^5 rad per hour. The specimen was
then heated for 1 minute to 140°C. in a drying cabinet. The weight increase was about 15%.

The fabric so treated showed an excellent water-repellent effect. The
sprinkling test gave the following values: water absorption: about 1 to 2%.
Water-repellent effect: 5 (very good). The test for the oil-repellent effect
by the 3 M-test gave a wetting resistance of 70.

Example 2:   15 g. of a fabric of polyethylene teraphthalate as that described
in Example 1 were wetted with a mixture of 6.5 g. of n-heptofluoropro-
pyl-vinyl ether and 2 g. of trifluoroacetic acid and irradiated with a dose
of 12 Mrad of electron rays (3 mev electron accelerator). The specimen
was then heated for 1 hour at 80°C. in a drying cabinet. The weight increase was about 12%. The water-repellent effect corresponded to the result obtained in Example 1.

Example 3:   25 g. of a polyethylene teraphthalate fabric as that described
in Example 1 were wetted with a mixture of 10 g. of n-heptafluoropropyl-
vinyl ether and 3.5 g. of trifluoroacetic acid and irradiated with 4 Mrad
of gamma-rays (Co 60) under exclusion of the oxygen of the air. The

specimen was then heated for 1 minute to 120°C. The weight increase was about 15%. The test of the finish obtained, which was effected after having subjected the specimen 5 times to dry cleaning with perchloro-ethylene, still showed a good water-repellent effect, the value of the water-absorption was unchanged within the error limit and the oil-repellent effect had not changed. The test results are shown in the following table under (a).

Another specimen of 25 g. of the same fabric was wetted, for comparison, without simultaneous use of trifluoroacetic acid, solely with 10 g. of n-heptafluoropropyl-vinyl ether and then further treated as described above. The values (b) determined in a test of the fabric thus treated were distinctly poorer than the values obtained with the finishes produced with simultaneous use of trifluoroacetic acid. The superiority of the finish of the process was more evident when subjecting the specimen to a test having subjected it 5 times to dry cleaning.

		Finished			
Test	Untreated	(a)	(b)	(a)*	(b)
Water absorption (%)	10.0	1.0	2.0	0.8	5.5
Water throughput (ml.)	425	143	220	212	330
Water-repellent effect	1	4-5	3	3	2
3 M-test	**	80	70	70-80	50

   *After 5 dry cleaning operations
  **Below 50

Silicone Graft Copolymerized Fiber or Cloth

In a process described by T. Chitani, S. Yokoyama and S. Nishide; U.S. Patent 3,617,187; November 2, 1971; assigned to Hakuyosha Co., Ltd., Japan silicone graft copolymerized fiber or cloth is prepared by first dyeing the cloth with a dyestuff containing metal in the form of a complex salt, and then applying a silicone having a general formula of

$$\left[\begin{array}{c} R \\ | \\ -Si-O- \\ | \\ R' \end{array} \right]_n$$

where R and R' are hydrogen or alkyl radicals up to 5 carbon atoms and R and R' can be the same or different, and n is an integer, to the dyed cloth and subjecting the silicone coated dyed cloth to an ionizing radiation where the dosage of ionizing radiation is within a range of from 10^4 rad to 10^7 rad.

Any of mono-azo dyestuffs, poly-azo dyestuffs and phthalocyanine dyestuffs are utilizable in this process, provided that they are metal-containing dyestuffs. In short, any dyestuff containing metal in its molecular structure can be used. It is preferable that the dyestuff contains the metal in the form of complex salt, since in this form the dyestuff has good efficiency even if a low irradiation dose is used. Further, any of dyeing processes generally used in the art can be used in this process. The degree of grafting, of course, varies in accordance with the concentration of the dyestuff, but in any event, the grafting rate is increased by use of this process. The process is illustrated by the following examples.

Example 1: No. 60 cotton broadcloth was dyed using 0.6% and 2.0% of Solar Rubinol B (a kind of copper complex salt of diazo dyestuff). Silicone was applied to the cloth thus treated in an amount of 5% (silicone to fiber) and the cloth was subjected to irradiation of Co 60 gamma-ray in the presence of air and dried 5 hours at 80°C. The cloth thus treated was subjected to an extraction with petroleum benzine for 20 hours and successively rinsed sufficiently with methanol and water and dried.

Thereafter, the weight of the cloth was measured and the degree of grafting was calculated from the data. The results are tabulated in the table below. As the controls, nondyed cloth and the cloth dyed with Chryeamine G (a kind of diazo dyestuff) were treated in the same manner and the degree of grafting obtained are also tabulated.

Sample number	Dyestuff	Conc. of dyestuff (percent)	Irradiation total dose (rad.)	Degree of grafting (percent)	Grafting yield (percent)
1	Solar Rubinol	2	5.2×10^6	4.9	97
2	do	2	1.3×10^6	4.1	92
3	do	2	0.7×10^6	4.3	85
4	do	2	0	1.9	37
5	do	0.6	5.2×10^6	4.9	98
6	do	0.6	1.3×10^6	4.2	84
7	do	0.6	0.7×10^6	4.2	84
8	do	0.6	0	0.1	2
9	Chryeamine	2	4.7×10^6	0.7	13
10	do	0.6	4.7×10^6	0.7	13
11	Not dyed	0	4.7×10^6	0.7	14
12	do	0	0	0	0

Samples No. 9–12 are given for reference purposes only.

Example 2: No. 60 cotton broadcloth was dyed using 0.6% and 2% of Direct Brown BRS (cupro triazo dyestuff). The cloth thus dyed was treated as in Example 1. The results are tabulated below.

Sample number	Dyestuff	Conc. of dyestuff (percent)	Irradiation total dose (rad.)	Degree of silicone grafting (percent)	Grafting yield (percent)
13	Direct Fast Brown BRS	2	5.2×10^6	43	85
14	do	2	0	0.1	2
15	do	0.6	5.2×10^6	4.6	92
16	do	0.6	1.3×10^6	9.0	18
17	do	0.6	0.7×10^6	5.0	10
18	do	0.6	0	0.2	3

Triallyl Phosphate and Acrylamide Mixture for Flameproofing

T.D. Miles and A.C. Delasanta; U.S. Patent 3,592,683; July 13, 1971; assigned to the U.S. Secretary of the Army describe a process for flame-proofing materials, more particularly cotton textile fabrics, which com-prises impregnating the material with a phosphorus compound, such as tri-allyl phosphate, and a comonomer, such as N-methylol acrylamide, and exposing the material to ionizing radiation at a dose level sufficient to cause production of a solid copolymer of the phosphorus compound and the comonomer in situ in the material. The preferred triallyl phosphate and N-methylol acrylamide are present in the impregnating bath in a ratio and concentration such that flame resistance imparted to the material is retained through at least 15 launderings under standardized conditions.

Example 1: A 9 x 12 inch sample of 8.2 oz. cotton sateen fabric was im-mersed in a solution containing 125 cc of triallyl phosphate and 25 cc of N-methylol acrylamide (60% aqueous solution by weight) In this propor-tion, 8.3 to 1 of triallyl phosphate to N-methylol acrylamide by weight, the two chemicals are miscible. The sample was squeezed to a wet pick-up of 100 to 125% by weight, placed in a sealed 4 mil polyethylene bag and exposed to 2 megarads of irradiation in air at room temperature using a 24 mev, 18 kw. electron LINAC as the source of radiation.

Electron radiation consisted of scanning the sample with an electron beam approximately 3 to 4 cm.^{2} in area as the sample moved through the beam path on a conveyor. The scan width was 16 inches and the dose rate, while the pulsed beam was on, was 10^7 rads/sec. The repetition rate was 60 pulses/sec. and the pulse duration was 5 microsec. The sample was held in a vertical position during irradiation. The sample was then rinsed in hot water (40°C.) for 2 minutes and dried at 100°C. in an oven.

The add-on of dry solids on the fabric was 15% by weight. Flame tests were made using the Vertical Bunsen test (Method 5903 of Federal Test Method Std. No. 191) both before and after laundering according to Method 5556 of the same Test Method Standard. Initially the after-flame was 0 sec. and the char length was 5.8 inches. After 15 launderings, the after-flame was 0 seconds and the char length was 5.0 inches.

Example 2: The same type of fabric of the same sample size as in Example 1 was treated with the same ratio of chemicals as used in Example 1 and then exposed to 4 megarads of irradiation in air at room temperature using a 1.25 x 10^6 Co^{60} Co isotope source. Gamma irradiation was under am-bient conditions with a dose rate of 2.84 to 3.86 x 10^4 rads/min. After the radiation exposure, the sample was rinsed in hot water (40°C.) for 2 minutes and dried at 100°C. in an oven. The add-on of dry solids on the fabric was 26% by weight. Flame tests initially were 0 sec. after-flame and 5.0 in. char length; after 15 launderings 2 sec. after-flame and 5.5 in. char length.

Example 3:   The same type of fabric of the same sample size as in Example 1 was immersed in a solution of 10 g. of acrylamide, 10 cc of triallyl phosphate, 5 cc of water and 10 cc of methanol. The sample was exposed in the same manner as in Example 2, but to 6 megarads of irradiation. After rinsing and drying as in Example 1, the flame resistance of the treated fabric was 0 sec. after-flame and 3.9 in. char length.

Example 4:   The same type of fabric of the same sample size as in Example 1 was immersed in a solution containing 10 cc acrylonitrile (70% by weight aqueous solution) 10 cc of triallyl phosphate, 5 cc of water and 5 cc of methanol. The sample was exposed in the same manner as in Example 2, but to 6 megarads of irradiation. After rinsing and drying, as in Example 2, the flame resistance of the treated fabric was 0 sec. after-flame and 7.1 in. char length.

Example 5:   The same type of fabric of the same sample size as in Example 1 was immersed in a solution comprising 20 cc of triallyl phosphate and 20 cc of acrylic acid. The sample was exposed in the same manner as in Example 2 to 4 megarads of irradiation. After rinsing and drying as in Example 2, the flame resistance of the treated fabric was 0 sec. after-flame and 4.6 in char length.

Dye Receptive Graft Copolymer on Polyamides and Polyesters

In a process described by E.E. Magat and D. Tanner; U.S. Patent 3,670,048; June 13, 1972; assigned to E.I. du Pont de Nemours and Co. a shaped article produced from a synthetic organic condensation polymer, in intimate contact with an organic compound, is subjected to bombardment by ionizing radiation to produce chemical bonds between the shaped article and the organic compound. In one case, an organic modifier is applied to the surface of a shaped article produced from a synthetic organic condensation polymer and the shaped article is irradiated with ionizing radiation to induce chemical bonding.

For deep seated modification, the organic compound is permitted to diffuse into the substrate prior to the irradiation. Alternatively, the organic modifier, especially when it is of high molecular weight, may remain upon the surface of the polymer substrate during the irradiation step, thus producing a uniform coating chemically grafted to the polymer substrate. The organic compound employed as modifier may be a nonpolymerizable organic compound or it may be polymerizable; either form is chemically grafted to the shaped article formed from an organic condensation polymer.

Figure 6.4a is a target arrangement for sample bombardment with proton, deuteron, and alpha particles. In the illustration the particle is accelerated in cyclotron (1) and following emergent particle path (2) passes through window and beam defocusing arm (3) where the beam is spread. Thereafter the spread beam is passed through carbon shutter impinging on sample (5), the sample being enclosed in wrapper (6). Electrometer (7) measures the beam-out current at the carbon shutter.

Figure 6.4b shows a section of the defocused beam pattern. The rectangular checker area (8) represents the irradiated area of sample with curves (9) and (10) denoting intensity distribution along the x and y coordinates of the sample, respectively. Figure 6.4c is a typical target arrangement for fast neutron bombardment. Beryllium target (11) is bombarded with (24) mev deuterons generated in cyclotron(1). The neutrons produced are impinged on target (12) disposed along the emergent neutron beam (13).

FIGURE 6.4:   DYE RECEPTIVE GRAFT COPOLYMERS ON POLYAMIDE
AND POLYESTER SUBSTRATES

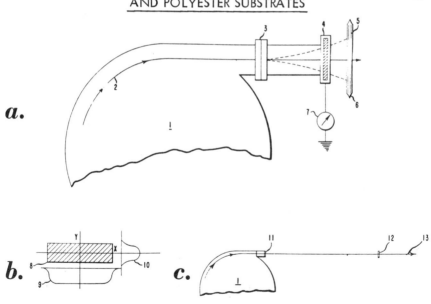

(a)-(c) Electron Beam Arrangement

Source: E.E. Magat and D. Tanner; U.S. Patent 3,670,048; June 13, 1972

Unless otherwise noted, "66 nylon fabric" employed in the examples is a taffeta fabric, woven from 70 denier polyhexamethylene adipamide continuous filament yarn having a denier per filament of 2.0. The polyamide is produced from hexamethylenediamine and adipic acid (ergo "66"), and has a relative viscosity of 37, 39 equivalents of $-NH_2$ ends and 92 equivalent of $-COOH$ ends per 10^6 grams of polymer (referred to hereinafter as 39 amine ends and 92 carboxyl ends, respectively). The polymer is prepared using 0.34 mol percent acetic acid stabilizer (which ends are, of course, not titratable), which is equivalent to 15 amine ends. From these data, following the method of G.B. Taylor and J.E. Waltz (Analytical Chemistry, Vol. 19, p. 448; 1942) the molecular weight (number average) is calculated to be about 13,700.

The "standard washing" to which samples are subjected consists of a 30-minute immersion in 18 liters of 70°C. water contained in a 20 liter agitation washer. The wash solution contains 0.5% of detergent. The detergent employed is known as Tide. This detergent contains, in addition to the active ingredient, well over 50% (sodium) phosphates. Analysis shows the composition to be substantially as follows:

16% sodium lauryl sulfate
6% alkyl alcohol sulfate
30% sodium polyphosphate
17% sodium pyrophosphate
31% sodium silicates and
sodium sulfate

The static propensity of the fabric is indicated in terms of direct current resistance in ohms per square, measured parallel to the fabric surface, at 78°F. in a 50% relative humidity atmosphere. High values, reported as the logarithm (to the base 10) of the resistivity (log R) indicate a tendency to acquire and retain a static charge. It should be noted that highly hydrophobic unmodified polymer substrates have such a high resistivity that the log R determined may depend somewhat on the sensitivity of the meter employed; log R values of 13 to over 15 have been observed, using the same fabric and different meters.

However, these differences substantially disappear when a satisfactory antistatic modification is produced, e.g., for log R values of 11 or less. Moreover, data reported in any given example are consistent, i.e., all measurements were made on the same instruments under the same conditions. A meter suitable for this determination is described by Hayek and Chromey, American Dyestuff Reporter, 40, 225 (1951). Wickability as measured in the examples is determined by placing a drop of water upon the fabric, and measuring the diameter of the wet spot after a standard time interval, e.g., 60 seconds. Alternatively, especially useful where decreased wickability is obtained, is a determination of the length of time required for a drop placed upon the fabric to disappear by soaking into the fabric. Discrepancies observed between control fabrics in the different examples are thought to be due to different preparation techniques. Data within each example are comparable.

Where quantitative values for hole melting are presented, they are measured by dropping heated glass beads of constant weight and diameter from a fixed height from a constant temperature oven onto the fabric. The temperature at which the fabric is stained is called the first damage temperature, and the temperature at which the glass bead melts completely through the fabric is referred to as hole-melting temperature. Where the hole-melting tendency is presented in qualitative terms, the designation "poor" (referring to polyamides) denotes a quantitative rating of about 300°C.; "fair", a rating of about 400° to about 500°C.; "good", a rating of about 600°C. or slightly better; and "excellent", a rating well over 600°C.

The fiber melt temperature reported in some examples is determined by placing a thread, unraveled from a fabric if necessary, upon an electrically heated tube and observing the tube temperature at which visible melting, fusing of filaments to the tube, or instantaneous decomposition occurs.

Post-formability is evaluated by contacting a sample yarn with a tube heated to about 225°C. A fiber which can be drawn in contact with the tube and without substantially fusing the filaments, to two or three times its original length is designated "elastic". When the stretch is retained without restraint after cooling, it is designated "post-formable".

Crease recovery is evaluated by crumpling a fabric in the hand, and observing the rate at which it recovers from this treatment. Wet crease recovery indicates the rate and extent of disappearance of creases from the crumpled fabric when it is wetted. Numerical values are obtained using the Monsanto Crease Recovery Method, described as the "vertical strip crease recovery test" in the American Society for Testing Materials Test No. D1295-53T.

In determining wet crease recovery by this method, the specimens are soaked for at least 16 hours in distilled water containing 0.5% by weight of Tween 20, a polyoxyalkylene derivative of sorbitan monolaurate, a wetting agent. Immediately prior to testing, excess water is removed from the test fabrics by blotting between layers of a paper towel. Results are reported as percent recovery from a standard crease in 300 seconds.

Example 1: A sample of "66" nylon fabric is soaked in liquid methoxydecaethyleneoxy methacrylate. After removal of excess liquid by wring-
ing, but while still wet, it is enclosed in an aluminum foil wrapper and
subjected to electron irradiation in a 1 Mev resonant transformer with a
beam-out current of 560 microamperes. The sample is placed on a conveyor
belt which carries it through the electron beam at a rate of 16 inches per
minute. At the sample location, the beam supplies an irradiation dose,
for textile samples, of 5.6×10^6 rad (5.6 Mrad) per pass. The sample is
traversed back and forth across the beam until a total dose of 17 Mrad is
attained. The sample is given the standard wash, rinsed in distilled water
and dried.

Its direct current resistance in ohms is then measured. It logarithm is 9.8.
After 5 washings, the value rises only to 10.7. After 10 additional washings, the value increases only to 10.9. This compares favorably with cotton (a material with but little tendency to accumulate static charges) which
has a value of 10.8. The product has a softening point of 239°C., and
except for a trace, is soluble in 98% formic acid. A control sample of
the original fabric has a log resistivity of 13.2, a softening point of 247°C.,
and is completely soluble in formic acid. When exposure of the soaked,
wrapped sample to irradiation is increased to 67 Mrad, although the product displays good antistatic properties (a log resistivity of 10.7), it is insoluble in 98% formic acid and is infusible, indicating a high degree of
cross-linking.

Example 2: In order to test the penetration of electron radiation into relatively thick samples, 60 samples of polyhexamethylene adipamide fabric
are individually padded with liquid methoxydecaethyleneoxy methacrylate
and thereafter stacked into a flat package of a thickness equal to the sum
of the thicknesses of the 60 pieces, a total of about 0.24 inch. The package is wrapped in aluminum foil and is irradiated from one side only in the
equipment and under the conditions of Example 1 to a total irradiation
dose of 33 Mrad.

The samples, numbered from 1 to 60, beginning at the top (nearest the electron source) are then subjected to a series of various treatments during which
the log resistivity of selected samples, after being rinsed and dried, is
measured and is reported in Table 1. The first treatment is a series of 15
consecutive standard washings. Column A is the observation taken after
the second washing, while column B is taken after the fifteenth. The 15
consecutive washings are followed by a sodium chlorite bleach and another
standard washing, the subsequent observation being shown in column C.
The samples are then washed 14 hours in synthetic detergent. Column D re-
ports the log resistivity observed. Finally the samples are given 5 consecutive washings in hot soapy water, these values being shown in Column E.

TABLE 1

Sample No.	A	B	C	D	E
2	9.2	10.1	10.1	10.1	10.3
8	9.1	10.0	10.0	9.9	10.3
12	9.5	10.1	10.1	10.3	10.5
16	9.7	10.5	10.5	10.2	10.7
20	9.8	10.7	10.7	10.7	11.0
24	10.1	10.6	10.6	10.8	11.1
26	10.6	12.2	12.2	12.3	13.2
30	12.9	13.3	13.3	13.1	13.3
35	13.5	13.1	13.1	—	—
40	12.8	13.3	13.3	—	—
45	13.3	—	—	—	—
50	13.1	—	—	—	—
60	13.1	—	—	—	—

These results show that the electron beam penetrates the fabric pile far enough to induce appreciable modification in about the top 30 fabric layers. The total thickness effectively penetrated by the 1 Mev electrons is about 0.1 inch (0.25 cm.).

Example 3:   A series of nylon taffeta fabrics are scoured in a solution containing 0.5% olive oil soap flakes and 0.46% trisodium phosphate. The scoured fabrics are soaked 8 hours in an aqueous solution containing 25% acrylic acid. Excess liquid is removed from the samples, and they are packaged in 5 mil polyethylene film packets and are then irradiated. The radiations used in this example are produced in a cyclotron, arranged to bombard the samples with high speed protons, deuterons, alpha particles, or neutrons. Duplicate samples for each test are provided.

The arrangement of the samples with respect to the cyclotron, for charged particle irradiation is shown in Figure 6.4a. In this example a distance from window to carbon shutter of 24 inches and from carbon shutter to sample of 5 inches is used, the beam width at the sample being 7 cm.

Figure 6.4b shows schematically the distribution of the charged particles as they impinge upon the fabric samples. Ionized hydrogen molecules are accelerated in the cyclotron, but are dissociated to form a proton beam on passing through the carbon shutter. Fast neutrons are produced by bombarding a beryllium target with 24 Mev deuterons. The emergent beam from the beryllium target impinges upon the sample at a distance of 30 inches from the target, arranged as shown in Figure 6.4c. These neutrons at this position have an average energy of 10 Mev.

After irradiation, the ungrafted acrylic and polyacrylic acid is removed by rinsing the fabrics in distilled water, boiling them for 1 hour in a 5%

aqueous solution of sodium carbonate, rinsing again in distilled water, followed by boiling in a 2% aqueous solution of acetic acid followed by a second distilled water rinse. The solution-to-fabric weight ratio is 500:1 in each operation. The exposure conditions and the results of the tests given to each sample are indicated in Table 2. Prior to testing, duplicate swatches to samples 3A, 3B, 3C and 3D are dyed with a basic dye. The location of the irradiated portion is clearly seen because it is deeply dyed by the basic dye, due to the grafted polyacrylic acid. Fabric tests subsequently made on duplicate samples 3A to 3D are carried out on the areas corresponding to those which were deeply dyed.

The weight gain due to acrylic acid grafted to each sample is indicated in Table 2, as well as the total of original carboxyl groups plus those attached to the nylon via grafting, as determined by titration as explained hereinabove. Carboxyl group concentration found in a typical control nylon sample (3E) is shown for comparison. Following the carboxyl group determination, a portion of each sample is boiled for 1 hour in 5% aqueous sodium carbonate, forming the sodium salt of the grafted acrylic acid. The log R values are indicated in Table 2. In addition, the wickability of the sodium salt of the grafted acrylic acid modification is determined; its resistance to hole-melting is estimated by dropping hot ashes from a burning cigarette upon the fabric.

TABLE 2: IRRADIATION CONDITIONS AND PROPERTIES OF THE MODIFIED SAMPLES

Sample Number	3A	3B	3C	3D	3E
Particles	12 Mev H^+	48 Mev He^{++}	24 Mev D^+	10 Mev neutrons	Control
Current (mμ amp.)	20	80	40	---	---
Total beam exposure (μ amp.-hr.)	0.0005	0.0020	0.0010	---	---
Integrated current flux (μ amp.-sec./cm.2)	0.12	0.48	0.24	---	---
Accumulated exposure (μ amp.-hrs.)	----	----	----	100	---
Dosage mrad	1	1	1	---	---
% weight gain of sample exposed to beam	13.9	17	14.5	3.5	0
COOH/10^6 g.	1,900	2,350	1,950	500	90
Log R	7.5	7.5	7.5	10.1	13.3
Wickability of sodium salt (time in sec. for disappearance of a drop of water)	<3 sec.	<3 sec.	<3 sec.	3 sec.	>60 sec.
Hole-melting resistance	Very good	Very good	Very good	Half-way between control and samples 3A, 3B, 3C	Very poor

H^+ = proton
He^{++} = α particles
D^+ = deuteron

Example 4: A sample of "66" nylon fabric is immersed in liquid acrylonitrile. It is then wrapped in aluminum foil and irradiated with 2 Mev x-rays as described below, until a dose of 23 Mrad is attained. The sample is exposed to x-radiation using a resonant transformer x-ray machine known as a Two Million Volt Mobile X-Ray Unit. The packaged sample is placed in an open top box made from 1/16-inch sheet lead, and positioned so that

the sample is 8 cm. from the tungsten tube target. At this location, using
a tube voltage of 2 Mev, and a tube current of 1.5 milliamperes, the irradiation rate for the sample in question is 1.2 Mrad per hour. The beam
irradiates a circle about 3 inches in diameter; all fabric tests are made on
the irradiated portion.

Following the irradiation, ungrafted polymer is removed by washing with
dimethylformamide. After 15 standard Tide washings, the dried nylon fabric has a superior crease recovery and greater resilience than before treatment by the process. A second sample is immersed in liquid acrylonitrile.
It is then wrapped in aluminum foil and irradiated as before to a dose of
5 Mrads. After thorough rinsing, the weight gain is 12%. Larger irradiation doses produce larger weight gains.

It is shown that a bulk modification has been obtained by hydrolyzing the
nylon-acrylonitrile graft by a 30-minute boil-off in 3% sodium hydroxide.
The fabric, which now contains a large number of additional carboxyl groups
due to hydrolysis of the nitrile groups, is then dyed with a basic dye (DuPont
Brilliant Green). Cross sections of the filaments taken from these fabrics
are deeply dyed throughout the modified filament, whereas only light shades
are observed in cross sections of filaments taken from control fabrics which
had received the same caustic boil-off and dyeing treatment without irradiation. The hydrolyzed test fabric has a log R of 9.4 vs. 13.3 for control.

Example 5: The process is readily carried out using gamma-rays, for example, derived from cobalt 60 as shown in this example. A nylon taffeta
sample is soaked for 16 hours in a 15% aqueous solution of maleic acid and
is then wrapped in aluminum foil while soaking wet. The foil package is
wrapped around a 3/4 inch glass tube which in turn is inserted in a 1 1/2
inch glass tube. This combination is lowered into a cobalt 60 source of
gamma radiation (1.3 Mev gamma-rays). The dose rate available from this
source is 7,200 Mrads per minute.

After a total dose of 27.5 Mrad, the sample is removed and scoured in hot
water. After extracting ungrafted maleic acid, a weight gain of 5% is observed. When a portion of the fabric is dyed with a basic dye, the dyed
cross section shows that the maleic acid has penetrated the fiber and is
grafted throughout its interior. When the maleic acid-grafted nylon is
converted to the calcium salt by boiling in 1% calcium acetate solution,
the resulting fabric is found to have excellent resistance to hole-melting.

Example 6: A sample coded 6A of "66" nylon fabric is immersed in liquid
acrylonitrile. It is then wrapped in aluminum foil and irradiated in the
apparatus and under the conditions of Example 1 until a dose of 17 Mrad
is attained. The product softens at 240°C. and is almost completely soluble

in formic acid. It is observed to possess a higher crease resistance and greater resiliency than the original sample. This improved resiliency is retained even after 15 standard washings following a washing in dimethyl-formamide (a solvent for polyacrylonitrile).

The test is repeated with nylon samples 6C and 6D, which are soaked in solutions of acrylonitrile, water and methanol as indicated in Table 3, for 24 hours at 25°C. Each sample (6C, 6D) is enclosed in a polyethylene bag with excess solution and irradiated, using a Van de Graaff generator under the conditions listed below.

Voltage Mev	2
Tube current, microamps	290
Conveyor speed, in./min.	40
Dose per pass, Mrad	1
Number of passes	2
Total dose, Mrad	2

After a hold-up time of 2 hours, the samples are thoroughly rinsed in di-methylformamide at 70°C., followed by acetone and then water. The weight gain of each is determined and listed in Table 3. The breaking strength of representative yarn samples from each fabric is determined after 0 and 500 hours exposure to ultraviolet light in a Weatherometer. A control, 6B, is not exposed to the high energy electrons. The grafted acrylonitrile greatly increases the light durability of the nylon.

TABLE 3

			Yarn Breaking Strength, g.	
Sample	Grafting Solution	% Weight Gain	0 hour	500 hour, Weatherometer
6B	none	none	386	30
6C	5/22/22*	10.1	393	58
6D	25/12/22	36.1	402	100

*Solution composition = ml. acrylonitrile/ml. H2O/ml. CH3OH

OTHER PROCESSES

Hydrophilic Silicones for Contact Lenses

J. Laizier and G. Wajs; U.S. Patent 3,700,573; October 24, 1972;

describe a method of preparation of hydrophilic silicones by radiochemical grafting. The compound of the silicone family is first exposed to ionizing radiation in an oxidizing medium. The irradiated body is then put in the presence of a monomer at a temperature within the range of 120° to 160°C. and thus endowed with an absorptive capacity for water. This method is of particular interest for the preparation of silicone contact lenses. The following examples illustrate the process.

Example 1: As plate of Si 182 which is a siloxanic chain having a vinylic group at the end on which a reticulation agent is reacted and having a length of 5 cm., a width of 1 cm. and a thickness of 1.5 mm. was subjected in the presence of air to a radiation dose of 10 mrads at an intensity of 1 mrad/hr. After putting in the presence of N-vinylpyrrolidone, and after vacuum degassing and introduction in a vacuum-sealed glass ampoule, the plate was heated to a temperature of 80°C. for a period of 4 hours. The grafted plate was washed with distilled water, then brought to the boil for a further period of 30 minutes and finally oven-dried at a temperature of 100°C. The weight of grafted polyvinylpyrrolidone was then 24.8 mg. (2.7%) and the water absorption was 17.2 mg. (1.9%) after immersion in distilled water for a period of 24 hours.

Example 2: A plate of RTV 10341 manufactured by the Societe Rhone-Poulenc which is a siloxanic chain having a vinylic group at the end, and having a weight of 0.7363 g. was subjected in the presence of air to a radiation dose of 12 mrads at an intensity of 0.75 mrad/hr. After putting in the presence of N-vinylpyrrolidone and after vacuum degassing and introduction in a vacuum-sealed glass ampoule, the plate was heated to a temperature of 80°C. for a period of 4 hours. The grafted plate was washed with distilled water, then brought to the boil for a further period of 30 minutes and finally oven-dried at a temperature of 100°C. The weight of grafted polyvinylpyrrolidone was then 23 mg. (3.1%) and a water absorption of 32.3 mg. (4.4%) was noted after immersion in distilled water for a period of 24 hours.

Example 3: A plate of RTV 10341 having a weight of 0.4984 g. and a length of 5 cm., a width of 1 cm. and a thickness of 0.5 mm. was subjected in the presence of air to a radiation dose of 0.75 mrad at an intensity of 0.75 mrad/hr. The irradiated plate was put in the presence of N-vinylpyrrolidone, degassed in vacuo and introduced in a vacuum-sealed glass ampoule, then heated to a temperature of 130°C. for a period of 1 hour. The grafted plate was washed with distilled water, then brought to the boil for a period of 30 minutes and finally oven-dried at a temperature of 100°C. The weight of grafted polyvinylpyrrolidone was then 27.9 mg. (5.6%) and the water absorption was 27.4 mg. (5.5%) after immersion in distilled water for a period of 24 hours.

Plastic Impregnated Wood

J. Miettinen, T. Autio and J. Stromberg; U.S. Patent 3,663,261; May 16, 1972 describe a method for preparing plastic impregnated wood, in which dry wood (moisture content below 10%) is impregnated with liquid resin material in a manner prior known per se, e.g., the wood is put into a vacuum impregnation vessel, the air is evacuated, the vessel flushed with nitrogen, the liquid resin is introduced in the vessel until the wood is immersed, an atmospheric nitrogen pressure or an overpressure of 1 to 8 atmospheres gauge is applied.

The wood is transferred into bags or containers filled with nitrogen, then hermetically sealed, then the impregnated wood is cured utilizing radioactive radiation optionally in the presence of chemical catalyst addition. The resin utilized for the wood impregnation contains commercial unsaturated polyester between 35 and 95% and styrene and/or methyl methacrylate. The amount of radiation is between 0.5 and 1.5 Mrad in the case of gamma-radiation and between 1 and 5 Mrad if high energy electron radiation is used. As chemical catalyst is used an organic peroxide and the curing is carried out at a temperature between 40° and 60°C. utilizing as additives when necessary chemical inhibitors such as pinene in an amount of less than 5%.

Bonding Wood Veneer to Coated Substrate

L.S. Miller and F. Shafizadeh; U.S. Patent 3,616,028; October 26, 1971; assigned to Weyerhaeuser Company describe a method of simultaneously bonding and curing an impregnated overlay material of, particularly wood veneer, to a substrate material by thoroughly impregnating the overlay material with a polymerizable monomer or polymer dissolved in monomer, bringing the overlay into contact with the substrate material previously coated on the surface to be contacted with a bonding material, and exposing the resulting product to high-energy radiation. The following example illustrates the process.

Example: A strip of birch veneer, 1.5 x 3 x 1/16 inch, was completely immersed in tetraethylene glycol dimethacrylate for 2 hours, then placed on a piece of 1/2 inch Douglas fir plywood with sufficient monomer to wet the plywood surface. The veneer was covered with a sheet of 5 mils thick polyethylene terephthalate film then irradiated at a dose of 4 megarads by passing it under the electron beam of a Van de Graaff accelerator set at 1 mev., 0.1 ma. After polymerization and curing the polyester film was stripped away. The veneer had a clear, glossy plastic surface and was so tightly bonded to the plywood that it could not be wedged off without tearing the wood.

Monomer Impregnated Rice Hulls

In a process described by S.L. Casalina; U.S. Patent 3,660,223; May 2, 1972 discrete, cellulose containing, rice hull particles are impregnated with a monomer, usually a liquid, having a radiation activatable reaction group, molded under pressure to place the particles in substantially touching relation and form an article, and subjected to high-intensity ionizing radiation to polymerize the monomer and bind the particles into an object of stable shape. Additionally, the monomer may be selected to produce objects that are rigid and hard or flexible and resilient. The following example illustrates the process.

Example: Relatively thin sheets of the product of the process were prepared by casting and irradiation. One set of sheets was formed from a mixture of rice hulls and ethyl acrylate monomer having 9% by weight carbon tetrachloride. The ratio of monomer to rice hulls was approximately 0.25 to 1 or about 20 to 21% monomer. The sheets were cast to have the following dimensions: 2 x 3 x 1/4 inch thick.

A second set of sheets was cast having the same dimensions and using methyl methacrylate. A third set of sheets was cast using parboiled rice hulls and ethyl acrylate. All sheets were irradiated by a 33,000 Curie source having a dose rate of 6 Mrad/hr. for 25 min. The sheets were then clamped or secured along one of the 2 in. sides and a bending force applied at the opposite side approximately 2 1/2 in. away. The table below sets forth the force required, in grams, to bend or displace the sheets through 90°. As indicated in the table, the flexibility of the sheet produced by ethyl acrylate is much greater than the sheets produced by methyl methacrylate. It will be seen that the use of parboiled rice hulls results in a further increase in flexibility.

Material	Monomer	Samples		
		No. 1	No. 2	No. 3
Rice Hulls	Methyl Methacrylate	No Measurable Displacement at 2000 grams		
Rice Hulls	Ethyl Acrylate	890	897	881
Rice Hulls (Parboiled)	Ethyl Acrylate	770	771	

COMPANY INDEX

INVENTOR INDEX

U.S. PATENT NUMBER INDEX

NOTICE

Nothing contained in this Review shall be construed to constitute a permission or recommendation to practice any invention covered by any patent without a license from the patent owners. Further, neither the author nor the publisher assumes any liability with respect to the use of, or for damages resulting from the use of, any information, apparatus, method or process described in this Review.

PRINTING INKS 1972
by A. Williams

Printing ink production in the U.S.A. now exceeds one billion pounds per year, supplying an extremely complex market. The printing industry is one of the largest consumers of pigments due to an increasing demand for color.

The mounting emphasis on pollution control has required many formulation changes and particularly the use of water-based systems in flexographic inks. Solvent-free systems are being developed, and "curing" by high energy radiation and ultraviolet is being studied.

Specialty inks (magnetic inks, conductive and fluorescent inks, etc.) constitute ca. 20% of the U.S. market, while the bulk of the supply goes into newsprint (letterpress), lithographic, rotogravure, and flexographic inks.

159 Processes based on 184 U.S. patents. Numbers in ( ) indicate numbers of processes per heading in the partial table of contents that follows.

293 pages

POWDER COATINGS AND
FLUIDIZED BED TECHNIQUES 1971

by Dr. M. W. Ranney

It is now potentially possible to paint auto bodies and other large items in a completely automatic plant with no manual labor and no environmental pollution. This is accomplished by solventless powder coating using fluidized bed, electrostatic spray, and other techniques which can be applied to metal, glass, wood and plastic surfaces. This book describes 161 processes based on U.S. patents. Due to the actuality of the subject, most patents are of very recent date, although some older, basic patents were included to give a complete technological picture of the state of the art.

Partial contents:

249 pages

FATTY ACIDS 1973

Synthesis and Applications

by N. E. Bednarcyk and W. L. Erickson

Chemical Technology Review No. 9

This book describes practical syntheses and applications of those fatty acids that are used by the rubber and synthetic resin industries, and to some lesser extent by manufacturers of paints, printing inks, adhesives, and allied products.

Naturally occurring fatty acids are higher straight-chain unsubstituted carboxylic acids containing an even number of carbon atoms. They may be saturated or unsaturated, i.e., they may contain double bonds. Since synthetic modifications are countless, the definition is broadened to include odd- and even-numbered compounds containing six or more carbon atoms. They may have various substituents along the chain, they may be branched, and even contain additional short side chains. They may be oxidized to yield dicarboxylic or aldehydo acids.

Short chain, water-soluble acids, such as maleic acid, do not fall into this definition. Exempted also are fatty acids used by the food industry, because they are the subjects of future Noyes monographs.

138 patent-based processes. Numbers in ( ) indicate a plurality of processes per topic. Chapter headings are given and some of the more important subheadings are mentioned.

ISBN 0-8155-0485-3

353 pages

ANTIFOULING MARINE COATINGS 1973

by A. Williams

Coatings Technology Review No. 1

One of the earliest needs for performance-oriented coatings was in the marine environment. Very early formulas were designed around known toxins such as copper and mercury compounds, and the earlier patent literature is replete with hundreds of directions for using these materials in creosote and natural drying oil formulations. Later information indicated in this book involves more sophisticated materials.

The two areas of a ship requiring specialty coatings are the bottom and the boot-topping area. The boot-topping area, intermittently exposed to air, sunshine, and water, represents a surface particularly difficult to protect from the elements.

For ships' bottoms, antifouling compounds, based on copper, mercury, and tin, are incorporated into somewhat water-sensitive binders to afford gradual breakdown of the film to allow for a sustained release of the poison. This required self-erosion necessitates frequent repainting of the ship's bottom, depending on geographical location and severity of exposure. By contrast, boot-topping paints are designed to provide a high level of resistance to both salt water and weather. Typically phenolic resin-tung oil, vinyl resin combinations are used.

This book describes many patented processes which provide high performance antifouling coatings based on metal compounds as well as organic coating compositions.

A partial and abbreviated table of contents is given here. Numbers in ( ) indicate the nos. of patents or application techniques discussed under a given heading. Chapter headings are given, followed by examples of important subtitles.

ISBN 0-8155-0464-0

271 pages